MATHEMATICS:
AN INTEGRATED VIEW

Merrill Mathematics Series
Erwin Kleinfeld, Editor

MATHEMATICS: AN INTEGRATED VIEW

ROLAND F. SMITH
RUSSELL SAGE COLLEGE

Charles E. Merrill Publishing Company
A Bell & Howell Company
Columbus, Ohio

Published by
Charles E. Merrill Publishing Co.
A Bell & Howell Company
Columbus, Ohio 43216

International Standard Book Number: 0-675-08814-3
Library of Congress Catalog Card Number: 73-92583

1-2-3-4-5-6-7-8 — 81-80-79-78-77-76-75-74
Printed in the United States of America

PREFACE

This book is an attempt to select some of the important ideas of mathematics, interweave them to make a unified pattern, yet leave some colorful strands protruding to encourage the reader to extend the pattern. Highly technical prerequisites would defeat these purposes. Though the text presupposes two or three years of high school mathematics, or the equivalent, it demands little specific knowledge or technique beyond elementary arithmetic. The need, instead, is for careful reading, active thinking, and the willingness to experiment with ideas and examples.

The choice and organization of the material is designed to provide an integrated introduction to or extension of some important aspects of modern mathematics with which an educated person should be familiar. Thus the book may prove helpful to able high school students, liberal arts students in junior and four-year colleges, elementary education majors and teachers, and others wishing to investigate mathematical ideas on their own.

The central core is a sequential treatment of sets, logic, and number systems. The book, however, stresses interrelations among these topics and also ties in many additional mathematical concepts. As a text, it contains enough material for a three-hour college course for nonmajors. It is tightly organized in two respects. First, the chapters and sections (except for the starred sections) should be studied in order, though the pace and the amount of emphasis on proving theorems may be varied. Second, the exercises form an integral part of the presentation, illustrating and extending the preceding concepts and questions and occasionally even introducing new ones. In this way the reader is able to become actively involved in understanding and applying the ideas presented, almost as if he or she were using a self-instruction program.

To allow for individual differences, most exercise sets contain a few starred, more challenging problems. In addition, five starred (optional) sections are included; they may be omitted or deferred without loss of continuity. Each chapter ends with a broad list of Suggestions for Investigation and a working bibliography of books and articles to encourage further exploration. The section immediately following the text contains selected answers to about half the unstarred problems, and is followed by a general bibliography and a reference list of all definitions and theorems by number and page.

Like most expository books, this one is the product of study, classroom experience, and personal contact with teachers, colleagues, students, and others to such an extent that a complete list of acknowledgements is impossible. Generous thanks are due to the students in my general mathematics classes, in particular to those who lugged around heavy preliminary editions, pointed out numerous errors and unclear passages, and made valuable suggestions for improvement. A mathematics major, Ms. Belinda Phillips, helped considerably in working out answers for many exercises. I am also grateful to the administration of Russell Sage College and to my colleagues for their encouragement in writing and using the materials; and to the typists and staff who produced them. Finally, I appreciate greatly the advice and technical assistance of the editors and staff of the Charles E. Merrill Publishing Company in converting the manuscript into a book.

Roland F. Smith

CONTENTS

7 THE INTEGERS AND FURTHER EXTENSIONS 263

1

The Nature of Mathematics

How old is mathematics? Is any mathematics really *new? What* is *mathematics, anyway? And what do mathematicians do besides adding numbers and drawing circles?*

Section 1.1 Mathematics: New or Old?

What comes immediately to mind when the word *mathematics* is mentioned? The response is likely to be in terms of personal experience; and it probably reveals more about the responder than about the subject of the inquiry. For some, mathematics connotes order, precision, logical reasoning, exciting challenges and discoveries, even beauty. But for many others it evokes the tasks of adding long columns of numbers, manipulating letters, solving equations, memorizing dull lists of definitions and proofs — at most mastering an unchanging body of more or less useful facts inherited from the Greeks and Arabians. In fact, the latter impression is not only quite uncomplimentary to mathematics, but it is also largely false: it is as much a travesty as to picture nursing as simply taking patients' temperatures or swimming as managing somehow to stay afloat in water.

The idea that mathematics is a cut-and-dried accretion of facts which one learns once and for all is epitomized in the phrase "as certain as two and two make four." (We shall see later that even this bit of arithmetic is not certain; in some circumstances one may claim that $2 + 2 = 10$, or $2 + 2 = 0$, or $2 + 2 = 1$.) It is true, of course, that the Greeks and the Arabians made signal contributions to mathematics that are relevant today. Indeed, evidences of written mathematics predating 3000 B.C. occur in Egypt, and a rudimentary sense of number probably goes back to primitive times.

In this connection, it is interesting to discuss the truth or falsity of the following two statements:

1. More *physics* has been discovered (or invented) since the death of Newton (1727) than in all previous recorded history.
2. More *mathematics* has been discovered (or invented) since the death of Newton (1727) than in all previous recorded history.

When college students who are not majoring in science confront these statements, they are likely to agree by a small majority that the first is true. As for the second statement, however, they are likely uncertain, and at best evenly divided when judging it true or false. Many of those who mark both statements true, moreover, admit that they are guessing or that they cannot justify their answers. Both statements *are* true, of course, and plenty of evidence is at hand to document them. As for mathematics, virtually any measure one chooses to apply shows that it is participating fully in the current "knowledge explosion."

One measure of the growth of mathematics is the volume of published materials, especially that concerned with new results. For example, the *Proceedings* of the American Mathematical Society, which principally contains newly created mathematics, appears monthly. The issue of April 1973 contained 222 pages with 44 articles; some issues have been considerably larger. This is *one* research publication of *one American* research organization: the Society publishes other research periodicals; several other American research societies publish journals; and many other journals are published in England, Germany, France, Russia, Japan, China, and elsewhere.

A related statistic is the number of papers presented at professional meetings of mathematicians. The American Mathematical Society is *one* of the *American* organizations which has sponsored regular conferences, the largest of which has been the annual meeting. The program of the 1973 meeting contained 733 short papers and 12 major addresses; in addition, 109 papers were listed "by title" and summarized in writing.[1] We underline the point that the above data concern only one meeting, and note that the January, February, April, and June, 1973, issues of the *Notices* of the American Mathematical Society gave announcements of a total of 50 regular and special conferences on mathematics scheduled in the United States during the calendar year 1973, in addition to 58 meetings in Canada, Mexico, and overseas.

---

[1] *Notices*, American Mathematical Society, vol. 20, no. 1, January, 1973.

If one wishes to judge by the number of *new* fields or areas of mathematics since Newton's day, it is interesting to examine a recently devised classification scheme. The classification list contains 63 major headings, some with 80 or more subheadings. Of the major headings, 42 (67%) refer to fields whose development has occurred almost entirely since Newton's time. Of the remaining 21 fields, virtually all include a large proportion of subheadings which belong to the post-Newtonian era. This scheme is used in connection with the computerized "Mathematical Offprint Service"[2] to help mathematicians keep abreast of research activities in their particular specialties. Through this service, a mathematician can obtain weekly lists of all published articles in his field and reprints of those which especially interest him.

In 1966, a well-known mathematician and historian, K. O. May, presented a paper on the growth of mathematical literature during the past hundred years.[3] His conclusion was that mathematical literature is growing at an annual rate of about $2\frac{1}{2}\%$. Put another way, the body of mathematical literature doubles about every 28 years. This is a *continuous* rate of growth, comparable with the rate at which money would grow in a bank which compounds interest moment by moment, rather than two or four times a year. Assuming that the volume of published articles in mathematics is a reliable index of the growth of mathematical knowledge, we must conclude that the statement concerning post-Newtonian mathematics is much too conservative. Indeed, it might justifiably be replaced by one such as "More mathematics has been created within the past generation than in all previously recorded history."

From a different point of view, we could ask how many mathematicians are active in the United States now, or how many are entering mathematical professions annually. A crude measure of the latter is the number of persons who earn degrees in mathematics each year. A few statistics are presented in Table 1.1.

TABLE 1.1

Earned Degrees in Mathematics and Statistics

	Bachelor's		*Master's*		*Doctor's*	
Year	*Number*	*Percent**	*Number*	*Percent**	*Number*	*Percent**
1959–60	11,399	9.9	1757	9.6	303	6.5
1964–65	19,547	14.3	4290	14.3	688	8.3
1969–70	28,986	15.7	7095	16.8	1343	9.7
1970–71**	29,940	15.9	7770	17.6	1480	10.1

SOURCE: *Projections of Educational Statistics to 1980–81*, 1971 ed., Office of Education, U. S. Department of Health, Education and Welfare (Washington, D.C.: U. S. Government Printing Office, 1972), excerpts from pp. 47, 50, 53.

*The base is the number of earned degrees in all the natural sciences.

**Estimated.

[2] *AMS (MOS) Subject Classification Scheme* (1970). Mathematical Offprint Service, American Mathematical Society, P. O. Box 6248, Providence, Rhode Island 02904.

[3] *Notices*, vol. 13, no. 5, August, 1966, pp. 579, 636–15.

The source of Table 1.1 also gives projections through 1980–81, based on past trends. Since, however, there is considerable evidence that an employment plateau has been reached in mathematics and many of the other sciences, it is unlikely that the trend will continue, at least in the near future. Projections, therefore, are not given in Table 1.1, for they might be misleading.

A related qualitative trend, nevertheless, is clearly discernible. Mathematics, which traditionally has been applied chiefly in the fields of physical science and engineering, is increasingly being utilized in the biological, social, and managerial sciences. These new applications, in turn, are leading to the creation of new mathematical techniques. Though this trend is as yet hard to document statistically, its implications for mathematics and many other fields of human endeavor are extremely significant. Though the chapter bibliography includes a few illustrative references, many articles demand a great deal of knowledge of mathematics or of the field of applications.

A "straw in the wind" was the Conference on the Application of Undergraduate Mathematics in the Engineering, Life, Managerial and Social Sciences.[4] A dozen speakers, each with advanced training in both mathematics and another specialty, presented examples of applications of mathematics in political science, psychology, sociology, physiology and medicine, ecology, computer science, industrial management, national security, and national educational policy. Even so, many applications were not included: a few are architecture, econometrics, communications and languages, and the catchall, operations research. The scope of mathematics used, moreover, was largely confined to what is expected of an undergraduate mathematics major. Whether or not one classifies these applications strictly as mathematics, it is clear that a professional working in almost any field will find a substantial background in mathematics a decided advantage, and, more and more frequently, a requirement.

We trust that you are by now properly convinced, if not overwhelmed, by the evidence that mathematics is a living, growing force in our culture. If your aim, moreover, is really to know "modern mathematics," you are attempting the impossible: even the fully trained mathematician cannot keep up with developments in all fields of mathematics, to say nothing of their applications. Fortunately, however, even a person with a limited background in the subject can profitably explore selected elementary portions of relatively recent mathematics, especially if the new material is well organized within itself and is related to his or her previous knowledge and experience. Accordingly, the tasks of this book are to weave strands of the "new" and the "old" into a coherent whole, and, more than this, to develop a modern point of view and an appreciation of the nature of mathematics today.

Suggestion for Investigation

1. It used to be assumed that the principal vocational opportunity for college mathematics majors was teaching at the secondary or college levels. While

[4]This conference, partially supported by the National Science Foundation, was held at Georgia Institute of Technology June 13–15, 1973.

this assumption has never been strictly true, it is certainly false now. To check on the current career opportunities in mathematics, consult references such as the frequently revised pamphlet, *Professional Opportunities in Mathematics*.[5]

Section 1.2 What Are the Characteristics of Mathematics?

If a person is asked, "What is mathematics?," the answer will doubtless be influenced strongly by the number and quality of the mathematics courses in his or her formal education and their applications to other courses or to his or her occupation. For example, one whose chief experience has been three or four years of high school mathematics, plus a physical science, and perhaps a course involving simple statistics, may think of mathematics primarily as a study of quantity and measurement. Numbers, and symbols which represent numbers (e.g., x, y, z), are, to be sure, important components of mathematics; and the theory and techniques of manipulating them comprise a large part of most elementary mathematics courses.

An obvious next step in answering the question "What is mathematics?" is to consult a dictionary. One dictionary in wide use defines mathematics as "a science that deals with the relationship and symbolism of numbers and magnitudes and that includes quantitative operations and the solution of quantitative problems."[6] This quotation neatly supports the ideas of the preceding paragraph. It is clear, then, that we should list as *one* characteristic

1. Mathematics includes the study of number and quantity.

Persons with a high school background, however, know that much of the geometry they studied is concerned more with shape and spatial relations than with number. Although numbers are used in plane geometry for simple counting and measuring, the fundamental ideas of congruence and similarity, as usually developed, are nonnumerical. Other nonnumerical properties are symmetry, order (one point between two others on a line), and incidence (intersection of lines and/or planes). At a more advanced level, the field of projective geometry was elaborated in the nineteenth century with little reference to number or quantity. Certainly, then, another characteristic is that

2. Mathematics includes the study of spatial configurations and properties.

A popular conception, noted earlier, is that mathematics is a set of techniques and rules of computation, or, worse still, a "bag of tricks." One important objective of this chapter is to disabuse you of this narrow view. It does not follow, however, that techniques are to be ignored. On the contrary, the

---

[5]*Professional Opportunities in Mathematics*, 8th ed., 1971. Mathematical Association of America, 1225 Connecticut Avenue, N.W., Washington, D.C. 20036.

[6]*Webster's Third New International Dictionary* (Springfield, Mass.: G. & C. Merriam Co., 1961), p. 1393.

dramatic rise of computer science owes its origin to a combination of mathematics and engineering; and computers provide a powerful tool for hitherto impractical research. It is well to recognize, however, that the computational aspect of mathematics leads to theoretical questions, such as the efficiency of a given computer program in terms of accuracy, speed, and generality, and to the development of new methods of calculation — usually approximate in nature — which exploit the advantages of the electronic computer. With this understanding, it is only fair to claim as another characteristic that

3. Mathematics includes techniques for computing and manipulating symbols.

It is often said that mathematics is a language. Obviously it has a set of symbols that can be combined according to certain rules of syntax to form expressions or sentences. The use of these symbols really begins in the early elementary grades, or before, with the recognition and writing of Hindu-Arabic and Roman numerals. On graduation from high school a student knows many additional symbols and a considerable number of rules for combining them. In solving "word problems" in algebra, he or she has even had experience in "translating" English sentences into algebraic equations. Much of the symbolism, moreover, is universally employed by mathematicians, whatever their native tongues; in this sense mathematics is a kind of international language. Thus another characteristic is that

4. Mathematics constitutes a symbolic language.

In the teaching of high school geometry — and increasingly also arithmetic and algebra — logical reasoning is stressed. Reasoning, as distinct from mere guesswork or a rough sketch or drawing, is first used informally to organize a set of statements into a coherent whole. Next, the system is formalized by being cast in a deductive mold: assumptions (axioms or postulates) and definitions are clearly stated, and from them, by the application of logical rules, theorems are proved. Increasingly the emphasis is on reasoning with abstract concepts, such as "ideal" triangles and circles, rather than on applications to approximately triangular or circular objects, such as house roofs or wheels. So another characteristic is that

5. Mathematics utilizes the study and application of reasoning to both abstract and concrete objects.

From a slightly different standpoint, mathematics can be considered as a study of abstract structures. In high school geometry, for example, instead of emphasizing the processes of reasoning and the methods of proof, one can focus on the result: the abstract system of Euclidean plane geometry. In a similar fashion, one can speak of the system of ordinary arithmetic, in which abstract numbers like 2 and 3 can be combined by an abstract operation called addition to produce the number 5. In fact, a substantial portion of this text will examine this very system and its extensions to include negative integers, fractions, and other kinds of numbers. Moreover, by changing one or several fundamental assumptions of an abstract system, mathematicians have created different geometrical and algebraic systems, some of which turn out to have unexpected beauty or utility. Hence,

6. Mathematics includes the study of abstract systems.

One aspect of mathematics which, though present from the beginning of history, has become increasingly explicit in modern times is its role as a reservoir of theoretical structures (or *models*, as they are often called) which can be applied in a variety of practical situations. The Euclidean geometry of high school, for example, provides an accurate description of most of the spatial phenomena of the world in which we live, and is therefore a useful model. For instance, if a surveyor stakes out a large triangle on a flat piece of ground and measures the angles at the three stakes, he finds that the sum of the angles is 180°, within the limits of accuracy of his measuring instruments. As a trivial example, the abstract arithmetic sum $15 + 25 = 40$, when applied to the transaction of buying a 15¢ candy bar and a 25¢ drink at a snack bar, correctly predicts that the cost of the two articles together is 40¢. So we say that the geometric and arithmetic models apply in the respective physical situations just described. (Curiously, these models, familiar though they are, may fail to apply under other circumstances. Euclidean geometry, for instance, is not an accurate model for describing the dynamics of the atom. And even the arithmetic sentence $15 + 25 = 40$ does not correctly predict the volume of liquid obtained by mixing 15 cm^3 of ethyl alcohol and 25 cm^3 of water; the actual volume is measurably less than 40 cm^3.)

Historically, much of mathematics has been created specifically to solve important concrete physical problems. In a number of cases, however, the order has been reversed: an abstract system created for theoretical purposes, with no thought of applications, was later found to fit a set of experimental data. Often, indeed, the order has been mixed: a practical problem would suggest a model which, in turn, would be developed theoretically far beyond the scope of the original problem; then the extended model would find application in an entirely different physical situation, which would again inspire another abstract model, etc.[7] A most important characteristic, therefore, is that

7. Mathematics provides abstract models which can be applied to concrete phenomena.

Suggestions for Investigation

1. Look up the word *mathematics* in several dictionaries. Do you consider the definitions adequate? Why or why not? (You may learn more about dic-

[7] As documented near the end of Section 1.1, these problems and situations are not restricted to the physical sciences, but occur also in many fields formerly not associated with mathematics.

tionaries than about mathematics! However, consider, in fairness, the dictionaries' monumental task of defining every word in the language within reasonable limitations of space.)

2. By contrast, peruse the article on *mathematics* in a good, many-volume encyclopedia. Compare it with the discussion of Section 1.2, noting any major omissions or differences in emphasis.

3. If you can find an elementary mathematics text written in a foreign language, try picking out and reading explanations and exercises on familiar topics. A plane geometry text, including figures for the proofs and exercises, might be especially interesting.

Section 1.3 An Operational Definition of Mathematics

An *operational definition* identifies a term by describing the observations or actions associated with it. In this vein, golf is the game the professional golfer plays. Similarly, mathematics is what the professional mathematician does. Putting the matter this way helps, to be sure, only those who are acquainted with mathematicians. But even a cursory observation indicates that a mathematician does not spend significant amounts of time computing like a machine. Moreover, *as a mathematician* a person does not teach formally in a classroom, although he or she may perform extremely valuable service to the profession and to students by teaching. But as a mathematician, a person creates new abstract concepts or relates old ones in new ways or experiments with and reasons upon concepts, either at the abstract level or at the concrete level of applications.

If the preceding, admittedly vague description seems to cover a whole spectrum of activities, this is indeed the case. These activities actually range over at least two spectra. In the first place, some mathematicians are attracted to working with abstract concepts and solving problems mentally, or symbolically on paper; others are interested in investigating concrete applications and solving problems experimentally in the real world. There is even a certain amount of rivalry between so-called pure and applied mathematicians. But the contrast is, for most people, a matter of degree or emphasis rather than of separate categories. Although some mathematicians prefer to work principally toward one or the other end of this continuous spectrum, others, including many famous ones, work happily in both and benefit from the way in which the abstract and the concrete complement and stimulate each other.

In the second place, much of a mathematician's work involves reasoning which can be informal or formal, bold and imaginative or cautious and analytical, creatively inductive or powerfully deductive. But here again we have a continuous spectrum from inductive to deductive approaches rather than a dichotomy. Some mathematicians have a flair for originating and exploring conjectures, some of which may turn out to be useless or even wrong; others have a penchant for clarifying concepts and testing the validity of the new discoveries. But all must recognize both the inductive and deductive aspects of mathematics. Typically a new idea is discovered or invented, tested, refined, and finally established by deductive reasoning. Sometimes this process is completed quickly by one person; frequently it continues for decades or even centuries, with contributions by many persons.

Although in this text the emphasis will be weighted somewhat toward the deductive approach, induction will not be neglected (see page 99). As an elementary example of how one approach shades into the other, you are invited to participate in the following exercise, which concerns summing the first two, three, four, etc., *odd* numbers and observing the sums. Thus $1 + 3 = 4$; $1 + 3 + 5 = 9$; $1 + 3 + 5 + 7 = 16$; etc. If you extend the process once or twice more, you will observe an apparent pattern in the sums. On the basis of these examples you may then guess that $1 + 3 + 5 + 7 + \ldots + 19$ has the sum 100, which simple addition shows to be correct. You may similarly explore other guesses; and as each checks out, your confidence in the observed pattern will increase. Thus, from an examination of particular cases you may reach a general conclusion or pattern. This is the essence of inductive reasoning: to generalize from specific facts. Yet — and this is a most important point — confirmation of this pattern in any number of individual cases, *no matter how large*, can never establish the pattern with certainty. This is so because the number of cases is infinite, so that it is physically impossible to verify each one and thus establish a *completely* general rule. This is the key difficulty of the *inductive* approach: though a powerful and valuable "suggester" (occasionally even of patterns which turn out to be false!), it can never yield certainty. Thus inductive reasoning can never produce a *proof* of the *general* statement,

$$1 + 3 + 5 + 7 + \cdots + (2n - 1) = n^2$$

where $2n - 1$ represents the last odd number in the sum. Instead, such a proof must utilize *deductive* reasoning, based on certain properties of the positive integers. Once we accept these properties, deductive reasoning logically forces us to accept the resulting equation: thus we gain the certainty we desire. We must, however, defer detailed discussion of deductive reasoning to Chapter 3 on Logic.

EXERCISES 1.3

1. (a) Express in your own words, as concisely as possible, the pattern concerning the sums of odd numbers observed in this section.

   (b) The following arrays of dots may help confirm the pattern of part (a). Study the three arrays, then draw the next two. Use them to make an informal argument which justifies the pattern.

$1 \qquad 1 + 3 \qquad (1 + 3) + 5$

*(c) If you have had experience with the principle of mathematical induction or with the sum of an arithmetic progression, try proving that the pattern always holds.

2. Polls are frequently used as devices for assessing opinions. A public official, for example, may wish to ascertain constituents' views on an important issue. If the official draws conclusions on the basis of a poll of these constituents, is he or she using inductive reasoning, deductive reasoning, or neither? Explain your answer.

3. One line lying in a plane separates the remaining points of the plane into two regions, one on each side of the line. Two *parallel* lines separate the plane in which they lie into three regions; but if the lines *intersect*, into four regions. So the *maximum* number of regions into which two lines separate a plane containing them is four. Continue this investigation to determine the maximum number of regions into which 3, 4, 5, . . . lines separate a plane. Attempt to find a pattern or formula which gives the maximum number of regions for n lines. Try to present at least an informal argument in support of your conjecture.

4. A student, amusing himself during a boring lecture, notices that when he squares an odd integer and then subtracts the number 1, the remainder is divisible by 8.

   (a) Check to see whether the student's conjecture appears to be correct. Does your checking involve inductive or deductive reasoning?

   *(b) In order to be sure that the conjecture is true for all odd integers, a proof is necessary. Try to discover one and write it up. Does the proof involve inductive or deductive reasoning?

Suggestions for Investigation

1. Drawing upon various courses of study (not only mathematics) which you have had during the past few years, give examples of topics or activities illustrating primarily "pure" mathematics, and others illustrating primarily "applied" mathematics. Explain your examples.

2. Compare the point of view of this chapter with that of the article by P. R. Halmos, a prominent contemporary mathematician, listed in the chapter bibliography.

Bibliography

Bell, E. T. *Development of Mathematics.* New York: McGraw-Hill Book Co., 1945.

______. *Men of Mathematics.* New York: Simon and Schuster, 1937.

---

*An asterisk identifies a (usually) more difficult exercise, not required in the main development, which extends the discussion or provides recreation.

Bell, M. S., ed. *Some Uses of Mathematics: A Source Book for Teachers and Students of School Mathematics*, SMSG *Studies in Mathematics*, XVI. Pasadena, Calif.: A. C. Vroman, Inc., 1967.

Boyer, C. B. *A History of Mathematics*. New York: John Wiley & Sons, Inc., 1968.

Coxeter, H. S. M. "Music and Mathematics." *Mathematics Teacher*, LXI (1968), 312–20.

Eves, Howard. *An Introduction to the History of Mathematics*, 3rd ed. New York: Holt, Rinehart and Winston, 1969.

Fischer, Irene. "The Shape and Size of the Earth." *Mathematics Teacher*, LX (1967), 508–16.

Gardner, Martin. *Mathematics, Magic and Mystery*. New York: Dover Publications, Inc., 1956.

Hadamard, J. S. *The Psychology of Invention in the Mathematical Field*. Princeton: Princeton University Press, 1949.

Halmos, P. R. "Mathematics as a Creative Art." *American Scientist*, 56:4 (1968), 375–89.

Hardy, G. H. *A Mathematician's Apology*. London: Cambridge University Press, 1940.

Hsiung, Chuan-Chi and Daniel Gorenstein. "Two Reports on the 1970 International Congress of Mathematics in Nice." *New York State Mathematics Teachers' Journal*, 21:2 (1971), 68–72.

Huff, Darrell. *How To Lie with Statistics*. New York: W. W. Norton & Co., Inc., 1954.

Huntley, H. E. *The Divine Proportion*. New York: Dover Publications, Inc., 1970.

Kemeny, J. G. and J. L. Snell. *Mathematical Models in the Social Sciences*. Boston: Ginn and Co., 1962. Chapters 1 and 2.

Kline, Morris. *Mathematics in Western Culture*. New York: Oxford University Press, 1953.

Lieber, L. R. *The Education of T. C. Mits*. New York: W. W. Norton & Co., Inc., 1944.

Mathematical Thinking in Behavioral Sciences. Readings from the *Scientific American*. San Francisco: W. H. Freeman and Co., 1968.

Mathematics in the Modern World. Readings from the *Scientific American*. San Francisco: W. H. Freeman and Co., 1968.

Mizrahi, Abe and Michael Sullivan. "Mathematical Models and Applications: Suggestions for the High School Classroom." *Mathematics Teacher*, LXVI (1973), 394–402.

Newman, J. R., ed. *The World of Mathematics*. 4 vols. New York: Simon and Schuster, 1956.

Polya, George. *How to Solve It*. New York: Doubleday & Co., Inc., 1957.

Professional Opportunities in Mathematics, 8th ed. Washington: Mathematical Association of America, 1971.

Rademacher, Hans and Otto Toplitz. *The Enjoyment of Mathematics.* Princeton: Princeton University Press, 1957.

Rice, J. A. "The Affinity of Mathematics to Music." *Mathematics Teacher*, LXI (1968), 268–75.

Saaty, T. L. "On Mathematical Structures in Some Problems of Politics." *Mathematics Teacher*, LXI (1968), 677–82.

———. *Mathematical Models of Arms Control and Disarmament.* New York: John Wiley & Sons, Inc., 1968. Chapter 1 and Sections 3.7, 3.8.

Schaaf, W. L., ed. *What is Contemporary Mathematics?*, SMSG *Reprint Series*, 3. Pasadena, Calif.: A. C. Vroman, Inc., 1966.

———. *Mathematics and Music.* SMSG *Reprint Series*, 8. Pasadena, Calif.: A. C. Vroman, Inc., 1967.

Scientific American, 211 (Sep. 1964). Eleven articles on mathematics.

———, 215 (Sep. 1966). Twelve articles on computing.

Sedgewich, C. H. W. *Apportionment of the U.S. House of Representatives.* Privately printed, 1964.

Siemens, D. F. "The Mathematics of the Honeycomb." *Mathematics Teacher*, LVIII (1965), 334–37.

Singh, Jagjit. *Great Ideas in Information Theory, Language, and Cybernetics.* New York: Dover Publications, Inc., 1966.

Steiner, H. G. "An Example of the Axiomatic Method: The Mathematization of a Political Structure." *Mathematics Teacher*, LX (1967), 520–28.

Steinhaus, Hugo. *Mathematical Snapshots*, 3rd American ed. New York: Oxford University Press, 1969.

Summers, G. J. *New Puzzles in Logical Deduction.* New York: Dover Publications, Inc., 1968.

Williams, H. E. "Some Mathematical Models in Plastic Surgery." *Mathematics Teacher*, LXIV (1971), 423–26.

Williams, J. D. *The Compleat Strategyst.* New York: McGraw-Hill Book Co., 1954.

2

Sets

Why all the fuss about sets? How can sets be usefully combined? In what ways are sets related to numbers? How do infinite sets behave?

Section 2.1 What Is a Set?

It is inevitable that a text dealing with "modern mathematics" should mention prominently the word *set*. For a while, indeed, it became a fad in some circles to make a great fuss about sets and, then, having dutifully pronounced the shibboleth, to relapse into a traditional treatment without any further reference to them. We shall try to avoid both errors by developing the set concept to the degree that it can clarify important parts of elementary mathematics and then by using it without fanfare for just this purpose.

The idea of a set is neither modern nor complicated: it is very old, simple, and pervasive. A set may be described as an assemblage or collection of objects.[1] These objects, or *elements*, may be physical or mental, concrete or

[1]Although good enough for our purposes, this description is inadequate for the formal theory of sets, because it permits certain embarrassing logical paradoxes to occur. See Exercises 2.1, Problems *7 and *8.

abstract. For example, the chairs in a classroom form a set, with each chair an element. So also do the people in the classroom — even though they might resent being called "objects." More abstract are the set of letters of the English alphabet, the set of whole numbers, and the set of theories of the structure of the universe.

An important requirement for a set is that there be some rule or characteristic by which we can determine whether a particular object does or does not belong to the set. Generally the characteristic is quite obvious; we may even be unaware of it, but it must be there. Thus, if the set is the letters of the English alphabet, m and n are elements of the set, but % and ? are not. The set of whole numbers is also clearly defined, as a later section will show. The set of theories of the structure of the universe is not as clearly defined. Although most people would have some notion of which theories would qualify as elements, they would, if pressed, find it hard to draw the line between a theory and a legend or religious belief and to decide the meaning or limits of the term *universe*.

In addition, it is not necessary for a set to contain elements all of the same kind. The set consisting of the ampersand symbol (&), Halley's comet, and the middle name of the current President of the United States is well defined, since an object belongs to this set if and only if it occurs in the list.

As an interesting aside, it should be noted that the word *set* is not the only English noun which expresses the described concept. Other terms are *collection*, *group*, and *class*. We shall, however, reserve *group* for a more precise, technical use which will be exploited in our study of number systems. The English language, moreover, is full of terms associated with sets of a particular kind. Thus we have a *flock* of sheep, a *herd* of elephants, a *school* of fish, a married *couple*, a string *quartet*, a *slate* of candidates, and even a *parliament* of owls and a *charm* of finches. James Lipton made a special study of such words, many connected with hunting, and cited over 300 of them.[2]

Though a particular set must be well defined, the term *set* itself has not been explicitly defined. The reason is that this concept is so fundamental that any attempt to define it introduces more complicated words and ideas which would, in turn, require definition.[3] For example, to define a set as a collection at once invites the question, "What is a collection?" A similar difficulty discourages a definition of the word *element*. We therefore accept these two words, *set* and *element*, as well as the phrase "is an element of the set," without attempting to define them. Instead, we gather their meaning intuitively on the basis of usage.

We may specify a set in several ways. First, we may use a descriptive phrase, such as "the states of the United States washed by the Pacific Ocean." This method is usually formalized by enclosing the descriptive phrase within a pair of braces, using x or some other symbol as a variable. Thus we may write $\{x \mid x$ is a state of the United States washed by the Pacific Ocean$\}$. This *set-*

---

[2]See the bibliography for this chapter.

[3]It *is* possible to define *set*. This is done, for example, in the Suppes reference (p. 19, Definition 1) after six pages of preliminaries. But such an abstract and detailed development subverts the purposes of this text.

builder notation is read, "the set of all elements x such that x is a state of the United States washed by the Pacific Ocean."

Secondly, if the set is small, we may simply list the individual elements, again between braces. So the example becomes {Alaska, California, Hawaii, Oregon, Washington}. Here the order of listing is inconsequential. In fact, it was done alphabetically; but it would have been equally correct to write {California, Oregon, Washington, Alaska, Hawaii}.

Finally, if frequent reference to a set is necessary, we may designate it by a single *capital* letter: for instance, P for "Pacific states." Note that we use lowercase letters, such as x, to refer to elements of a set.

If a set has many elements, however, it is not practical to list them all. For a well-known set we may use an incomplete listing, giving only the first few and the last element, indicating by three dots that some elements are not being listed. Thus the set of letters of the alphabet may be written: $\{a,b,c, \ldots, z\}$. If a set is infinite, it has no last element; so we use set-builder notation or incomplete listing without a last element. Hence for the set of whole numbers we write

$$W = \{n \mid n \text{ is a whole number}\}$$

or

$$W = \{0,1,2,3, \ldots\}$$

Since this set is so frequently used, we shall consistently designate it by the letter W.

For readers unfamiliar with these notations, we point out that a certain amount of formal symbolism is not only a convenience but may also lead to better understanding of the ideas represented. In any case, new notation is nothing mysterious, nor need it be frightening, if introduced slowly and in an orderly fashion. As an example, sentences like "California is an element of the set P" or "5 is an element of the set W" occur frequently. We abbreviate these as California $\in P$ or $5 \in W$, respectively. Note that the symbol $\in$ is read, "is an element of." Sometimes we wish to state the opposite, as "1/2 is *not* an element of W." We then make a slash through the symbol, writing $1/2 \notin W$. Which is correct: Ohio $\in P$ or Ohio $\notin P$?

EXERCISES 2.1

1. Use the listing or incomplete listing method to specify each of the following sets:
   (a) the set of New England states.
   (b) the set of states of the United States which have a point or boundary in common with your state.
   (c) the set of days of the week.

(d) the set of odd (whole) numbers.

(e) the set of all leap years from 1892 to 1916, inclusive.

(f) the set of whole numbers which satisfy the equation $x + 3 = 7$.

2. Rewrite each part of Problem 1 in the set-builder notation.

3. Some of the following sets are not well defined. Point out those which are not, and explain why they are not.

   (a) the original 13 states of the United States.

   (b) the southern states of the United States.

   (c) the set of married women currently residing in Indiana.

   (d) the set of vowels of the English alphabet.

   (e) the set of women Presidents of the United States from 1789 to 1969.

   (f) the set of whole numbers which satisfy the equation $x + 5 = 2$.

4. Insert the appropriate symbol, $\in$ or $\notin$, between the first and second expressions of the following:

   (a) Oregon, P (as defined in Section 2.1).

   (b) 3, W (as defined in Section 2.1).

   (c) 2.6, W

   (d) 0, $\{1,2,3,4,5\}$

   (e) h, $\{x \mid x$ is a letter of the English alphabet$\}$.

   (f) m, $\{v \mid v$ is a vowel of the English alphabet$\}$.

   (g) 1,000,000, W

   (h) 78, $\{n \mid n$ is an odd number$\}$.

5. Does $\{2,4,6, \ldots\}$ unambiguously define a set of numbers? Before answering, consider the set $\{2,4,6,10,16,26,42, \ldots\}$, in which each number after the second is the sum of the two numbers preceding it.

6. A *circle* may be defined as the set of points in a plane which are at a fixed distance from a fixed point in the plane. How would you define a sphere using the set concept?

*7. Despite the intuitive simplicity and generality of the undefined concept of sets, unrestrained use of it leads to logical difficulties. Some of these can be readily eliminated, but the embarrassing appearance of others has shaken the foundations of mathematics. A simple example of the first type is the *paradox* of the barber. It seems that the barber of a village shaves all the men and only the men of the village who do not shave themselves. We take as given that the barber is a man and lives in the village, and that no man in the village is permitted to go unshaven. The question now is, "Who shaves the barber?" There are only two alternatives: either he shaves himself or he does not. Explain why each alternative leads to a contradiction. Actually these contradictions prove, perfectly logically, that no such barber exists: so denying his existence removes the paradox.

*8. An example of a more serious difficulty is the following instance of Russell's paradox, which shook the mathematical world when he announced it in 1901. Consider the set B of all sets described in these Exercises 2.1. Since B is such a set, B is an element of itself. Now consider, by contrast, A, the set of United States astronauts through 1970. Although the individual elements of A are astronauts, the *set* A itself is not an astronaut, since a set is abstract and not a human being. So A is a set which is *not* an element of itself. It is clear, then, that there are two types of sets: those that are elements of themselves, and those that are not. Now let us denote by X the set of all sets that are not members of themselves. Is it true that $B \in X$? $A \in X$? $X \in X$? Explain why $X \in X$ is a contradiction. Again the only logical way out is to deny the existence of the set X, even though it appears to be a perfectly reasonable collection of objects. But rejecting X is much more troublesome to abstract set theory and to the foundations of mathematics than is denying the existence of an admittedly fictitious barber. For further reading on types of paradoxes and their consequences, read the Cohen and Quine references in the chapter bibliography.

Section 2.2 Subsets and Equal Sets

Sets may be related to each other in several ways. In particular, it may happen that all the elements of one set are also elements of another. For illustrations we shall use W, the set of whole numbers, and two other sets: D, the set of odd whole numbers, and E, the set of even whole numbers. Thus

$$D = \{1,3,5,\ldots\} \quad \text{and} \quad E = \{0,2,4,\ldots\}$$

Let us compare D and E with W. Since every element of D is also in W, we call D a *subset* of W. Is E a subset of W? Is E a subset of D?

We can discuss the subset relation more efficiently if we have a symbol for it. The symbol we shall adopt is $\subseteq$.[4] Thus we write $D \subseteq W$ and $E \subseteq W$. The following statements are also true: $E \not\subseteq D$ and $W \not\subseteq D$. The meaning of the slash (/) in $\not\subseteq$ is, of course, the same as in the symbol $\notin$. Hereafter, we shall use a slash without further comment to negate any relation.

Although the subset relation has been described and illustrated, terms used in mathematics should, if possible, be defined formally and precisely. For this reason, we introduce

---

[4]Be warned that some writers on set theory use different symbols (e.g., $\subset$ instead of $\subseteq$); so in studying another book, you should ascertain which symbols are employed, and with which meanings.

Definition 2.1 Subset If A and B are sets, A is said to be a *subset* of B (in symbols, $A \subseteq B$) if and only if every element of A is an element of B.

The phrase "if and only if" enables us to write two conditional statements very compactly. For Definition 2.1 these two statements are:

1. If A is a subset of B, then every element of A is an element of B.
2. If every element of A is an element of B, then A is a subset of B.

As a matter of fact, all formal definitions, whether in mathematics or in other fields, are to be understood in this way, even when the phrase "if and only if" is omitted. Thus in Exercises 2.1, Problem 6, one of the two conditionals is: "If a geometric figure is a circle, then it is a set of points in a plane which. . . ." You should complete this conditional and also write the second one.

Returning to the example of odd and even numbers, you should apply the definition to show that $D \subseteq W$ and $E \subseteq W$ are true. It is also important to show how the false statement $E \subseteq D$ violates Definition 2.1.

In discussing particular sets we usually have in mind a larger, more general set, of which all the others are subsets. In the preceding paragraphs, for instance, W is the general set and D and E are subsets. Put another way, W is the basic or *universal* set, or, as some authors call it, the *universe of discourse*. We can continue our discussion by considering still other subsets of W, such as the set of the ten smallest whole numbers, the set of all whole numbers divisible by 5, the set of page numbers of this text, etc. The common symbol for the universal set is U. For the preceding paragraphs, then, we have taken $U = W$.

Let us consider another example of a universal set. We earlier defined the set P of states of the United States washed by the Pacific Ocean. We might also wish to discuss the set of New England states or the set of states that voted Republican in the 1972 election. Here the universal set is clearly

$$U = \{x \mid x \text{ is a state of the United States}\}$$

Frequently the universal set is obvious; when it is not, or when ambiguity arises, U must be specified clearly.

Let us return to the discourse where $U = W$ and examine the following sets:

$$V = \{1,2,3,4\} \qquad X = \{4,2\} \qquad Y = \{1,3,2,4\} \qquad Z = \{1,3\}$$

It is easy to see that V, X, Y, Z are subsets of W. We can also apply Definition 2.1 to pairs of these sets. Thus we may write $X \subseteq V$ because the elements of X, namely 4 and 2, are also in V. Which of the following is (are) true? $X \subseteq Y$, $X \subseteq Z$, $V \subseteq X$.

It is worth noting that although $X \subseteq V$ is true, $V \subseteq X$ is not. An interesting question comes to mind: Is it ever possible for *each* of two sets to be a subset of the other? To answer this, we must carefully apply Definition 2.1. Let us consider V and Y. Is V a subset of Y? Yes, because each of the four elements of V

is an element of Y. But in the same way, Y is a subset of V. So we have both $V \subseteq Y$ and $Y \subseteq V$. This result is not surprising since V and Y have exactly the same elements. For this reason, we call V and Y *equal* sets. We do so formally by means of

Definition 2.2 Equal Sets If A and B are sets, A is said to be *equal* to B (in symbols, $A = B$) iff[5] $A \subseteq B$ and $B \subseteq A$.

A special case of the two definitions answers the question "Is a set a subset of itself?" That is, do we have $X \subseteq X$, $Y \subseteq Y$, etc.? Surely Definition 2.1 is satisfied here quite easily. Now we can apply Definition 2.2, with both A and B replaced by X, to get $X = X$. Thus every set is equal to itself — a rather transparent fact which, however, we have essentially proved in an informal way.

In the case of the sets X and V, on the other hand, Definition 2.2 fails, since $V \not\subseteq X$. But $V \not\subseteq X$ requires, according to the denial of Definition 2.2, that V contain at least one element which is not in X. Either 1 or 3 is such an element. You should similarly compare the sets Z and V.

Since, for the pair X and V, $X \subseteq V$ is true, but $V \subseteq X$ and $V = X$ are false (Why?), we call X a *proper subset* of V. More formally, we have

Definition 2.3 Proper Subset If A and B are sets, A is said to be a *proper subset* of B (in symbols, $A \subset B$) iff $A \subseteq B$ but $A \neq B$.

Thus, for the preceding example we may write $X \subset V$. As further exercises, is $Z \subset V$ true? Is $Z \subset X$ true? Other pairs from the sets V, X, Y, Z should be checked. Note that the symbol for a proper subset, $\subset$, is simply the earlier subset symbol, $\subseteq$, with the horizontal dash, —, which is part of the symbol for equality, removed. This symbolism accords neatly with that for numbers, in which $3 \leq 4$ means "3 is less than or equal to 4," but $3 < 4$ means "3 is strictly less than 4."

A further question is: "For any two pairs of sets, *must* one be a subset of the other?" An investigation of the pair X and V and the pair X and Z yields the answer. As we have seen, $X \subseteq V$, though $V \not\subseteq X$. For the second pair, by contrast, we have both $X \not\subseteq Z$ and $Z \not\subseteq X$. Yet the pair X and Z do have some special relation; for they have the same *number* of elements, though not the *same* elements. Because of this relation, we may say that X and Z are *equivalent* sets (in a sense to be defined formally in Section 2.7), although they are not equal.

In conclusion, we call attention to the distinction between the meanings of the symbols $\in$ and $\subseteq$ (or $\subset$). The first asserts set *membership:* it is a relation between an *element* and a *set*. The second (or third), on the other hand, asserts set *inclusion*: it relates *two sets*. Thus we write $4 \in W$ and $1, 2, 3, 4 \in W$, but not $\{1,2,3,4\} \in W$, since the braces make clear that $\{1,2,3,4\}$ is a set and not a list of individual elements. Instead we write $\{1,2,3,4\} \subseteq W$ or $\{1,2,3,4\} \subset W$,

---

[5]"iff" is *not* a misspelling; it is an abbreviation for the phrase "if and only if" and is becoming well established in mathematical literature.

where the symbols $\subseteq$ and $\subset$ connect two sets. Why is it incorrect to write $4 \subseteq W$? Is $\{4\} \subseteq W$ correct? In answering these questions, we note that the braces show that $\{4\}$ is a set, not an element: it just happens that this set contains only one element. We must be careful to distinguish the *element* 4 from the *set* $\{4\}$. Perhaps a more convincing example of this distinction is the following: Franklin D. Roosevelt, who was elected President of the United States in 1932, 1936, 1940, and 1944, was an individual, a person. But the set of presidents of the United States who have been elected four times, namely, {Franklin D. Roosevelt}, is an abstract collection which, as a matter of historical fact, contains only one element.

EXERCISES 2.2

1. Each part below describes several sets. In each case, determine a suitable universal set.

   (a) Chevrolet convertibles, Ford station wagons, Plymouth hardtops, Rambler sedans

   (b) $\{1,2,3\}$, $\{2,3,5,7\}$, $\{8\}$, $\{0,5,10\}$, $\{4,6,9\}$

   (c) {Mercury, Venus}, {Jupiter, Mars, Mercury, Saturn, Venus}, {Neptune, Pluto}, {Saturn, Uranus}

   (d) squares, parallelograms, rectangles

2. Let U be the set of the first ten letters of the alphabet: $U = \{a,b,c,d,e,f,g,h,i,j\}$.

   (a) Of the following sets, which (if any) are subsets of U? Which are proper subsets of U? (Use the symbols $\subseteq$ and $\subset$.)

   $A = \{a,d,c,f,g\}$ $\quad B = \{b\}$ $\quad C = \{f\}$

   $D = \{g,c,e,a\}$ $\quad E = \{g,h,a,i,j\}$ $\quad F = \{f,a,c,e\}$

   $G = \{c,a,g,e\}$ $\quad H = \{j,g,c,h,d,b,a,e,i,f\}$

   (b) Some of the sets are subsets of others. Name all such pairs: $C \subseteq A$, etc.

   (c) Repeat part (b) for *proper* subsets.

   (d) Similarly, write down all pairs of *equal* sets.

3. Referring to the sets of Problem 2, mark each of the following statements T (true) or F (false):

   $$a \in A, f \notin B, f \in D, D \subseteq H, d \subseteq H, B \subset H, B \in H,$$
   $$H \not\subset B, E \subseteq F, F \subset F, F \in F, \{c,d\} \subset A, \{c,d\} \in H$$

4. Let M and T be two sets from the set of whole numbers (i.e., let $U = W$).

   (a) Is it possible to choose sets M and T so that

   (i) $M \subseteq T$? $\quad$ (ii) $M \subseteq T$ and $T \subseteq M$?

   (iii) $M \subset T$? $\quad$ (iv) $M \subset T$ and $T \subset M$?

(b) For each part of (a) answered Yes, provide a specific example; if you answered No, explain why such a choice is not possible.

5. Let $U = W$, and let M be any set. Under which circumstances, if any, is it possible to have (a) $M \subseteq M$? (b) $M \subset M$? In each case justify your answer.

6. (a) The following are familiar kinds of living creatures: snakes, vertebrates, reptiles, cobras.
   (i) List them in an order so that each kind is a subset of the immediately following one.
   (ii) Is the first kind you listed in part (i) a subset of the last kind? Explain.

   (b) The following are familiar types of figures in plane geometry: rectangles, quadrilaterals, squares, parallelograms.
   (i) List them in an order so that each type is a subset of the immediately following one.
   (ii) Is the first type you listed in part (i) a subset of the last type? Explain.

7. Let $U = W$, and let M, R, T be any three sets.
   (a) Suppose that $M \subseteq R$ and $R \subseteq T$. Provide an example, citing three specific sets; then explain the relation, if any, between M and T.
   (b) Repeat part (a), using $\subset$ instead of $\subseteq$.

8. (a) Give a careful argument (really an informal proof) that if A, B, C are any three subsets of a universal set U and if $A \subseteq B$ and $B \subseteq C$, then $A \subseteq C$.
   (b) How, if at all, would your argument have to be modified if $\subset$ were substituted for $\subseteq$ in the statement of part (a)?

Section 2.3 Complementation: A Unary Operation on Sets

Suppose we wish to analyze in some way the full-time student body of Ivy College, a four-year liberal arts undergraduate institution. Clearly, $U = \{x \mid x$ is a student currently enrolled full time at Ivy College$\}$. One of the subsets of U is the class of freshmen, F. The set F is well defined, since the college catalog states that a freshman is any full-time student who has accumulated fewer than 32 semester credits. Associated with the sets U and F is another set consisting of all the students at Ivy College who are *not* freshmen. This set, called the *complement of F*, is the set of all sophomores, juniors, and seniors. We shall denote this complement by F' (read, "F-prime"); thus $F' = \{x \mid x$ is not a freshman$\}$. We do not need to mention in the braces that x is a student at Ivy, since the set U guarantees that. On the other hand, the set F'

would not have a definite meaning unless U had already been given. Thus U plays a crucial role in this section. As usual, we proceed to the formal

Definition 2.4 Complement If U is given and A is any set in U, then the *complement of A* (in symbols, A') is the set of all elements of U that are not elements of A. Briefly, $A' = \{x \mid x \in U \text{ and } x \notin A\}$.

Since determining A' is an operation on a single set, A, we may refer to complementation of a set as a *unary* operation. A similar example from numbers is "taking the reciprocal of" a fraction. By operating on a single fraction, say 2/3, we obtain another fraction, 3/2, called its reciprocal. By contrast, there are *binary* operations by which *two* elements are combined to form a third. An example is addition in the set W: we combine 2 and 3 to get 5. In later sections of this chapter we shall study three binary operations on sets.

Understanding of sets is greatly aided by a device attributed to John Venn (1834–1923), an English logician. A *Venn diagram* is a method of representing sets geometrically. The universal set, U, is represented by the interior of a rectangle; the subsets of U are depicted by the interiors or the exteriors of circles lying within the rectangle. (Points on the boundaries of the rectangle and circles are disregarded.) Note that the rectangle in Figure 2.1 is labeled U and the interior of the circle is labeled F. Points inside the rectangle represent students at Ivy College. What do the points inside the circle represent? In particular, the point x inside the circle indicates a freshman. The region outside the circle but inside the rectangle depicts the set F'. Thus the point y indicates an upperclassman (nonfreshman). It is correct to write $y \in F'$ or $y \notin F$. Two similar correct statements can be made about x. What are they?

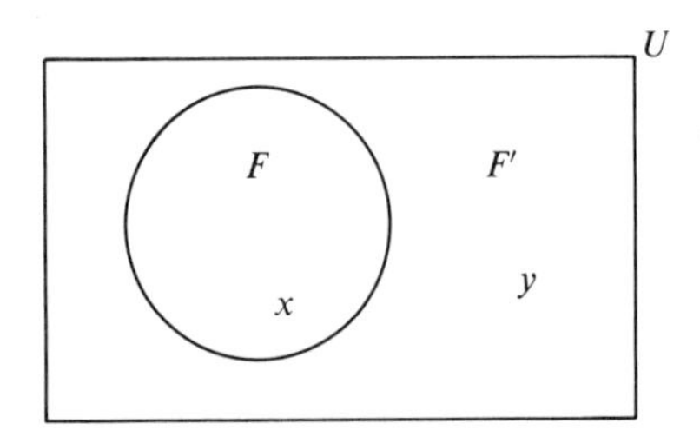

Figure 2.1

To take another illustration with the same set U, we may let M be the set of men students. The Venn diagram of this situation is shown in Figure 2.2. How should the set M' be described? Also, which elements would the set M' contain if Ivy College were a men's college? In that case, $M = U$, and M' would have no elements at all! Would M' be a set?

This question may be raised in a more general form. Since complementation is an operation on sets, we would like to have it applicable to *every* set, without any awkward exceptions. (It is a propensity of mathematicians to wish to generalize as much as possible and to try to fit any apparent exceptions into a

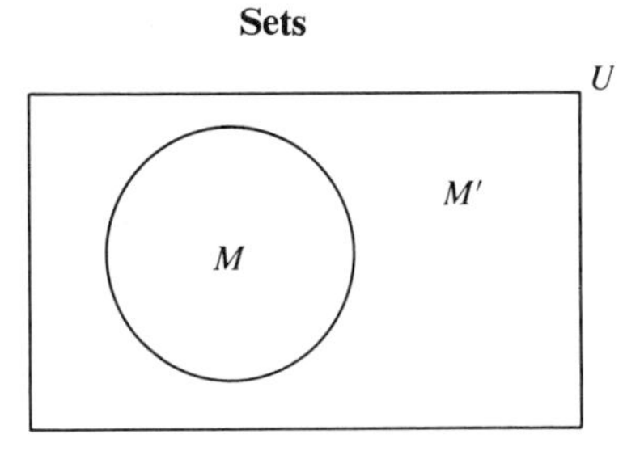

Figure 2.2

larger framework.) Since U is a set in U (i.e., $U \subseteq U$), we naturally inquire about U'. Now if U, as the universal set, contains *all* the elements under discussion, then none are left for U'! Can a set without any elements exist? Certainly it can exist *abstractly*, in our minds, if we agree to permit it. We call such a set an *empty*, or *null*, set, and designate it by the symbol $\varnothing$. (Typesetters often use the Greek letter *phi* for this purpose.) Another symbol in common use is a pair of braces with no elements listed inside ($\{\ \}$). Then $U' = \varnothing = \{\ \}$. What is M' in the preceding illustration?

Since any null set is indistinguishable from any other (there are no elements by which to tell them apart!), it is proper to speak of *the null* set. Thus we have the formal

Definition 2.5 Null Set The *null*, or *empty*, set (in symbols, $\varnothing$, or $\{\ \}$) is the set containing no elements.

Our motivation for introducing the null set was to find a complement for U. We shall soon find, moreover, that the null set fills other needs. But its introduction leads to four further questions:

1. What is the relation of $\varnothing$ to U? Since U is the universal set, we would like to have $\varnothing \subseteq U$. Let us see whether Definition 2.1 is satisfied. It requires that every element of $\varnothing$ (of which there are none!) be in U. Certainly this requirement is not contradicted: $\varnothing$ does not contain any elements which fail to be in U. We therefore agree that Definition 2.1 is satisfied *vacuously*. More generally, if A is any set, $\varnothing \subseteq A$. What is the reasoning for this statement? Are the following true or false?: $\varnothing \subseteq \varnothing$? $\varnothing \subset \varnothing$? $\varnothing \subset A$, where A is any set in U?

2. Our definition of equal sets (Definition 2.2) was made before we introduced the special sets U and $\varnothing$. Can it be extended to them? We need only show that $U \subseteq U$, with U playing the roles of both A and B in Definition 2.2; similarly, we must show that $\varnothing \subseteq \varnothing$. But $U \subseteq U$ follows from Definition 2.1 and the fact that the universal set includes all the elements under discussion; and $\varnothing \subseteq \varnothing$ follows from the preceding paragraph. So these special sets, too, are equal to themselves.

3. What is the meaning of $\varnothing'$? Since $\varnothing \subseteq U$, we would hope that $\varnothing$ has a complement in U. Does $\varnothing' = U$ satisfy Definition 2.4? We are now in a posi-

tion to state that *every* set in U (including $\varnothing$ and U itself) has a complement in U. In fact, it is not hard to see that *each set in U has one and only one complement in U*. This last statement, indeed, expresses an important characteristic of the unary operation of complementation of sets: the collection of sets of U forms a *closed system* when subjected to complementation in the sense that, for any set A of U, A' is a *unique set of U*. A briefer way to refer to this property is to say:

Complementation of sets is a closed operation in U.

4. How should the set $\varnothing$ be diagrammed? Since it contains no elements, our first thought might be that it should be omitted from the Venn diagram. This may be done; but we cannot then meaningfully say that $\varnothing$ has been diagrammed. Another method, which goes back to Venn, is to represent the null set by *shading* a circle which represents it, as in Figure 2.3. As another example, for an all-male Ivy College with $U = M$, we would have the Venn diagram shown in Figure 2.4.

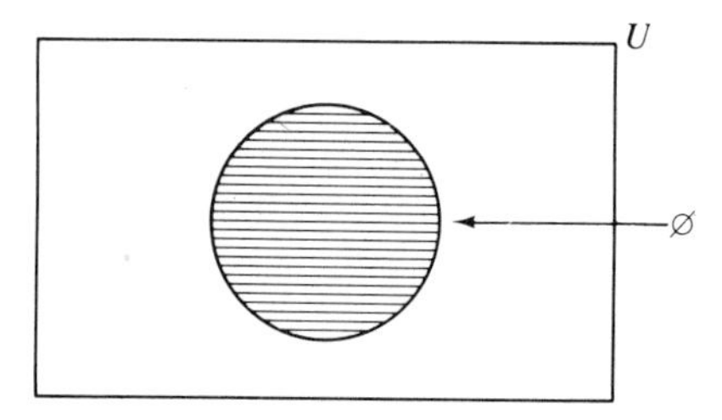

Figure 2.3

We are now able to explore a question which is interesting in itself and rather significant for higher mathematics. If U is a set with a fixed number of elements, how many subsets does it have? For instance, let $U = \{a,b\}$ be a set with two elements. Then $\{a\}$ and $\{b\}$ are subsets of U; in addition, U is a subset of itself; and $\varnothing \subseteq U$, since the null set is a subset of every set. Thus $U = \{a,b\}$ has four subsets in all. In the same manner, we may explore the cases where U contains three elements or one element or none ($U = \varnothing$). In fact, an excellent exercise in inductive reasoning is to discover a formula for the number of subsets of a set containing n elements, and a parallel exercise in deductive reasoning is to prove that the formula is correct.

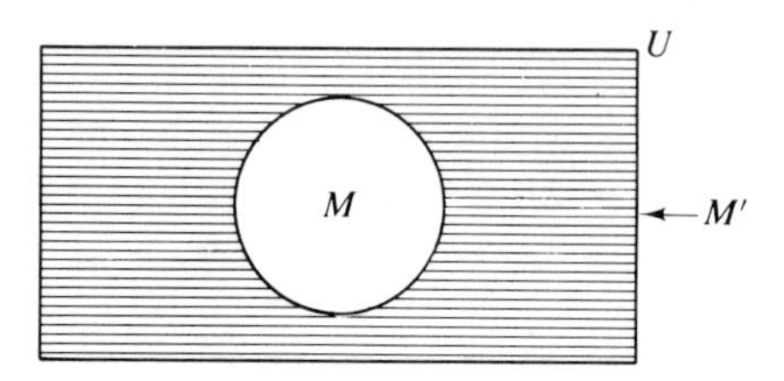

Figure 2.4

EXERCISES 2.3

1. Let $U = W$, the set of whole numbers, and D be the set of odd numbers. Then $D \subseteq W$.

   (a) List several elements of D', the complement of D.

   (b) Describe D' in words.

   (c) Indicate the relations among U, D, and D' by means of a Venn diagram.

2. Let U be the set of states of the United States, and let S be the subset of these states which lie in the southern hemisphere. Depict U, S, and S' in a Venn diagram, using shading where appropriate.

3. Let $U = \{a,b,c,d,e,f,g,h,i,j\}$ and define subsets as follows:

$$A = \{a,d,c,f,g\} \qquad B = \{b\} \qquad C = \{f\}$$

$$D = \{g,c,e,a\} \qquad E = \{g,h,a,i,j\} \qquad F = \{f,a,c,e\}$$

$$G = \{c,a,g,e\} \qquad H = \{j,g,c,h,d,b,a,e,i,f\}$$

   (a) List the elements of each of the following sets: A',F',C',H',U'.

   (b) Use a Venn diagram with appropriate shading to indicate the relations among H, H', and U.

   (c) Mark each of the following statements T (true) or F (false):

$$d \in A,\ d \in A',\ d \notin A',\ d \in B,\ d \notin B,\ d \in B',\ D \subseteq H,$$

$$D' \subseteq H,\ D \subseteq H',\ H' \subseteq D',\ B \subseteq F,\ B \subseteq F',$$

$$B' \subseteq F,\ B' \subseteq F',\ C \subseteq F,\ F' \subseteq C'$$

4. (a) Using the information of Problem 3, determine and list the elements of $(F')'$. $(F')'$ is the complement of the complement of F (i.e., the operation of complementation is applied twice in succession).

   (b) Similarly determine and list the elements of $(C')'$.

   (c) Let X be any set in U. How is $(X')'$ related to X? Give an informal argument to support your answer, using one Venn diagram to show X and X' and another to show X' and $(X')'$.

5. We have stated that $\varnothing$ is a subset of any set A, because Definition 2.1 is satisfied vacuously.

   (a) If $A = \varnothing$, is $\varnothing \subseteq A$ still true? Explain.

   (b) For at least some sets A it is true that $\varnothing \subset A$. Give an example, taking A to be a set from the whole numbers (i.e., $A \subseteq W$).

   (c) Does $\varnothing \subset A$ hold for *every* set A? If so, explain why; if not, provide a counterexample (i.e., an example showing that it fails). (*Hint:* What if $A = \varnothing$?)

6. (a) In Exercises 2.2, Problem 8(a), the following statement occurs: "If A, B, $C \subseteq U$ and if $A \subseteq B$ and $B \subseteq C$, then $A \subseteq C$."

    (i) Does this statement remain true if $A = \varnothing$? Explain.

    (ii) Under which circumstances, if any, would the statement be true if $B = \varnothing$? if $C = \varnothing$?

   (b) Answer part (a) with the relation $\subset$ substituted for $\subseteq$, as was done in Exercises 2.2, Problem 8(b).

*7. The definition of a null set (Definition 2.5) is rather informal, depending on an intuitive understanding of the phrase "containing no elements." A more precise definition is

$$\varnothing = \{x \mid x \neq x\}$$

Explain why $\{x \mid x \neq x\}$ contains no elements.

*8. (a) Set up the following table:

Number of Elements in U	Number of Subsets in U
0	
1	
2	4
3	
4	

   (b) If U has no elements ($U = \varnothing$), does U have any subsets? (*Hint:* recall that $\varnothing \subseteq \varnothing$.) Record the number of these subsets in the first line of the right-hand column of the table.

   (c) Suppose U has one element, e.g., $U = \{a\}$. Determine the number of subsets that U has and record the results in the table.

   (d) Continue the experiment, taking $U = \{a,b\}$ (already done and recorded), $U = \{a,b,c\}$, etc., until you recognize a pattern in the entries of the right column. Describe the pattern in words and/or by an algebraic formula for the general case where U has n elements.

   (e) Each time U is increased by one element, *which* new subsets are created? How many such "new" subsets are there? Have any "old" subsets been eliminated in this process? See if you can use these ideas to give at least an informal argument establishing the pattern which you discovered in part (d). If you are acquainted with the method of mathematical induction, you can present a formal proof.

Section 2.4 Intersection: A Binary Operation on Sets

Let us return to Ivy College, with U the set of all its students and F the set of freshmen. Since Ivy is located in a small city, it has both boarding students (set B) and commuters (set C). Note that B, C, and F are all subsets of U. What are the relationships among these three subsets?

We first investigate how C and F are related. Is either a subset of the other? If we suppose that there are both boarders and commuters among the freshmen, and similarly among the upperclassmen, we cannot have $C \subseteq F$ or $F \subseteq C$. (Why not?) The two sets, nevertheless, overlap: for some freshmen commute, and they are members of both C and F. The Venn diagram in Figure 2.5 depicts this relationship. The heavily outlined region in which the two circles intersect contains exactly the points which represent freshman commuters. These persons, indeed, constitute a new set, derived from C and F, called the *intersection* of C and F and symbolized by $C \cap F$. The symbol $\cap$ is read "intersection" or "cap." The set $C \cap F$ is labeled in Figure 2.5. More generally, we have

Definition 2.6 Intersection If A and B are any two sets in U, then the *intersection* of A and B (in symbols, $A \cap B$) is the set of elements which are in both A and B.

Since according to this definition intersection is an operation on two sets rather than on one, it is a *binary* operation. Making a seemingly reasonable definition, however, is only the beginning of our work. We still face the question of whether Definition 2.6 applies to and is consistent with our development of set theory. We recall that the definition of complementation led to the problem of interpreting U'; the solution was to invent the null set. A similar difficulty emerges from Definition 2.6: do two sets in U always intersect in another set? Consider, for example, the sets F and S (sophomores) of Ivy College. We may draw a Venn diagram, as before, and indicate $F \cap S$, putting a heavy boundary around it. But which students are members of both F and S? None! According to collegiate practice, no student is classified as both a freshman and a sophomore. Is $F \cap S$, then, a set? Fortunately, the reply is, "Yes, it is the null set." Thus $F \cap S = \varnothing$. If we had not already introduced the concept of

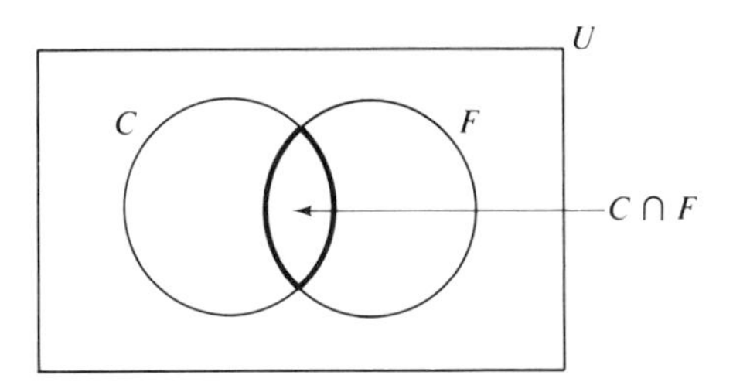

Figure 2.5

an empty set, we would have needed it here. In order to indicate the fact that $F \cap S$ is empty, we shall shade the intersection, just as we shaded null sets in Section 2.3. The diagram that results is shown in Figure 2.6.

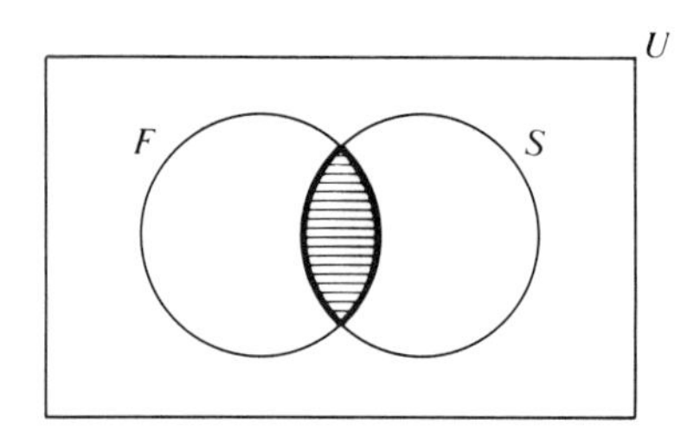

Figure 2.6

As a third example, it would be useful to diagram $B \cap F$. You should do so. Is $B \cap F = \varnothing$?

The preceding examples show that any two sets, X and Y, in U either have elements in common or else they do not. In the latter case, we write $X \cap Y = \varnothing$ and say that X and Y are *disjoint*. Since in all cases $X \cap Y$ is a set (empty or not), we observe that the operation of intersection can be performed on any two sets and will always yield one and only one set. Briefly, we have noted the property:

The intersection of sets is a closed operation in U.

At this point we sound a warning about the manner in which we are developing set theory. After accepting *set*, *element*, and *is an element of* as undefined terms, we are formally defining the remaining concepts. For psychological reasons, however, we are relying heavily on examples and appeals to intuition, both in defining and in studying the properties of set concepts. But in so doing, we must emphasize that we have not attempted any mathematical proofs. Instead, we are discussing and using, without proof, results which a number of mathematicians have rigorously established. This we shall continue to do in order to save time and to concentrate on the understanding and application of these concepts.

Besides closure, the binary operation of intersection has three properties that are not only significant for sets, but that also provide a recurrent theme for much of mathematics. The first property can be illustrated by comparing the sets $C \cap F$ and $F \cap C$ of Ivy College. Apparently these two sets contain exactly the same members, namely, the commuting freshmen. So we may write $C \cap F = F \cap C$. Examination of Definition 2.6 reveals that this property is perfectly general:

For any two sets X and Y, $X \cap Y = Y \cap X$.

In words, we state that intersection of sets is a *commutative* operation.

The commutative property occurs in so many parts of mathematics, especially in connection with the number systems in this text, that it *and its correct spelling* should become a part of your active vocabulary as quickly as possible. The word *commutative*, of course, has no logical connection with *commuting students*. But it may help the memory to note that the commutative property concerns the effect of reversing the order of performing an operation; and a commuting student, in coming from and returning to his home, reverses the direction of his travel.

A second important property of intersection is the grouping, or *associative*, property. It can arise from the attempt to intersect three sets at once. Suppose that X, Y, Z are sets in U. We wish to assign a meaning to the expression $X \cap Y \cap Z$. Definition 2.6 does not help directly, for it pertains to *two* sets. But we can apply it in two stages, first combining X and Y to get $X \cap Y$, which is also a set (Why?); then combining the set $(X \cap Y)$ with Z to get $(X \cap Y) \cap Z$. We surround $X \cap Y$ with parentheses, as $(X \cap Y)$, when we wish to consider it as a single set rather than as two. On the other hand, we could have combined Y and Z first, obtaining $(Y \cap Z)$, and then intersected X with $(Y \cap Z)$ to get $X \cap (Y \cap Z)$. But will the two methods produce the same result? You are invited to find the Ivy College sets $(B \cap C) \cap F$ and $B \cap (C \cap F)$. What emerges from this discussion is the associative property, which may be stated:

For any three sets X, Y, Z, $(X \cap Y) \cap Z = X \cap (Y \cap Z)$.

Note that in this equation the *order* of the sets, X, Y, Z, is the same on both sides, so that there is no commuting: what changes is the way in which the sets are *grouped*. Since either grouping yields the same result, we may consider either to be the meaning of the expression $X \cap Y \cap Z$.

A third property of intersection of sets is of a somewhat different nature. So far, when one set was intersected by another, as $C \cap F$, the resulting set was different from either original set. The multiplication of whole numbers behaves similarly: the product of two numbers is generally different from either factor, as $3 \cdot 4 = 12$. But the number 1 has the special characteristic that, when it is multiplied by any other number, the product is identical with the other number: thus $1 \cdot 3 = 3$, $4 \cdot 1 = 4$, and, in general, $1 \cdot x = x \cdot 1 = x$. For this reason we call 1 the *identity* for the operation of multiplication in W. By analogy let us seek a special set in U which will serve as an identity for the operation of intersection of sets. Let us try $F \cap U$ and $U \cap C$. Their respective Venn diagrams (see Figure 2.7) suggest that $F \cap U = F$ and $U \cap C = C$. Further examples and informal reasoning indicate that, in general:

For any set X *in* U, $X \cap U = X$ *and* $U \cap X = X$.

Hence U is the *identity for intersection* of sets.

You may be wondering how the special set $\varnothing$ behaves under intersection. If so, you are invited to experiment and then turn to the exercises.

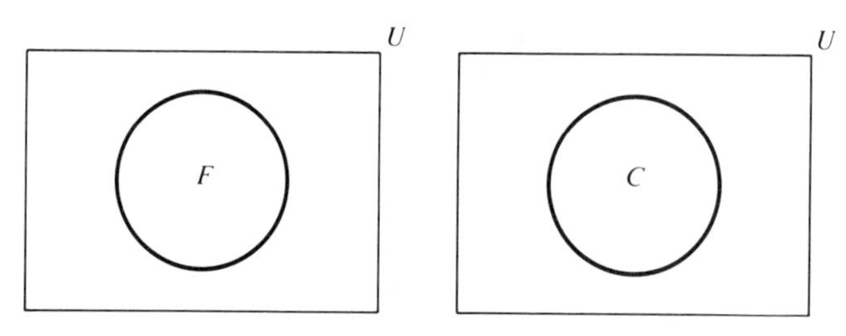

Figure 2.7

EXERCISES 2.4

1. Let U = {Mercury, Venus, Earth, Mars, Jupiter, Saturn, Uranus, Neptune, Pluto}

   and L = {Mars, Jupiter, Saturn, Uranus, Neptune, Pluto}

   S = {Mercury, Venus, Earth}

   M = {Earth, Mars, Jupiter, Saturn, Uranus, Neptune}

   V = {Earth, Mars, Venus, Jupiter, Saturn}

   X = {Venus, Mars}

   (Note that U is the set of planets of the solar system: some (L) have orbits larger than that of the earth; others (S) do not; some (M) have moons; some (V) are visible easily to the naked eye; some (X) are nearest to the earth.)

   List the elements of each of the following sets:

   $S \cap X$, $M \cap V$, $L \cap S$, M', X', $M' \cap S$, L', $S \cap L'$, $M' \cap V'$ (*first* complement M and V individually; then intersect the result), $(M \cap V)'$ (*first* intersect M and V; then complement the result)

2. (a) Using the sets of Problem 1, verify that

   (i) $M \cap V = V \cap M$

   (ii) $(L \cap M) \cap V = L \cap (M \cap V)$

   (iii) $U \cap V = V$; $S \cap U = S$

   (b) For each part of (a) name the property of set intersection which is illustrated.

   (c) Determine $M \cap \varnothing$ and $\varnothing \cap U$.

   (d) What, if anything, can be said about the intersection of any set with $\varnothing$? Explain.

(e) If A and B are any two sets in U, is it true that $A' \cap B' = (A \cap B)'$? Explain, using results of Problem 1.

3. Let U be the set of Ivy College students, B, the set of boarders, F, the set of freshmen, and E, the set of students majoring in English. Describe in words each of the following sets:

(a) $E \cap F$ (*Answer:* The set of freshman English majors)

(b) B' (c) F' (d) $(B \cap F) \cap E$ (e) $F \cap B'$

(f) $F' \cap B'$ (g) $(F \cap B)'$

4. We can represent the four sets of Problem 3 by means of the Venn diagram in Figure 2.8.

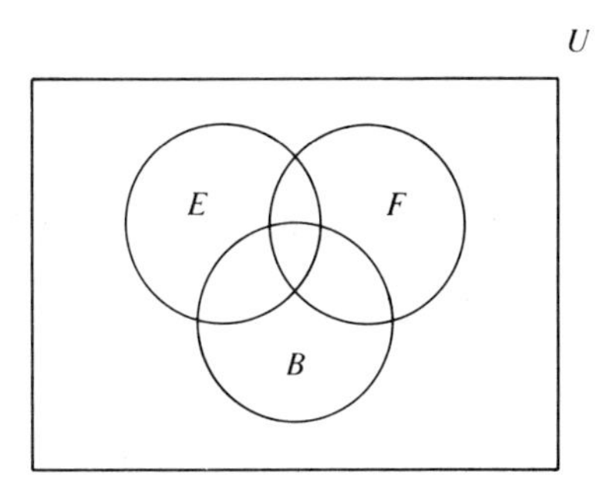

Figure 2.8

(a) For *each* of the following parts, copy Figure 2.8 and draw a heavy line around the boundary of the designated set:

(i) $E \cap F$ (ii) $F \cap B$ (iii) $E \cap U$

(iv) $(E \cap F) \cap B$ (v) $E \cap (F \cap B)$

(b) Note that a comparison of your drawings for parts (iv) and (v) makes evident a property of intersection of sets. Which property is it?

5. (a) Returning to Problem 1, note that $X \subseteq V$. Now find $X \cap V$ and describe its relation to X or to V.

(b) Write a statement generalizing your discovery in part (a). Begin: "If A and B are any two sets in U and if $A \subseteq B$, then. . . ."

(c) Is the general statement of part (b) true?

(d) In part (b), replace $\subseteq$ by $\subset$. Is the resulting statement true?

(e) The relationships discussed in parts (a)–(d) can be shown by Venn diagrams. Thus $A \subseteq B$ is indicated in Figure 2.9, where the shaded region is empty, so that all the elements of A are in the unshaded portion of the circle labeled "A." Explain how this diagram justifies parts (c) and (d).

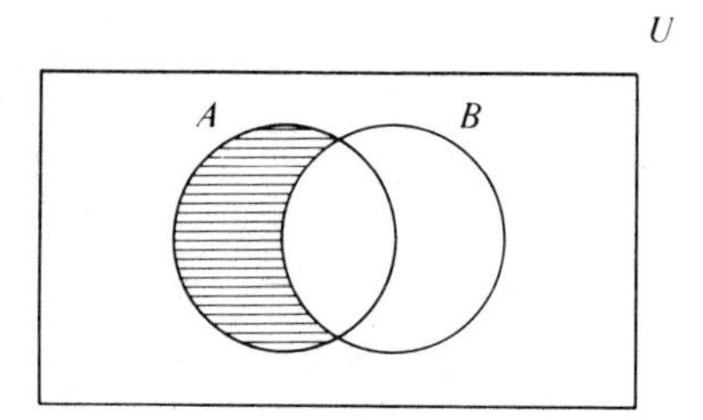

Figure 2.9

(f) By means of a suitable Venn diagram, determine whether the following statement is true: "If Y and Z are any two sets in U and if $Y \cap Z = Y$, then $Y \subseteq Z$."

*6. If $A, B \subseteq U$, the *difference* $A - B$ is defined by the equation

$$A - B = A \cap B'$$

(a) Using Problem 1, list the elements of $V - S$; of $L - S$.

(b) Using Problem 3, describe in words the set $F - E$; also construct a Venn diagram and draw a heavy line around the boundary of the set $F - E$.

(c) Is subtraction of sets commutative? Justify your answer.

(d) If $A, B \subseteq U$, what can be said about the set $(A - B) \cap (B - A)$? Explain your answer.

Section 2.5 Union: Another Binary Operation on Sets

In the Ivy College example we combined two sets, C and F, to get $C \cap F$, a set containing all the elements common to both sets. Sometimes it is desirable to refer to sets whose elements are in *at least one* of two given sets, but not necessarily in both. To continue our example, the set of Ivy students who are commuters or freshmen (including those who are both) is called the *union* of the two given sets. This set is shown as the heavily outlined region in Figure 2.10. The union of C and F is symbolized by $C \cup F$, read, "C union F" or "C cup F." The symbol $\cup$ is to be distinguished from the letter U (for the universal set). The appropriate definition is

Definition 2.7 Union If A and B are any two sets in U, then the *union* of A and B (in symbols, $A \cup B$) is the set of elements which are in A or in B or in both A and B.

Since both union and intersection are binary operations on sets, we might expect them to have somewhat similar properties. In particular, we shall ask whether union is a closed operation. In other words, given any two sets A and B, is $A \cup B$ a *unique set?* The expression $A \cup B$ *is* a well-defined set in U,

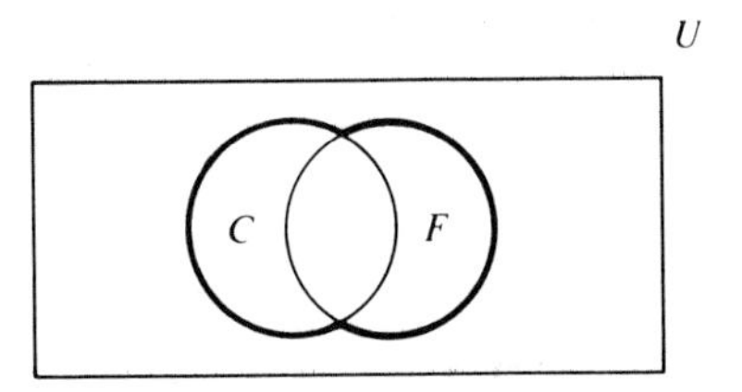

Figure 2.10

because any element of U is either in A or in B or in both — in which case(s), it is an element of $A \cup B$ by Definition 2.7 — or else it is in neither A nor B, and therefore is not in $A \cup B$. Do the special cases where $A = U$ or $B = U$ or both present any difficulties? What if A or B or both are null sets?

Intersection, we found, has three properties in addition to closure. Does union have these properties? If so, which set serves as the identity for the operation of union? Discovery of these matters is left largely for the exercises. It is timely, however, to present one technique utilizing Venn diagrams which may be helpful in exploring such properties. Let us use it to check the associative property for union.

Let A, B, C be any three sets in U. We construct a Venn diagram (see Figure 2.11) showing the intersection of each set with the others in U. By this means, the interior of U is divided into eight distinct, nonoverlapping regions. For easy

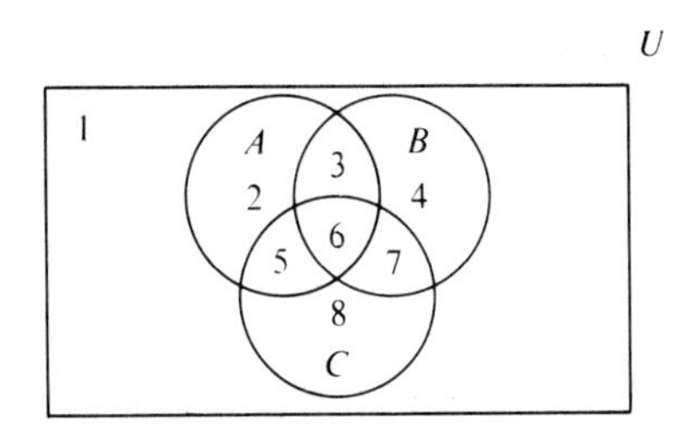

Figure 2.11

reference these regions have been identified by numbers 1–8. For example, "1" designates the portion of U which lies outside all three circles; similarly, "6" designates the portion of U which lies inside all three circles. We are now ready to check the associative property for union:

$$(A \cup B) \cup C = A \cup (B \cup C)$$

We do this by showing that the numbered regions contained in $(A \cup B) \cup C$ are exactly the same ones that are contained in $A \cup (B \cup C)$. We begin by noting that the set A contains regions #2, #3, #5, and #6; briefly, $A = \{\#2, \#3, \#5, \#6\}$. (We use the # sign so that the reader will not think that A contains the numbers 2, 3, 5, 6 as elements; indeed, the elements of A might be letters, planets, students, etc.) Continuing, we list the regions for B, for C, for $(A \cup B)$, and, finally, for $(A \cup B) \cup C$. We obtain:

$$A = \{\#2,\#3,\#5,\#6\} \qquad B = \{\#3,\#4,\#6,\#7\} \qquad C = \{\#5,\#6,\#7,\#8\}$$

$$(A \cup B) = \{\#2,\#3,\#4,\#5,\#6,\#7\}$$

$$(A \cup B) \cup C = \{\#2,\#3,\#4,\#5,\#6,\#7,\#8\}$$

By a similar procedure, we may build up $(B \cup C)$ from B and C, then perform the union $A \cup (B \cup C)$. There should be no difficulty in obtaining

$$A \cup (B \cup C) = \{\#2,\#3,\#4,\#5,\#6,\#7,\#8\}$$

which is identical with $(A \cup B) \cup C$, thus confirming the associative property for union.

In a similar manner, it is easy to check the associative property for intersection. This task and others are left for the exercises. Note that the manner of placing the numbers is not important, provided that each nonoverlapping region is identified by one and only one number. We may, in fact, use any identification scheme we wish, provided we use the same scheme for both sides of the set equation.

Now that we have two binary operations on sets, we may look for relations between them. An analogy from arithmetic operations on whole numbers will be helpful here. We begin by asking you to perform a simple calculation *entirely* in your head, writing only the answer on paper:

$$(29 \cdot 36) + (29 \cdot 64)$$

Before struggling with the obvious, but difficult feat of finding the product of 29 and 36 and then trying to remember it while performing the other product, you would do well to examine the problem carefully to see if it can be replaced by a simpler one. Certainly so; an equivalent problem is

$$29 \cdot (36 + 64)$$

whose answer is immediate. The key fact used here is that

$$(29 \cdot 36) + (29 \cdot 64) = 29 \cdot (36 + 64)$$

More generally, addition and multiplication of whole numbers are related by the equation

$$ab + ac = a(b + c)$$

where a, b, c represent any three whole numbers.

This equation and the equivalent one, $a(b + c) = ab + ac$, state the *distributive property of multiplication over addition*. Note that the factor a in $a(b + c)$ is "distributed over" the two addends, b and c, to yield $ab + ac$.

It is natural to raise the question: "Does intersection of sets distribute over set union?" In symbols, is it true that

$$A \cap (B \cup C) = (A \cap B) \cup (A \cap C)?$$

We could "try out" this equation by using specific sets from previous exercises. A more general technique is to use the numbered regions of the Venn diagram for three sets in U. Referring back to Figure 2.11, we have

$$A = \{\#2,\#3,\#5,\#6\} \quad \text{and} \quad B \cup C = \{\#3,\#4,\#5,\#6,\#7,\#8\}$$

so

$$A \cap (B \cup C) = \{\#3,\#5,\#6\}$$

since we are now *intersecting* A and $(B \cup C)$. On the other side of the equation, we find that

$$(A \cap B) = \{\#3,\#6\} \quad \text{and} \quad (A \cap C) = \{\#5,\#6\}$$

so

$$(A \cap B) \cup (A \cap C) = \{\#3,\#5,\#6\}$$

which agrees with $A \cap (B \cup C)$.

EXERCISES 2.5

1. Let U = {Mercury, Venus, Earth, Mars, Jupiter, Saturn, Uranus, Neptune, Pluto}

   and L = {Mars, Jupiter, Saturn, Uranus, Neptune, Pluto}

   S = {Mercury, Venus, Earth}

   M = {Earth, Mars, Jupiter, Saturn, Uranus, Neptune}

   V = {Earth, Mars, Venus, Jupiter, Saturn}

   X = {Venus, Mars}

   List the elements of each of the following sets:

   $L \cup S$, $S \cup X$, $S \cup M$, $M' \cup S$, $M \cup S'$, $M' \cup V'$, $(M \cup V)'$

2. (a) Using the sets of Problem 1, verify that

   (i) $M \cup S = S \cup M$

   (ii) $(L \cup M) \cup S = L \cup (M \cup S)$

   (iii) $S \cap (L \cup M) = (S \cap L) \cup (S \cap M)$

   (iv) $S \cup (L \cap M) = (S \cup L) \cap (S \cup M)$

   (b) For each part of (a) name the property of sets which is illustrated.

   (c) Determine $M \cup U$, $U \cup S$, $M \cup \varnothing$, $\varnothing \cup S$.

   (d) What, if anything, can be said about the union of any set with U? With $\varnothing$? Explain in each case.

   (e) If A and B are any two sets in U, is it true that $A' \cup B' = (A \cup B)'$? Explain using results of Problem 1.

3. Let U be the set of Ivy College students, B, the set of boarders, F, the set of freshmen, and E, the set of students majoring in English. Describe in words each of the following sets:

   (a) $E \cup F$ (*Answer:* The set of students who are freshmen and/or English majors)

   (b) $B \cup C$ (c) $(B \cup F) \cup E$ (d) $F \cup B'$

   (e) $F' \cup B'$ (f) $(F \cap B)'$ (g) $F' \cap B'$

   (h) $(F \cup B)'$

4. We can represent the four sets of Problem 3 by a Venn diagram containing three overlapping circles within a rectangle representing U.

   (a) For each of the following, make such a Venn diagram, drawing a heavy line around the boundary of the designated set:

   (i) $E \cup F$ (ii) $F \cup B$ (iii) $F \cup \varnothing$

   (iv) $(E \cup F) \cup B$ (v) $E \cup (F \cup B)$

   (b) The drawings for (iii), (iv), and (v) illustrate two properties of the union of sets. Which are they?

5. (a) Number each of the four distinct, nonoverlapping regions into which the two circles divide the interior of the set U in Figure 2.12.

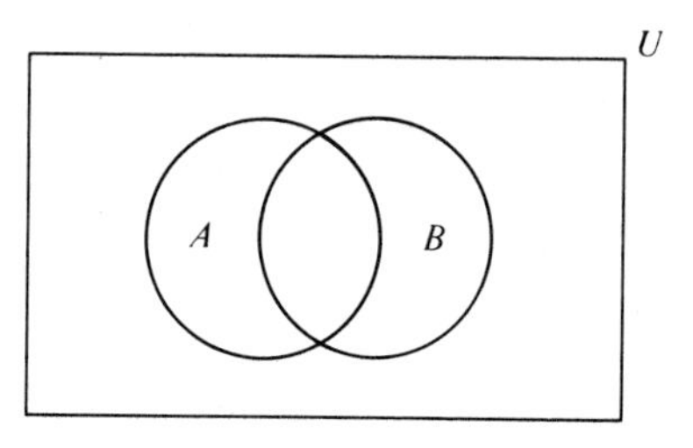

Figure 2.12

(b) Use this diagram to check the following two set equations, which are called the *DeMorgan Laws:*

(i) $(A \cup B)' = A' \cap B'$ (ii) $(A \cap B)' = A' \cup B'$

6. Use eight numbered regions on a Venn diagram for three sets to check each of the following set equations, where A, B, C are any sets in U:

(a) $(A \cap B) \cap C = A \cap (B \cap C)$

(b) $A \cup (B \cap C) = (A \cup B) \cap (A \cup C)$

(*Note:* Part (b) is a *second* distributive law. Which operation is distributed over which?)

7. (a) Returning to Problem 1, note that $X \subseteq V$. Now find $X \cup V$ and describe its relation to X or to V.

(b) Write a statement generalizing your discovery in part (a). Begin it: "If A and B are any two sets in U and if $A \subseteq B$, then. . . ."

(c) Is the general statement of part (b) true?

(d) In part (b), replace $\subseteq$ by $\subset$. Is the resulting statement true?

(e) The relationships discussed in parts (a)–(d) can be shown by shading a Venn diagram, as was done in Exercises 2.4, Problem 5. Use this method to justify parts (c) and (d).

(f) By means of a suitable Venn diagram, determine whether the following statement is true: "If Y and Z are any two sets in U and if $Y \cup Z = Z$, then $Y \subseteq Z$."

(*Note:* A comparison of part (f) of this problem with part (f) of Exercises 2.4, Problem 5, shows that

$$A \subseteq B \quad A \cap B = A \quad \text{and} \quad A \cup B = B$$

are equivalent statements in the sense that, if any one of them is true, the other two are also true.

*8. Write down all the statements of properties involving union which have occurred in this section. Then rewrite each statement, interchanging the symbols $\cup$ and $\cap$, and also the symbols U and $\varnothing$. For example,

$$A \cap (B \cup C) = (A \cap B) \cup (A \cap C)$$

becomes

$$A \cup (B \cap C) = (A \cup B) \cap (A \cup C)$$

Which of the resulting statements are true? Note that a striking *duality* between the two binary operations ($\cup$, $\cap$) and the two special sets (U, $\varnothing$) is thereby revealed.

*9 Solve the following logical puzzle, indicating your reasoning:

Mary's ideal man is tall, dark, and handsome. She knows four men: Alec, Bill, Carl, and Dave. Only one of the four men has all of the characteristics Mary requires.

1. Only three of the men are tall, only two are dark, and only one is handsome.
2. Each of the four men has at least one of the required traits.
3. Alec and Bill have the same complexion.
4. Bill and Carl are the same height.
5. Carl and Dave are different heights.

Which one of the four men satisfies all of Mary's requirements?[6]

*10. Set concepts are used increasingly in geometry, as the following sequence of definitions shows. Generally, *point* and *line* are taken as undefined terms. The line determined by two given points A and B is often designated by $\overleftrightarrow{AB}$ and is represented as in Figure 2.13(a).

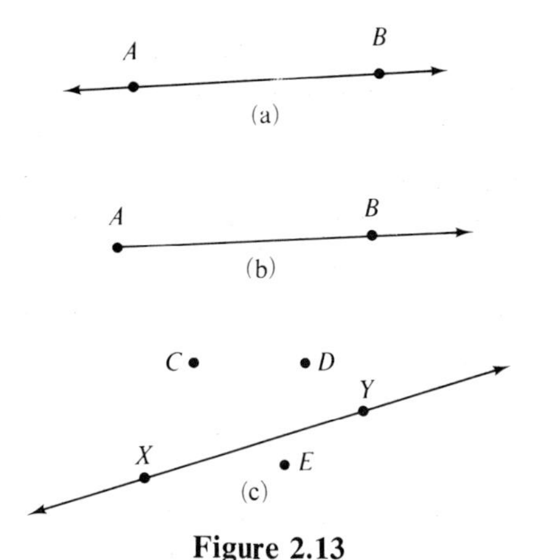

Figure 2.13

Later a *ray*, or *half-line*, is defined: thus ray $\overrightarrow{AB}$, starting at point A, is represented in Figure 2.13(b). Also, any line in a plane separates the plane into two *half-planes*. Thus the line $\overleftrightarrow{XY}$ in Figure 2.13(c) separates the plane of this page into two half-planes, one containing the two points C and D and the other containing the point E. We say that C and D are *on the same side of* $\overleftrightarrow{XY}$ since they lie in the same half-plane, and that C and E are *on opposite sides of* $\overleftrightarrow{XY}$.

(a) From a given point P draw two rays, $\overrightarrow{PQ}$ and $\overrightarrow{PR}$, choosing the points Q and R so that the rays do not lie on the same line.

---

[6]George J. Summers, *New Puzzles in Logical Deduction* (New York: Dover Publications, Inc., 1968), p. 7. Reprinted through permission of the publisher.

(b) Since the rays $\overrightarrow{PQ}$ and $\overrightarrow{PR}$ are sets of points, set operations can be performed on them.

(i) Find $\overrightarrow{PQ} \cup \overrightarrow{PR}$ and describe this set. The union of two noncollinear rays at the same point is called an *angle*. Do you see why?

(ii) Find $\overrightarrow{PQ} \cap \overrightarrow{PR}$ and describe this set. What is the *vertex* of an angle?

(c) Given an angle ($\angle$), such as $\angle QPR$ of part (b), a certain set of points is called the *interior* of the angle. A formal definition of the interior of $\angle QPR$ (with vertex at P) is: the intersection of the set S of points on the same side of $\overleftrightarrow{PQ}$ as is R and the set T of points on the same side of $\overleftrightarrow{PR}$ as is Q. Make a second drawing of $\angle QPR$ indicating S, T, and $S \cap T$ by different kinds or colors of shading. Does $S \cap T$ correspond to your intuitive notion of the interior of an angle?

*Section 2.6 The Formal Algebra of Sets

Mathematicians and amateur devotees of mathematics occasionally rejoice over a beautiful theorem, an exquisite geometric figure, or an elegant proof. Appreciation for the beauty of mathematics can be cultivated, even by those who are not proficient and knowledgeable in it. But many people, because of limited or unfortunate experiences with courses in the schools or because of prejudice instilled by another person, have been prevented from seeking such appreciation. Sometimes it helps to know what to look for.

Much of the beauty of mathematics inheres in its symmetry or balance. The visual symmetry of some common geometric figures, such as the circle, square, and ellipse, is an example of this beauty. More subtle is the sense of balance, or parallelism, in some abstract systems. One of these is the algebra of sets with its duality principle, illustrated in Exercises 2.5, Problem *8. (You are advised to spend some time with this exercise before proceeding.) A summary of some of the properties of sets under the operations of intersection and union is given in Table 2.1. Study it both for its informational content and for its aesthetic quality. Note that the duality principle holds without exception.

There is also a kind of austere beauty, or at least satisfaction, in a rigorous proof of a mathematical statement. This is true at least when a method of proof is applied for the first time; if it becomes routinized through frequent and mechanical use, the senses are dulled. At this point mathematicians tend to lose interest and transfer their energies to something more exciting.

We have not yet constructed any rigorous proofs. Instead, we have first illustrated properties of sets, then checked them by means of Venn diagrams. Now, the appearance of a diagram or figure, suggestive as it may be to one's intuition or imagination, is not an acceptable substitute for a carefully reasoned

TABLE 2.1

Comparison of Intersection and Union of Sets

Property	*Intersection*	*Union*
1. Closure	$A \cap B$ is a unique set.	$A \cup B$ is a unique set.
2. Commutativity	$A \cap B = B \cap A$	$A \cup B = B \cup A$
3. Associativity	$(A \cap B) \cap C = A \cap (B \cap C)$	$(A \cup B) \cup C = A \cup (B \cup C)$
4. Identity	$A \cap U = A$	$A \cup \varnothing = A$
5. Distributivity	$A \cap (B \cup C) = (A \cap B) \cup (A \cap C)$	$A \cup (B \cap C) = (A \cup B) \cap (A \cup C)$
6. Self-distributivity	$A \cap (B \cap C) = (A \cap B) \cap (A \cap C)$	$A \cup (B \cup C) = (A \cup B) \cup (A \cup C)$
7.	$A \cap \varnothing = \varnothing$	$A \cup U = U$
8. Idempotency	$A \cap A = A$	$A \cup A = A$
9. Complementation	$A \cap A' = \varnothing$	$A \cup A' = U$
10. DeMorgan laws	$(A \cap B)' = A' \cup B'$	$(A \cup B)' = A' \cap B'$
11.	$A \cap B \subseteq A$	$A \cup B \supseteq^* A$
12.	$A \subseteq B$ iff $A \cap B = A$	$A \supseteq B$ iff $A \cup B = A$

*The symbol $\supseteq$ means "is a *superset* of"; thus X is a superset of Y iff Y is a subset of X.

logical proof. To make a direct proof it is necessary to supply an orderly sequence of statements leading from the *hypothesis(es)* (what is "given") to the *conclusion* (what is "to be proved"), with each step justified carefully by a logically sound reason (or reasons) derived from the structure of the mathematical system under consideration. But if all the statements and reasons were written out in detail, even a simple theorem might well require a lengthy and complicated proof. Indeed, the state of mathematics has advanced so far that some of the proofs that famous mathematicians made a century or two ago are no longer considered acceptable by their successors. Although absolute rigor, by today's standards, is very difficult to achieve, we can, with care, provide examples of simple theorems from set theory which have acceptable proofs.

As an example, we shall prove the following theorem (one of the DeMorgan laws):

THEOREM If A and B are any two sets in U, then $(A \cup B)' = A' \cap B'$.

Our proof will not attempt to persuade through drawings or examples; instead, it will be based strictly on the definitions we have stated, especially Definition 2.2 for the equality of sets. This definition says, in effect, that two sets are equal whenever each is a subset of the other. We shall therefore separate our proof into two parts: in the first part we shall establish that $(A \cup B)' \subseteq$

$(A' \cap B')$, and, in the second, that $(A' \cap B') \subseteq (A \cup B)'$; then Definition 2.2 will provide the equality.

To prove $(A \cup B)' \subseteq (A' \cap B')$, we must use Definition 2.1 in the following way. We take x to be any element of $(A \cup B)'$: $x \in (A \cup B)'$. Then we show that $x \in (A' \cap B')$. When we have completed this task, we know that every element of $(A \cup B)'$ is in $A' \cap B'$, so that, by Definition 2.1, $(A \cup B)'$ is a subset of $A' \cap B'$. The details of this argument are as follows:

Let x be any element of $(A \cup B)'$: $x \in (A \cup B)'$. By the definition of set complement (Definition 2.4), since $x \in (A \cup B)'$, we must have $x \notin (A \cup B)$. Now, by the definition of set union (Definition 2.7), since x is not in the set $(A \cup B)$, it cannot be an element of A or an element of B: hence $x \notin A$ and $x \notin B$. But the last two statements, again by Definition 2.4, imply that $x \in A'$ and $x \in B'$. In turn, this means that x is an element common to the sets A' and B'; hence, by the definition of set intersection (Definition 2.6), $x \in (A' \cap B')$. Since any element (here, x) of $(A \cup B)'$ has been shown to be an element of $A' \cap B'$, by Definition 2.1, $(A \cup B)' \subseteq (A' \cap B')$.

In order to complete the proof, we must establish the second part, $(A' \cap B') \subseteq (A \cup B)'$; then we use Definition 2.2. The details, which are similar to those for the first part, are left for Exercises 2.6, Problem *1.

*EXERCISES 2.6

*1. Complete the proof of the preceding theorem by establishing $(A' \cap B') \subseteq (A \cup B)'$.

*2. Prove the second DeMorgan law, namely,

THEOREM If A and B are any two sets in U, then $(A \cap B)' = A' \cup B'$.

*3. Justify by means of Venn diagrams some of the properties in Table 2.1 that have not been mentioned previously (6,7,8,9,12).

*4. Make a careful proof of each of the following properties from Table 2.1: 3, 5, 9, 12.

Section 2.7 Equivalent Sets and Cardinal Numbers

Suppose two two-year-olds, playing with a set of blocks, start quarreling because each thinks the other has more than his share. For the harried parent to count the blocks and tell each child he has twenty is not very convincing, for the children are not old enough to count or understand numbers. How can they be convinced? One method might be to hand each child one block at a time: "One block for Billy"; "one block for Johnny" — until the entire pile has been distributed. (Let us hope that there is an even number of blocks!) In mathematical terms, by this procedure the parent is establishing a *one-to-one corre-*

spondence between the set of Billy's blocks and the set of Johnny's blocks. The parent then applies the mathematical concept that two sets which can be matched one-to-one contain the same number of elements. Whether the psychological appeal of this procedure to the children is strong enough to settle their quarrel is a question best left to your experience!

Also, in the history of the race, the idea of one-to-one correspondence considerably antedates the concept of number. Pebbles, notches in sticks, and tally marks were used by early man to keep track of quantity, according to the principle that one pebble or notch or tally mark corresponded to one physical object of the set under consideration. Moreover, fingers were (and still are) ready-made devices for ascertaining the size of a small set: a person made a one-to-one correspondence between a specific set of objects and a subset of the fingers of one or both hands. All this could be done without any concept of number as an abstract entity: primitive man observed the matching of a set of pebbles with a set of sheep, rather than counting *four* sheep; and the idea of a *number* four which could be applied to sheep or days or miles was beyond his ken.

As man began to develop his capacity to deal with quantity, there was a stage of identifying sets of different sizes with appropriate objects. Thus any collection of five objects was thought of as — and perhaps was called — a "hand"; a set of two objects might have been identified with the wings of a bird, a set of four objects with the legs of an animal, etc. Only when men began abstracting a certain quality of "fiveness" shared by *all* sets which matched the set of fingers of one hand could they be said to be inventing and using numbers.

A similar development occurs with children today: the first-grade teacher and first-grade workbooks devote considerable attention to helping pupils pick out sets that can be matched one-to-one with a given set, and supplying the number that should be associated with these sets.

In this course we shall accept without definition — just as we did with the terms *set* and *element* — the fundamental notion of *one-to-one correspondence* between two sets. We can now state the following:

Definition 2.8 Equivalent Sets Two nonempty sets A and B are said to be *equivalent* (in symbols, $A \sim B$) iff there exists a one-to-one correspondence between A and B. If A and B are both empty, they are equivalent.

The special provision for null sets is necessary, since they have no elements with which to set up a one-to-one correspondence. But since all null sets are equal (Why?), it is reasonable to define them to be equivalent to each other.

It is important to note that *equal sets* and *equivalent sets* are not synonymous terms. As an illustration, let U be the set of letters of the English alphabet and $A = \{a,b\}$, $B = \{b,a\}$, $C = \{c,d\}$, $D = \{a,b,c\}$. Now $A = B$ because they contain exactly the same elements; but $A \neq C$. (Why are A and C not equal?) However, $A \sim C$, because there is a one-to-one correspondence between them, e.g., a can be matched with c, and b with d. (This is not the *only* one-to-one correspondence between A and C. Find another.) Which of the above pairs of sets are equivalent? Is A equivalent to B? Are there pairs that are neither equal nor equivalent?

From the above discussion it turns out that $A \sim B \sim C$. Are there other sets equivalent to A? Certainly: some of these are $E = \{1,2\}$, $F = \{57,100\}$, the set consisting of a person's hands, the set of wheels on a bicycle, and the set consisting of a pair of dancers at a ball. (Of course, in mentioning the sets of this paragraph we have greatly enlarged the universe of discourse; it is no longer simply the letters of the alphabet.) It is easy to name or describe still other sets which are equivalent to A. The totality of all such sets forms what may be called a *collection*, or *class*,[7] of equivalent sets. Note that some of the sets in this class contain abstract elements, others physical objects, and still others persons; but they all have one common characteristic: they are equivalent to A and to each other, or, less formally, they "have the same number of elements." It is advantageous to give this class a name. In the English language the name we assign to this class is *two;* in the Hindu-Arabic numeration system it is *2*. Similarly, the set $D = \{a,b,c\}$ and the set $G = \{1,2,3\}$ belong to a different equivalence class, which we name *three* (*3*). Another way of expressing the content of this sentence is: "The *cardinal number* of set D is 3"; in symbols, $n(D) = 3$. More formally, we have

Definition 2.9 Cardinal Number The class containing a set A and all sets equivalent to A is called the *cardinal number of A* (in symbols, $n(A)$).

We shall assume that you have learned the English words and the Hindu-Arabic symbols for the finite cardinal numbers. As further examples, $n(C) = n(F) = 2$; $n(G) = 3$; $n(\{a,b,c,d\}) = 4$; $n(M) = 12$, where M is the set of months in a civil year; and, we note especially, $n(\varnothing) = 0$.

The concept of cardinal numbers seems so simple and familiar that there is danger of overlooking its significance. Yet, as mentioned before, it took the human race millenia, and it takes the kindergartener a year or more, to grasp, even in concrete terms, the idea that sets as different as a pair of marks on a wall and the President and Vice-President of the United States are equivalent sets and so have the same cardinal number, 2. However, there is one sequence of abstract sets whose cardinal numbers are very easy to determine. Note that $n(E) = n(\{1,2\}) = 2$; $n(\{1,2,3\}) = 3$; $n(\{1,2,3,4\}) = 4$; . . . . Evidently the cardinal number of a set of whole numbers, *in the usual order and beginning with 1*, is the same as the last element of that set: $n(\{1,2,3, \ldots ,k\}) = k$. We use this property when we *count* a set of objects; for counting is simply a matter of setting up a one-to-one correspondence between a particular finite subset of the whole numbers (starting with 1) and the set being counted, in such a way that the last-named element of the subset of W is the cardinal number of the given set.

The sets we have discussed so far have been *finite* sets, and their cardinal numbers have been whole numbers; but what about *infinite* sets? Are there

---

[7]It is perfectly proper to say, "*set* of sets"; but for variety and clarity we prefer a synonym.

classes consisting of infinite sets which are equivalent to each other? If so, what are the cardinal numbers of these sets? It is best to answer the second question first, for a special set, the set of whole numbers. On the assumption that the set W is one member of such a class of sets, we shall assign it a cardinal number. But since each of the whole numbers is a finite number, none is large enough to serve as a label for W itself. George Cantor[8] (1845–1918), sometimes called "the father of sets," met this problem head-on by inventing a symbol for the cardinal number of W. He chose *aleph* ($\aleph$), the first letter of the Hebrew alphabet, and attached to it a subscript, zero. Thus Cantor is responsible for the following:

Definition 2.10 Aleph-Null The cardinal number of the set W is *aleph-null* (in symbols, $\aleph_0$).

Having defined $n(W) = \aleph_0$, we return to the question of whether there are other infinite sets having the same cardinal number. The results are surprising, and to many people, even disconcerting. Another infinite set is E, the even numbers. By the same argument as for W, $n(E)$ cannot be a whole number. Could it be $\aleph_0$? Let us list the elements of W and those of E in the following array:

$$W = \{0,1,2,3,4,\ 5,\ldots,\ k,\ldots\}$$

$$E = \{0,2,4,6,8,10,\ldots,2k,\ldots\}$$

Note that below each element of W appears the element of E that is its double. To put the matter another way, corresponding to $0 \in W$ is $0 \in E$, and to $1 \in W$ is $2 \in E$ (more briefly, $0 \leftrightarrow 0$, $1 \leftrightarrow 2$, $2 \leftrightarrow 4$, $3 \leftrightarrow 6, \ldots, k \leftrightarrow 2k, \ldots$). Thus we have established a one-to-one correspondence between W and E. But then W and E are equivalent and have the same cardinal number, so that $n(W) = n(E) = \aleph_0$. Nevertheless, E is a *proper* subset of W; indeed, an infinite number of elements of W are missing from E. This seemingly bizarre situation in which two sets, one a proper subset of the other, have the same cardinal number never arises with finite sets.

In high school geometry you may have met or used Euclid's axiom that "the whole is greater than any of its parts."[9] Considering that E is a *proper* subset of W—and therefore a "part" of W—the fact that $n(E) = n(W)$ suggests that Euclid's axiom may be false for infinite sets. The example of E and W, moreover, is not a freak happening. To emphasize this point, we give a

---

[8]For the significant, but tragic life of this pioneer in mathematics, read Chapter 29, "Paradise Lost," in the Bell reference, whose remaining 28 chapters have just as intriguing titles and contents.

[9]Some texts use a similar algebraic statement, such as: "if $a = b + c$ and $c > 0$, then $a > b$."

second example; and others appear in the exercises or can be readily constructed. Let S be the set of *squares* of whole numbers. Then we have

$$W = \{0,1,2,3, 4, \ldots, k, \ldots\}$$
$$S = \{0,1,4,9,16, \ldots, k^2, \ldots\}$$

This array suggests the one-to-one correspondence $k \leftrightarrow k^2$, which shows that $S \sim W$, so that $n(S) = n(W) = \aleph_0$. But it is also clear that $S \subset W$. Again we face the disconcerting situation that, though relatively few of the whole numbers are perfect squares, the set of perfect squares is nevertheless the "same size" in terms of its cardinal number as the set of all the whole numbers! So also, it turns out, is the set of whole numbers which are *not* perfect squares, although it is more difficult mechanically to set up a confirming one-to-one correspondence.

Let us explore the contrast between finite and infinite sets a bit further. Intuitively we picture a finite set as "coming to an end": if we attempt to list all of its elements, we should eventually — at least theoretically — reach the last one. In fact, some writers define a finite set as one that is equivalent to the set $\{1,2,3, \ldots, k\}$ of the first k nonzero whole numbers, an idea that was illustrated following Definition 2.9. An infinite set, on the other hand, is so large as to be inexhaustible: we can never reach its "last" element, for it has none. For our later work, however, we shall need precise definitions of both types of sets. In view of the discussion of the preceding paragraphs, it is convenient to state the following:

Definition 2.11 Infinite Set A set is *infinite* iff it is equivalent to some proper subset of itself.

This definition suggests that the bizarre failure of Euclid's axiom concerning the whole and its parts is a reliable characteristic by which to distinguish infinite from finite sets. Indeed it is, though we shall not prove this. Immediately we have the associated

Definition 2.12 Finite Set A set is *finite* iff it is not infinite.

To round out the discussion, you should convince yourself intuitively that any set equivalent to the set $\{1,2,3, \ldots, k\}$, where k is a nonzero whole number, *cannot* be matched with any proper subset of itself, and is therefore finite. The only finite set omitted from consideration is the null set, $\varnothing$, which cannot be matched one-to-one even with the set $\{1\}$, for which $k = 1$. Yet $\varnothing$ certainly satisfies Definition 2.12, as you should check. In addition, as a set, $\varnothing$ must have a cardinal number: the obvious choice is $n(\varnothing) = 0$, which we shall announce formally in Chapter 4.

In summary, we begin with the basic idea of one-to-one correspondence to define equivalent sets. Then we introduce the concept of classes of equivalent sets and define these classes as cardinal numbers. With each set we thus associate a unique cardinal number. The whole numbers serve as cardinal numbers for all finite sets, including the null set. The cardinal numbers of infinite sets, which cannot be whole numbers, are called *transfinite* cardinals. The only such number mentioned so far is aleph-null, the cardinal number for the set of whole numbers. This is the smallest transfinite cardinal. There are an infinite number of transfinites larger than aleph-null.[10]

Though we have emphasized equivalent sets, we close this section with a brief reference to sets that are not equivalent. As an example, if $A = \{a,b\}$ and $H = \{x,y,z\}$, clearly $A \not\sim H$; moreover, A is equivalent to a proper subset of H, say $H_1 = \{x,y\}$. Since sets are equivalent iff they have the same cardinal number (by Definition 2.9), we know at once that $n(A) = n(H_1)$, but $n(A) \neq n(H)$. In fact, we should like to state more precisely that the cardinal number of A *is less than* that of H: $n(A) < n(H)$. Thus we make

Definition 2.13 Unequal Cardinal Numbers Let A and B be sets, *with A not equivalent to B*. Then the cardinal number of A *is less than* the cardinal number of B (in symbols, $n(A) < n(B)$) iff A is equivalent to a proper subset of B. Moreover, the cardinal number of B *is greater than* that of A (in symbols, $n(B) > n(A)$) iff $n(A) < n(B)$.

You should note that the language of Definition 2.13 does not restrict it to finite sets and should therefore try it out by comparing two finite sets, and also by comparing a finite with an infinite set. You should ponder, too, why the phrase "*with A not equivalent to B*" of Definition 2.13 is essential. Exercises explore these ideas further.

EXERCISES 2.7

1. Let $U = \{a,b,c,d,e,f,g,h,i,j\}$ and define subsets as follow:

$A = \{a,d,c,f,g\}$ $\quad B = \{b\}$ $\quad C = \{f\}$

$D = \{g,c,e,a\}$ $\quad E = \{g,h,a,i,j\}$ $\quad F = \{f,a,c,e\}$

$G = \{c,a,g,e\}$ $\quad H = \{j,g,c,h,d,b,a,e,i,f\}$

List all the pairs of equivalent sets. Are any of these pairs equal sets?

2. Which, if either, of the following statements is (are) true? In each case, justify your answer.
   (a) If two sets are equal, then they are equivalent.
   (b) If two sets are equivalent, then they are equal.

---

[10]See the Gardner reference and the third Suggestion for Investigation at the end of this chapter.

3. Let U = {Mercury, Venus, Earth, Mars, Jupiter, Saturn, Uranus, Neptune, Pluto}

and L = {Mars, Jupiter, Saturn, Uranus, Neptune, Pluto}
S = {Mercury, Venus, Earth}
M = {Earth, Mars, Jupiter, Saturn, Uranus, Neptune}
V = {Earth, Mars, Venus, Jupiter, Saturn}
X = {Venus, Mars}

Determine the cardinal number of each of these sets by making a one-to-one correspondence between each of them and an appropriate subset of W.

4. (a) At a certain concert, an auditorium containing 1500 seats had every seat occupied by one person, and everyone in the audience had a seat. Describe a one-to-one correspondence between two sets (specifying the sets) which enables you to determine the exact number of people in the audience without counting them. In your explanation, use correctly the word *equivalent*.

   (b) To locate a book in a library, use is made of the card catalog. Is this an example of a one-to-one correspondence between two sets? Explain your answer.

   (c) Make up a concrete example of your own of a one-to-one correspondence between two sets. Be sure to specify the two sets and describe the correspondence.

5. (a) Give examples of sets of physical objects whose cardinal numbers are 4, 2, 1, 5, 10, and 12, respectively. Use examples different from those of the text, trying to select ones that would be meaningful to an elementary school child.

   (b) What is the cardinal number of the set of all persons whom you have met who are over 12 feet tall?

   (c) Make up an example of your own of a set that is equivalent to the one of part (b). Are the two sets equal?

   (d) Explain why $\varnothing$ fails to satisfy Definition 2.11 for infinite sets, and so, by Definition 2.12, is a finite set.

   (e) Let C be the set of cubes of whole numbers, namely,

$$C = \{0,1,8,27, \ldots ,k^3, \ldots\}$$

   Using an appropriate one-to-one correspondence between W and C, show that $n(C) = \aleph_0$.

   (f) Using an appropriate one-to-one correspondence, show that $n(D) = \aleph_0$, where D is the set of odd numbers.

(g) Find $n(F)$, where $F = \{0,5,10,15, \ldots ,5k, \ldots\}$, the set of multiples of 5.

(h) How many different sets are there which are equivalent to W? Give reasons for your answer.

6. (a) Which, if any, of the sets in Problem 3 have the same cardinal number?

(b) Referring again to Problem 3, illustrate Definition 2.13 by comparing the sets (i) V and L (ii) V and S.

(c) Is Definition 2.13 satisfied by the following pairs of sets? In each case, explain your answer.

(i) T and W, where $T = \{0,1,2, \ldots ,999\}$, the first thousand whole numbers, and W is the set of whole numbers.

(ii) E and W, where E is the set of even numbers.

(d) (i) Explain why the phrase "*with A not equivalent to B*" in Definition 2.13 is essential if the definition is to apply to infinite sets.

(ii) Could this phrase be omitted if Definition 2.13 were restricted to finite sets? Explain.

*7. (a) Let $A = \{a,b\}$ and $C = \{c,d\}$, as in the text. We showed that $A \sim C$ by exhibiting the one-to-one correspondence: $a \leftrightarrow c$, $b \leftrightarrow d$. Supply a second one-to-one correspondence between A and C. Can you find any more one-to-one correspondences?

(b) In part (a), $n(A) = n(C) = 2$, and it is possible to specify two different one-to-one correspondences between A and C. Suppose X and Y are sets such that $n(X) = n(Y) = 3$. For example, let $X = \{a,b,c\}$ and $Y = \{d,e,f\}$; then surely $X \sim Y$, for *a specific* one-to-one correspondence is the following:

$$a \leftrightarrow d$$
$$b \leftrightarrow e$$
$$c \leftrightarrow f$$

Several other one-to-one correspondences are possible, such as

$$\begin{array}{ccc} a \leftrightarrow d & & a \leftrightarrow e \\ b \leftrightarrow f & \text{and} & b \leftrightarrow f \\ c \leftrightarrow e & & c \leftrightarrow d \end{array}$$

List the remaining one-to-one correspondences. How many are there in all between two sets whose cardinal number is 3?

(c) Determine the number of one-to-one correspondences between two sets whose cardinal number is 4.

(d) Develop a general formula giving the number of one-to-one correspondences between two sets whose cardinal number is k.

(e) Consider the dilemma of an instructor faced on the first day of classes with a group of 20 students, none of whom he knows. Each face in front of him matches a name on his enrollment list for the course. Use the formula of part (d) to determine how many possible one-to-one correspondences there are between two sets of 20 elements each. *Only one* of these possible correspondences, of course, is a completely correct matching of names and faces. The instructor has to learn this one and eliminate all the others if he or she wants to claim to know all the students. A class of 20 is small. How much greater is an instructor's task with a class of 40? of 200? (If you are really serious about calculating the answers to these questions, you will need special tables, logarithms, or other computing devices!)

Section 2.8 Cartesian Product: Ordered Pairs

A road map or atlas has a device for locating any town or city printed on it. Key letters ($A,B,C,\ldots$) are uniformly spaced down the margins and key numbers ($1,2,3,\ldots$) across the top. To locate a place on the map, one first consults an index to find the key letter and number of the place: for example, on a road map of New York State, the index gives Oswego the symbol pair $E8$. This means that Oswego is within the region opposite the letter E in the margin and below the number 8 at the top. If the reference system has 12 letters and 15 numbers, any pair of symbols in the form letter-number identifies one of 180 regions on the map. Note that the first member of this pair is an element of the set $Q = \{A,B,C,\ldots,L\}$ and the second member is an element of the set $R = \{1,2,3,\ldots,15\}$. Thus the symbol pair $E8$, for example, constitutes an *ordered pair* of elements from the two given sets. We call $E8$ an *ordered* pair because the element from Q is always placed first, followed by the element from R: for example, the map consistently uses $E8$ rather than $8E$.

We note that the ordered pairs themselves form a set, namely, $\{(A,1),(A,2),(A,3),\ldots,(A,15),(B,1),\ldots,(C,1),\ldots,(L,15)\}$. The elements of this set, however, are not simple elements like those of Q or R, but are *compound* elements of the form (q,r), with $q \in Q$ and $r \in R$. From two sets of simple elements, a new set, having ordered pairs as its elements, has been created. This new set is called the *cartesian product* of the sets Q and R and is denoted by the symbol $Q \times R$ (read, "Q cross R"). This yields the following:

Definition 2.14 Cartesian Product If A and B are sets, the *cartesian product* of A and B (in symbols, $A \times B$) is the set of ordered pairs of elements formed by taking the first element of the pair from A and the second element from B. Symbolically:

$$A \times B = \{(x,y) \mid x \in A \text{ and } y \in B\}$$

Let us take another example, this time with small sets, so that all the elements of the cartesian product can be conveniently listed. A certain mail-order catalog pictures mixed ensembles for misses: a blouse (b) or a pullover sweater (p) can be combined with a culotte (c), a skirt (s), or walking shorts (w). Symbolically expressed, two sets are depicted: $A = \{b,p\}$ and $B = \{c,s,w\}$. The available combinations are given by the cartesian product:

$$A \times B = \{(b,c),(b,s),(b,w),(p,c),(p,s),(p,w)\}$$

You should note the relations among $n(A)$, $n(B)$, and $n(A \times B)$. Also, it is possible to form the set $B \times A$; but it is considered to be a different set from $A \times B$ since the simple elements occur in a different order, e.g., $(b,c) \in A \times B$, but $(c,b) \in B \times A$.

A third example shows the pervasiveness of the concept of cartesian product. You have doubtless learned to locate on a two-dimensional graph points described by pairs of numbers (*coordinates*), such as (1,1), (1,2), (1,3), and (2,1). Figure 2.14 shows the location of these four points. These number pairs can be obtained as elements of a cartesian product of two sets. Let $A = \{1,2\}$ and $B = \{1,2,3\}$. Then $A \times B = \{(1,1),(1,2),(1,3),(2,1),(2,2),(2,3)\}$, a set which includes the points depicted in Figure 2.14. Does $A \times B = B \times A$? Clearly not, because $(1,3) \in A \times B$ but $(1,3) \notin B \times A$. Of course, $(3,1) \in B \times A$; but (3,1) is a different point from (1,3).

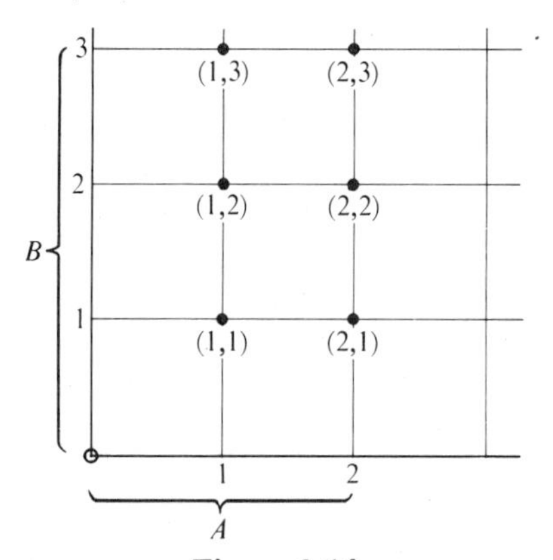

Figure 2.14

Since all our examples have started with two *different* sets, it is worth noting that Definition 2.14 also applies easily to equal sets. Thus, if $A = \{1,2\}$, we have

$$A \times A = \{1,2\} \times \{1,2\} = \{(1,1),(1,2),(2,1),(2,2)\}$$

What can be said about the cardinal number of the cartesian product of any set with itself?

Let us now use Definition 2.14 and the preceding examples to discover the properties of the binary operation of cartesian product. We have just seen that it is not commutative. Is it closed? We have to show first that whenever two sets A and B in U are combined by the operation of cartesian product, a unique set, $A \times B$, results. The only difficulty would seem to be with the null set. If in the mail-order illustration $A = \{b,p\}$ but $B = \varnothing$, of which pairs would $A \times B$ consist? One could not form any ordered pairs $A \times B$, since the second elements necessary to the pairs would be lacking. (In practical terms, if the catalog showed only blouses and pullover sweaters, a complete outfit could not be selected from it.) But a set containing no elements — whether no simple elements, no ordered pairs, or no dodos — is necessarily the null set. Thus $A \times \varnothing = \{\ \} = \varnothing$.

The second requirement of closure of an operation is that the result yielded by the operation on the original sets from U also be contained in U. But U is a collection of sets with simple elements, whereas $A \times B$ is a set of ordered pairs, and so is not in U. Hence, though $A \times B$ is a unique set, it is not the right kind of set; so we cannot call cartesian product a closed operation in U.

Because it is necessary later to use the distributive property of cartesian product over union, we shall illustrate it here. Let $L = \{1,2\}$, $Y = \{a,b,c\}$, and $Z = \{c,d\}$. Then

$$\begin{aligned} L \times (Y \cup Z) &= \{1,2\} \times (\{a,b,c\} \cup \{c,d\}) \\ &= \{1,2\} \times \{a,b,c,d\} \\ &= \{(1,a),(1,b),(1,c),(1,d),(2,a),(2,b),(2,c),(2,d)\} \end{aligned}$$

If the operation $\times$ is to distribute over $\cup$, we must have

$$L \times (Y \cup Z) = (L \times Y) \cup (L \times Z)$$

Let us build up the right side of our example:

$$L \times Y = \{1,2\} \times \{a,b,c\} = \{(1,a),(1,b),(1,c),(2,a),(2,b),(2,c)\}$$

and

$$L \times Z = \{1,2\} \times \{c,d\} = \{(1,c),(1,d),(2,c),(2,d)\}$$

In forming $(L \times Y) \cup (L \times Z)$ we write down the common elements $(1,c)$ and $(2,c)$ only once. Thus we obtain for the set $(L \times Y) \cup (L \times Z)$ exactly the same eight elements as we found for $L \times (Y \cup Z)$. More generally, we have the following:

THEOREM If A, B, C are any three sets of U, then

$$A \times (B \cup C) = (A \times B) \cup (A \times C)$$

The proof of this theorem is not given here, but is left as an optional exercise, since it requires an understanding of *Section 2.6.

EXERCISES 2.8

1. In a certain club (set U) of men and women, three men (set M) are eligible to be president and two women (set F) are eligible to be secretary. Let $M = \{\text{Albert, Bertrand, Charles}\}$ and $F = \{\text{Dorothy, Eve}\}$. List all possible slates of nominees — naming the presidential candidate *first* — that the club's nominating committee could present to the club business meeting. Which cartesian product do the slates illustrate?

2. Suppose that the sex discrimination implied in Problem 1 is eliminated by making all five persons eligible for both offices. Again list all possible slates, assuming that no person can be a candidate for both offices simultaneously. Does this example illustrate a cartesian product? Explain.

3. Let $N = \{1,3,5\}$ be a set of numerators and $D = \{2,7,11\}$ be a set of denominators. From N and D certain fractions can be formed (e.g., 1/2). List all such fractions. How, if at all, does this set differ from $N \times D$? How could the set $D \times N$ be interpreted?

4. Let $E = \{a,b\}$, $F = \{d,e,f\}$, $G = \{c,d,e\}$.
   (a) Show that $E \times (F \cup G) = (E \times F) \cup (E \times G)$.
   (b) According to the equation of part (a), which operation is distributed over which?
   (c) Is cartesian product distributed over intersection? Write the appropriate equation and then illustrate it with sets E, F, and G.

5. (a) In Problem 3, find $n(N)$, $n(D)$, and $n(N \times D)$.
   (b) In Problem 4, find $n(E)$, $n(F)$, and $n(E \times F)$.
   (c) Generalize the results of parts (a) and (b) so as to determine, if possible, a relation between the cardinal numbers of two nonempty, finite sets and that of their cartesian product.
   (d) Does the relation of part (c) continue to hold if one of the sets is the null set? Explain.
   (e) Is either of the following statements true? Explain.
   (i) $A \times B = B \times A$ (ii) $A \times B \sim B \times A$

*6. (a) Is cartesian product an associative operation? If it is, explain your answer; if it is not, provide a counterexample.
   (b) Does union distribute over cartesian product? Explain or justify.

*7. By the method of Section 2.6, prove formally that cartesian product distributes over union.

Suggestions for Investigation

1. For an amusing, though nonmathematical recreation, see the Lipton reference, which gives both historical and contemporary synonyms for *set* and encourages the coinage of more such terms.

2. Several other operations can be defined on sets. One is *subtraction:* $A - B = A \cap B'$. Another is the *symmetric difference:* $A \Delta B = (A - B) \cup (B - A)$. Explore these in the text and/or exercises of an elementary book or article on set theory, such as those by Christian, S. Smith, or Stoll, listed in the chapter bibliography.

3. Investigate further the concept of infinity. A superficially amusing, as well as a rewarding and mathematically honest way of doing this is to read the first few chapters of the Lieber reference, in which *potential* and *actual* infinities are discussed. It is more enlightening to continue into *hierarchy of infinities*. However, if you do not stop your reading then, you will be into calculus before you know it!

4. The theory of sets has a simple application to voting coalitions in a legislature or in a deliberative body such as the Security Council of the United Nations. Study the Kemeny (pp. 79–83) or Luce (pp. 37–45) references, which also contain problems.

5. Let U be the universal set and A and B be two subsets of U. A Venn diagram representing relationships among these three sets is given in Figure 2.15. Note that the circles representing sets A and B partition the interior of the rectangle for U into four distinct, nonoverlapping regions, numbered 1–4. Note further that the regions represent, in numerical order, subsets of U containing no elements of set A or set B, containing elements of set A only, containing elements of set B only, and containing elements in $A \cap B$. Thus these *four* regions together depict *all* the possibilities for an element x of U with respect to the sets A and B: x is in neither set, in A only, in B only, or in both.

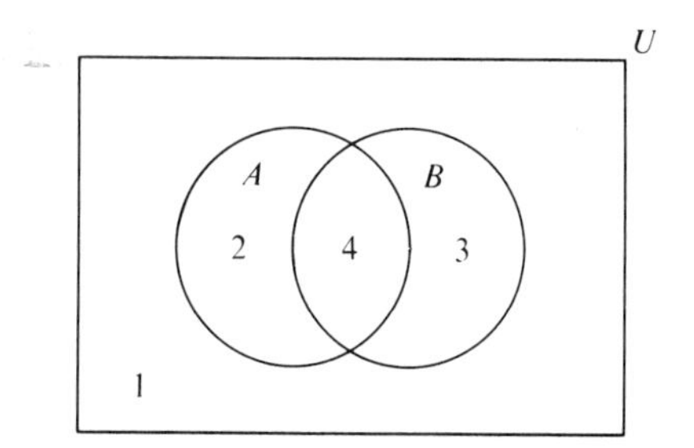

Figure 2.15

Now extend the above ideas to *three* subsets, A, B, and C of U, as depicted in Figure 2.16. Can you represent the three subsets by circles so positioned that they partition U into enough distinct, nonoverlapping regions so as to depict *all* the possibilities for an element x of U with respect to the three sets A, B, and C? To do this you would need a total of *eight* regions, representing, respectively, x in none of the sets, in A only, in B only, in C only, in A and B only, in A and C only, in B and C only, and in all three sets. In the above diagram, number in order from 1 to 8 the regions described in the preceding sentence.

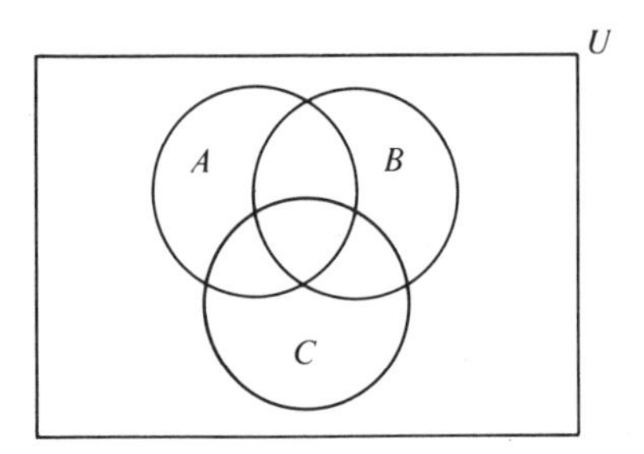

Figure 2.16

Now extend the above ideas to *four* subsets, A, B, C, and D of U. Ask again the equivalent question: this time, how many regions would be required to represent all the possibilities for an element x of U? List each of the possibilities. Then attempt to position four *circles* within the rectangle representing U in such a way as to obtain the required number of distinct, nonoverlapping regions, labeling them in order from 1 on, according to your list. You will discover that it is geometrically impossible to represent all these regions! What is the best you can do? Finally, experiment with representing the four sets by *ellipses* and show how this can be done successfully. After trying this, consult the Ranucci article, which shows that other closed figures may also be used and which extends the inquiry to *five* subsets.

Bibliography

Bell, E. T. *Men of Mathematics*. New York: Simon and Schuster, 1937.

Christian, R. R. *Introduction to Logic and Sets*, 2nd ed. Waltham, Mass.: Blaisdell Publishing Co., 1965.

Cohen, P. J. and R. Hersh. "Non-Cantorian Set Theory." *Scientific American*, 217 (Dec. 1967), 104–106 and 111–116. (This article has been reprinted in *Mathematics in the Modern World*, 212–20. San Francisco: W. H. Freeman and Co., 1968.)

Committee on the Undergraduate Program in Mathematics (CUPM). *Elementary Mathematics of Sets*. Washington: Mathematical Association of America, 1958.

Dinkines, Flora. *Elementary Theory of Sets*. New York: Appleton-Century-Crofts, 1964.

Dubisch, Roy. "Set Equality." *Arithmetic Teacher*, 13 (1966), 388–91.

Gardner, Martin. "The Orders of Infinity, the Topological Nature of Dimension and 'Supertasks,'" *Scientific American*, 224:3 (Mar. 1971), 106–108.

Geddes, Dorothy and S. I. Lipsey. "Sets: Natural, Necessary, Knowable?" *Arithmetic Teacher*, 15 (1968), 337–40.

———. "The Hazards of Sets." *Mathematics Teacher*, LXII (1969), 454–59.

Hamilton, N. T. and J. Landin. *Set Theory: The Structure of Arithmetic*, 2nd ed. Boston: Allyn and Bacon, Inc., 1961.

Kemeny, J. G., et al. *Introduction to Finite Mathematics*, 2nd ed. Englewood Cliffs, N.J.: Prentice-Hall, Inc., 1966.

Lieber, L. R. *Infinity*. New York: Rinehart & Co., Inc., 1953.

Lipschutz, Seymour. *Set Theory*. New York: Schaum Publishing Co., 1964.

Lipton, James. *An Exaltation of Larks, or the Venereal Game*. New York: Grossman Publishers, 1968.

Luce, R. D. *Some Basic Mathematical Concepts*, SMSG *Studies in Mathematics*, I. New Haven: School Mathematics Study Group, 1959.

McFadden, Myra. *Sets, Relations and Functions: Programmed Units in Modern Mathematics*. New York: McGraw-Hill Book Co., 1963.

Meder, A. E. "Sets, Sinners and Salvation." *Mathematics Teacher*, LIX (1966), 358–63.

Quine, W. V. "Paradox." *Scientific American*, 206:4 (Apr. 1962), 84–96. (This article has been reprinted in *Mathematics in the Modern World*, 200–208. San Francisco: W. H. Freeman and Co., 1968).

Ranucci, E. R. "Spatial Aspects of the Venn Diagram." *New York State Mathematics Teachers' Journal*, 18:2 (1968), 64–67.

Sanders, W. J. "Cardinal numbers and sets." *Arithmetic Teacher*, 13 (1966), 26–29.

Smith, C. W. "Intersection of Solution Sets." *Arithmetic Teacher*, 14 (1967), 504–506.

Smith, Sigmund. "Some Notes on Symmetric Difference." *New York State Mathematics Teachers' Journal*, 19:3 (1969), 111–13.

Stoll, R. R. *Sets, Logic and Axiomatic Theories*. San Francisco: W. H. Freeman and Co., 1961.

Summers, G. J. *New Puzzles in Logical Deduction*. New York: Dover Publications, Inc., 1968.

Suppes, Patrick. *Axiomatic Set Theory*. New York: D. Van Nostrand Co., Inc., 1960.

3

Logic

What is correct reasoning and how can it be identified? What are truth tables? What is the distinction between inductive and deductive reasoning? How are models used in science?

Section 3.1 What Is a Proposition?

Though the vocabulary of mathematics contains some terms which occur only in that field, mathematics also has a propensity for borrowing words from ordinary speech and giving them precise technical meanings. Examples are the words *set, intersection, union, group, field, ring, chain, open, continuous, limit, maximum, origin, tree* — to name a very few. One mathematician remarked facetiously some time ago that only the word *committee* has been left undisturbed by his colleagues; but by now, perhaps even this term has been pressed into service!

In mathematics, however, words are used much more carefully than they are in ordinary speech or in literature. Each word or symbol is supposed to mean exactly *one* thing: there should be no room for ambiguity. In poetry, by contrast, exactness of expression is not the primary goal: the poet uses a variety of words, whose meanings shade into one another, and whose connotations and ability to evoke images are of paramount importance. But poets and mathematicians can still understand and appreciate one another, if they make the effort.[1]

In the area of logic, one of the fundamental concepts is *proposition*, or *statement*. Its meaning is akin to the ordinary one for *statement*, particularly as used in courts of law, where the jury must decide whether the assertions of the various witnesses are true or false. But, as in the case of the term *set*, the concept of a proposition, or statement, is so basic that it is difficult to define: instead we take it and its two modifiers, *true* and *false*, as undefined terms. We can, however, give an intuitive description: a *proposition* is a sentence which has a definite *truth value*; it is either true (T) or false (F), but not both. Thus the proposition "Sacramento is the capital of California" is true; "Columbus discovered America in 1942" is false. However, in our study of classical logic we do not allow gradations of partial truth, or truth values dependent on vague circumstances.

Not every verbal or mathematical expression is a proposition. A few examples and counterexamples will clarify this point. The first of the following expressions, not being a complete sentence, is certainly not a proposition. Which of the others are, and which are not?

1. The daring young man on the flying trapeze.
2. What is your name?
3. What a storm we had last Friday!
4. Please fill out this form in triplicate.
5. The year 1968 had 366 days.
6. $3 + 5 = 9$.
7. If today is Thursday, then tomorrow is Friday.
8. The fourteenth state to become part of the United States was ______.

Of these expressions, only 5, 6, and 7 have truth values, for they are declarative sentences. Sentences 2, 3, and 4 are, in order, interrogative, exclamatory, and imperative; so they cannot be propositions. Sentences 5 and 6 are *simple* propositions, since they each contain only one subject and one predicate. Sentence 7 is a *compound* proposition, since it is composed of two simple propositions joined by the connectives, "if . . . then. . . ." Sentence 8, though it is in form a simple declarative sentence, is not a proposition, since its truth value depends on how the blank is filled. It would, however, *become* a true proposition if the blank were replaced by the name *Vermont*; the substitution of any other state's name would make it a false proposition.

We now restrict our attention to simple propositions, reserving discussion of

---

[1]See the D. E. Smith reference in the chapter bibliography.

compound ones for later sections. Among the various kinds of simple propositions, four special types occur frequently. These are characterized in English by use of the words *all*, *some*, or *no*. Examples of the four types are:

1. All cows eat grass.
2. Some hyenas do not eat grass.
3. Some 18-year-olds are eligible to vote.
4. No convicted criminals are eligible to vote.

The meanings of the first and last propositions are clear; but the middle two are likely to be misinterpreted, since *some* has a special meaning in logic. In common speech, *some* is frequently opposed to *all:* "Did you bring all your books?" "No, I left some at home." In logic, however, *some* means precisely "at least one." Thus Proposition 3 asserts that there is *at least one* 18-year-old who is eligible to vote. Beyond that, it makes no claim; it neither denies nor affirms the possibility that *all* 18-year-olds are eligible to vote. This means that *some* in logic is not opposed to *all;* on the contrary, *some* allows the possibility of *all*. Similarly, Proposition 2 merely asserts that there is at least one hyena which does not eat grass; it gives no information as to how many more do not eat grass.

In addition to the above types, a variant occurs. We discuss it only to reject it, for it is ambiguous and so has no place in mathematics. An example is, "All that glitters is not gold." What does this proverb mean? From the context or past association, it seems to mean, "Not everything that glitters is gold," or "Some things that glitter are not gold." The latter sentence is really of the form of Proposition 2, and its meaning is clear. But consider now another proposition that has the same form as "All that glitters is not gold," namely, "All numbers that are even are not odd." Does this mean merely that "Some even numbers are not odd"? No; from our previous knowledge we would probably interpret it as a stronger statement: "No even numbers are odd," which is of the form of Proposition 4. Thus apparently the type "All x is not y" should sometimes be translated into the form of Proposition 2, and sometimes into that of 4, depending on the context; but to avoid ambiguity, it should *not* be left unchanged. In conversation, to be sure, we may attempt to distinguish between these two interpretations by our tone of voice and emphasis. But without external knowledge of some kind it is impossible to choose between the two interpretations. Which, for example, is the correct interpretation of "All Arachnida are not Orthoptera"? Even biology majors might have to check that one!

There are, of course, many simple propositions which cannot be classified under one of the four special types. The following are examples:

1. The year 1968 had 366 days.
2. $3 + 5 = 9$.

3. Lincoln's birthday falls in January.
4. April 8, 1917, was Easter Sunday.
5. John was charged with speeding.

None of these begins with *all*, *some*, or *no*, yet they are declarative sentences with definite truth values, except possibly the fifth. The first is a true proposition. The second and third are false propositions. The fourth is certainly a proposition, though it may be necessary to consult an old calendar or do some figuring to determine its truth value: the important point is that an unequivocal determination can be made. The fifth example requires more discussion. If the name "John" refers to a specific individual, then its truth value can be established, and it is a proposition. If, on the other hand, "John" is used in a general way to denote an unspecified person (as "John Doe" is used in legal documents), the truth value of the sentence cannot be established, so it is not a proposition. In this case, we would do well to replace "John" by "______."

In view of the preceding paragraph, it is not safe to conclude that every declarative sentence is a proposition. Such sentences may fail to be propositions in two ways: their truth value may be indeterminate (e.g., when blanks occur), or they may be self-contradictory. Further instances of the first kind are "He kissed her" and "$x + 5 = 7$." We cannot assign a truth value to the first sentence unless we know to whom the pronouns refer. Similarly, the truth value of the second (algebraic) sentence depends on the meaning of the symbol x. Indeed, to "solve the equation $x + 5 = 7$" means to determine all the numbers which, when substituted for x, make the resulting sentence a *true* proposition. In both these cases, if the pronoun or letter or blank is replaced by an appropriate specific element, the resulting sentence is a proposition, true or false, depending on the replacement. For this reason, these sentences (before replacement by specific elements) are called *propositional functions* or *open sentences*. The latter term is especially used in a number of recent high school algebra texts.

The self-contradictory type of sentence is bizarre and significant for the foundations of logic, but does not occur in normal applications of logical reasoning. This type fails to be a proposition because it is simultaneously true and false. An example is, "This sentence is false." A moment's concentrated thought will reveal that, if the quoted expression is true, then it must be false, and vice versa. This kind of contradiction is related to certain paradoxes in set theory.[2]

EXERCISES 3.1

1. Determine which of the following expressions are simple propositions, which are propositional functions, and which are neither. For those that

---

[2]See Exercises 2.1, Problems *7 and *8.

are propositions, state whether they are true (T) or false (F). For the others, explain why they are not propositions.

(a) New York City is north of Philadelphia.

(b) Where have all the flowers gone?

(c) Augustus Caesar was born in 1000 B.C.

(d) All tricycles have three wheels.

(e) When in the course of human events it becomes necessary.

(f) The capital city of Massachusetts is ______.

(g) $2 + 9 = 29$

(h) Some giraffes have long necks.

(i) $x + 3 = 8$

(j) Some rectangles are not squares.

(k) Swing low, sweet chariot!

(l) No mushrooms are edible.

(m) In the U. S. census of 1970, Nevada had a greater population than Wyoming.

(n) I am lying.

2. Explain why each of the following sentences is logically ambiguous; then rewrite each in a way which unambiguously expresses your interpretation of it:

   (a) All cows are not purple.

   (b) Every day in August is not sultry.

   (c) A grade of "incomplete" means that all requirements of the course have not been met.[3]

3. Make up your own examples of:

   (a) a true proposition of each of the four special types.

   (b) a false proposition of each of the four special types.

*4. An advertisement for a logic game contained the following puzzle. Solve it, explaining your reasoning, using *only* the following information:

1. There are three numbered statements in this box.
2. Two of these numbered statements are not true.
3. This puzzle is a good exercise in logic.

Question: Is Statement #3 true?[4]

---

[3]Proposed faculty legislation at a certain college!

[4]Used by permission of WFF 'N PROOF Publishers.

Section 3.2 Negation: A Unary Operation on Propositions

Consider the simple proposition "July 20, 1969, was a Monday." (It was about 10 P.M. (EDST) on this date that Neil Armstrong became the first man to walk on the moon.) This sentence is a proposition since it has a truth value which is easily determined by consulting a calendar. We can also write a related proposition: "July 20, 1969, was *not* a Monday." The second proposition is called the *negation* of the first. Clearly, its truth value is opposite to that of the original one. As a matter of fact, the original is false, since July 20, 1969, was a Sunday; correspondingly, the negation is true.

Another example of a negation is the following. From the original proposition, "$3 + 4 = 7$," we can easily derive its negation. This can be expressed as "It is not the case that $3 + 4 = 7$," or, more briefly, as "$3 + 4 \neq 7$."

Before proceeding further, it will be convenient to invent a notation for referring to propositions. In Chapter 2 we designated specific sets by *capital* letters (A, B, C, . . .). In this chapter we shall again follow custom by designating simple propositions by lower-case letters (typically, p, q, r, . . .). Thus we may let p stand for "July 20, 1969, was a Monday" and q stand for $3 + 4 = 7$. We shall use $\sim p$ (read, "not p") and $\sim q$ to represent the negations of p and q, respectively. Thus $\sim p$ stands for "July 20, 1969, was not a Monday." What does $\sim q$ stand for?

As usual, examples lead to a general, but precise definition:

Definition 3.1 Negation If p is a proposition, the *negation* of p, "not p" (in symbols, $\sim p$), is the proposition which is false if p is true and true if p is false.

Definition 3.1 amounts to saying that, in all cases, the truth value of $\sim p$ is opposite that of p. It can be presented in the form shown in Table 3.1. This form, called a *truth table*, is to be read line by line. Thus the first line indicates that if p is true, then $\sim p$ is false. What does the second line indicate? Since p has only two possible truth values, the table describes completely the relationship between p and $\sim p$. Of course, any other lower-case letter could be used instead of p: for example, if the letter q were used instead of p, the body of the table would remain unchanged. Thus *our* proposition q ($3 + 4 = 7$), together with $\sim q$, illustrate the first line of the truth table. Which line of the table does *our* proposition p ("July 20, 1969, was a Monday") illustrate? The truth table is a handy device; we shall shortly exploit it more fully.

TABLE 3.1

p	$\sim p$
T	F
F	T

In order to illustrate Definition 3.1 and also to warn of some common pitfalls, we shall attempt to negate several other propositions, in particular the four special types discussed in Section 3.1. So far, experience indicates that a proposition is negated if it is preceded by phrases like "it is not the case that" or "it is false that." This procedure of negation is always correct, but it produces a rather cumbersome sentence and is so mechanical that it does not stimulate thought or provide insight into the nature of negation. A second method, to insert the word "not," is also mechanical and very tidy; but it is quite unreliable, since it frequently leads to wrong answers.

As a case in point, let p be the proposition "Some cats have tails." Merely inserting a *not* yields the proposition "Some cats do not have tails." But this result is not $\sim p$, since Definition 3.1 is not satisfied. For, checking the first part of the definition (or the first line of the truth table), if p is true, then the statement "Some cats do not have tails" is not necessarily false: it is quite possible that some cats have tails and some do not. (In fact, the Manx cat *is* tailless!) What, then, is $\sim p$? We can, of course, retreat to the mechanical circumlocution, "It is not the case that some cats have tails." But what does this mean in crisp English? It means "No cats have tails." Let us check that this proposition satisfies Definition 3.1. First, if p is true, we are guaranteed at least one cat with a tail, so that "No cats have tails" is false. So far, so good; but we must not stop here. Secondly, if p is false, then we cannot find even one cat with a tail, and "No cats have tails" is true. Thus we have checked both halves of the definition. Note that in the above argument we are not concerned with whether, in fact, p is true or whether it is false; we are instead in search of a proposition of *opposite* truth value to p, whichever we may later decide is its truth value.

As a second example we take q, "All trees are evergreens." In seeking $\sim q$, we must reject "All trees are not evergreens" out of hand, on account of its ambiguity (see Section 3.1). "Not all trees are evergreens" is a correct expression for $\sim q$, though mechanically derived. "No trees are evergreens" may sound plausible, but is actually incorrect. It is a good exercise to apply Definition 3.1 here and note where it fails. Specifically, if q is true, is "No trees are evergreens" required to be false? And if q is false, is the quoted statement necessarily true? Another exercise is to examine by means of Definition 3.1 the statement "Some trees are not evergreens," which turns out to be an expression for $\sim q$. It is of at least passing interest to note that this form of $\sim q$ is also one of the four special types of propositions, of which q is another. Is this also the case with p and $\sim p$ in the preceding paragraph?

By now you may be wondering if there is any connection between the *negation* of a proposition and the *complement* of a set. The following example suggests that there is. Let r be the propositional function, "x is an odd number." This function will become a proposition if the letter x is replaced by a whole number. But if x is replaced by "$4\frac{1}{2}$," "$\sqrt{3}$," "a triangle," or "the Sahara Desert," the resulting sentence will not make sense, since the definition of an odd number cannot be applied: so it will not be a proposition. In other words, associated with all replacements of x which make p a proposition is a set U; here $U = W$. Moreover, if $x \in W$, then p is true if $x = 1, 3, 5, \ldots$ (i.e., if $x \in D$) and p is false if $x \in E$. Thus, the replacements for x which make p true constitute a corresponding set $D \subseteq W$. What can be said about the replacements for x which make p false? A Venn diagram will depict these relationships. Figure 3.1 shows $U = W$, with D a subset of U. Since D is the set of replacements for x which make p true, D may be called the *truth set* for p. Further, $D' = E$ is the set of replacements for which p is false, so that D' is actually the truth set of $\sim p$. Thus the notion of truth set gives rise to a kind of parallelism between sets and propositions. Venn diagrams, moreover, seem useful in describing relations among propositions as well as among sets.

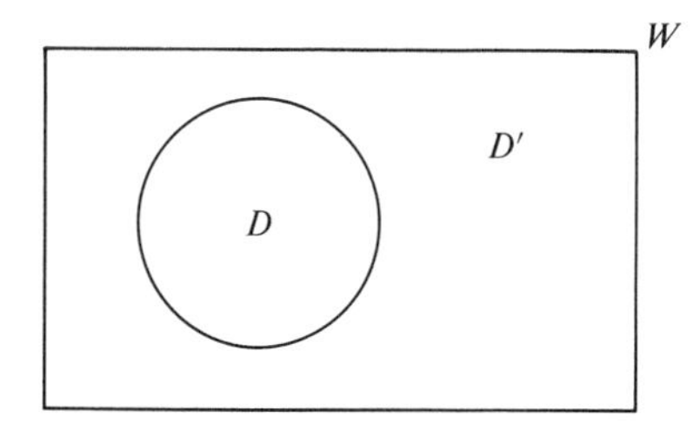

Figure 3.1

EXERCISES 3.2

1. Complete the text's discussion of the statement "All trees are evergreens" by applying Definition 3.1 to show that its negative is not "No trees are evergreens," but rather "Some trees are not evergreens."

2. (a) One form of the negation of "Some trees are not evergreens" is "It is not the case that some trees are not evergreens." Write this negation in simpler form, avoiding the double negative and the introductory clause. Note that, if we symbolize "Some trees are not evergreens" by $\sim q$ (as we did in the text), then the answer to this exercise can be written symbolically as $\sim(\sim q)$, the negation of the negation of q.

   (b) Similarly, write in simple form the negation of "No cats have tails" (which is called $\sim p$ in the text) and thus obtain $\sim(\sim p)$.

   (c) Write a general statement explaining the relation between *any* proposition r and $\sim(\sim r)$, the negation of its negation.

3. Some of the parts of Exercises 3.1, Problem 1, are propositions (T or F). For each of these parts write its negation in simplest form.

4. The four special types of propositions described in Section 3.1 are designated in classical logic as follows:

A-type: All x's are y's.

E-type: No x is a y.

I-type: Some x's are y's.

O-type: Some x's are not y's.

For each *type*, determine its negation in simplest form. Use examples as aids, if you wish, but state the result in general form. You will note that this system of four types is *closed* under the operation of negation, in the sense that the negation of any of the types is a unique one of the four types. Also, note that the answer of Problem 2(c) holds for this system.

*5. A soup, prepared by two cooks, may have been poisoned. The person to whom it is served wishes to determine whether it is safe to eat. He is permitted to ask *one* question of *one* of the cooks, a question answerable simply by Yes or No. He knows that one cook always tells the truth and the other always lies: but he doesn't know which is which. Which question should he ask?

*6. A computing firm personnel officer wishes to test the logical ability of three applicants for employment, all of whom are smart. She seats them around a table, blindfolds them, and explains that she is marking each of their foreheads with a black or a red cross. When the blindfolds are removed, each person who sees a *red* cross on another's forehead is to start tapping on the table; if someone sees only black crosses, he or she is not to tap. The first person who guesses correctly the color of the cross on his or her own forehead wins the job. The officer then marks red crosses on all three foreheads, removes the blindfolds, and all three persons begin tapping. Soon one announces, correctly, that he has a red cross on his forehead.

(a) Explain the reasoning of the successful candidate.

(b) Show that if the officer had marked the foreheads in any other way the test would have been considerably easier.

*7. Let p be the propositional function "The number of words in this sentence is x."

(a) What is the set, U, of all replacements for x which make the quoted sentence a proposition?

(b) Find P, the truth set of p.

(c) Write $\sim p$ and find its truth set.

(d) In particular, note that the truth value of p when $x = 9$ is T, and find the truth value of $\sim p$ when $x = 9$. Show that for these two propositions Definition 3.1 appears to be violated, i.e., we have a *paradox*.

(e) Similarly investigate the case where $x = 10$.

(f) The paradoxes of parts (d) and (e), if unexplained, destroy the concept of negation! Try to find a satisfactory explanation.[5]

Section 3.3 Conjunction: A Binary Operation on Propositions

In the preceding chapter, several binary operations on sets were introduced. In each case, two sets, combined according to some definition, yielded a unique set. In a similar way, it is possible to combine two simple propositions by means of a binary operation to produce a unique proposition. One such operation is *conjunction*: it is accomplished by connecting two propositions with the word *and*. As an example, if p is the statement "$2 + 3 = 5$" and q is "$7 - 6 = 1$," then the conjunction of p and q is the compound statement "$2 + 3 = 5$ and $7 - 6 = 1$." The customary symbol for conjunction is $\wedge$; thus the last statement is symbolized by $p \wedge q$.

Now if $p \wedge q$ is a proposition, it must have a truth value, which must depend on the individual truth values of p and q. In the preceding illustration p and q are both true. What, then, is the truth value of $p \wedge q$? Your intuition will probably suggest that, since both parts of the compound statement, with *and* as the connective, are true, so is the statement as a whole. For the present we shall thus be guided by intuition and accept that the conjunction of two true propositions is true.

A different case occurs when p is true and q is false. To see this, let p be the same proposition as before, but q be changed to "$7 - 6 = 2$." Now which truth value is it reasonable to assign to $p \wedge q$, "$2 + 3 = 5$ and $7 - 6 = 2$"? Since one of the parts (q) is false and the two parts are connected by *and*, it seems that $p \wedge q$ should be false.

There remain two other cases. What are they? It should be easy to supply examples and to determine from them on an intuitive basis the truth values of $p \wedge q$. The four cases are summarized in Table 3.2. Note that this summary includes all possible combinations of the individual truth values of p and q. Since the third column has an entry (T or F) in every line, it is clear that $p \wedge q$ is a proposition, for it has a unique truth value in every case.

TABLE 3.2

p	q	$p \wedge q$
T	T	T
T	F	F
F	T	F
F	F	F

[5]Adapted, by permission, from a paragraph in "Did You Know?," *New York State Mathematics Teachers' Journal*, 21:2 (April, 1971), 84.

We have come thus far with the aid of intuition. It is valuable in that it provides, for our abstract table, a natural connection with the language and reasoning practices of everyday life. But it cannot serve as a logical justification. The bold and quick approach is simply to accept the truth table as arbitrary "rules of the game" of logic, a way of defining what conjunction of propositions is. More formally, we make the following:

Definition 3.2 Conjunction If p and q are any two propositions, the *conjunction* of p and q, "*p and q*" (in symbols, $p \wedge q$), is a proposition which is true when p and q are both true, and false otherwise.

In addition to intuition and formal definition, a third approach to conjunction relates propositions to truth sets utilizing Venn diagrams. In invoking set theory, we start with the truth sets for two propositional functions, which we shall denote p and q. We assume, of course, that these truth sets are contained in a suitable universal set. For an example let us return to Ivy College,[6] with U the set of all its students; thus any proposition or propositional function has to do with Ivy College students. Now, let p be the propositional function "x is a commuting student" and q the function "x is a freshman." Then the truth set for p consists of all the replacements for x which make p a true statement — namely, the set C of commuting students. Similarly, the truth set for q is F, the set of freshmen. We now search for the truth set of $p \wedge q$. Since, by Definition 3.2, $p \wedge q$ is true if and only if both p and q are true, its truth set must consist of all replacements of x which are elements of both set C and set F. Thus its truth set is $C \cap F$, as shown in Figure 2.5. More generally, the truth set of the conjunction of any two propositional functions is the *intersection* of the truth sets of the separate functions. Indeed, this is not surprising, since the word *and* is the connective employed in both Definition 2.6 and Definition 3.2.

Fortunately, these three approaches to conjunction of propositions all lead to the same result, which is summarized in the truth table for $p \wedge q$. You should check this claim and then select whichever approach results in the best understanding of the concept of conjunction of propositions.

Associated with truth sets of propositions are several special situations of interest. To illustrate one, let q be again "x is a freshman" and r be "x is a sophomore." Associated with q and r are the truth sets F and S, respectively, where S is the set of sophomores. Now, what is the truth set for $q \wedge r$? To pose the question differently, which replacements for x will make q and r true simultaneously? In this case, x would have to represent students who are both freshmen and sophomores, which is impossible. Hence the truth set for $q \wedge r$ is $\varnothing$, the null set. This result parallels the fact that $F \cap S = \varnothing$, as shown in Figure 2.6.

In most of our examples propositional functions have been true for certain elements $x \in U$ and false for others: the former set is, of course, the truth set

[6]See Section 2.4.

of the given function. But functions occur which are *always* true, no matter what the replacements for x are, so long as $x \in U$. As an illustration, we take $U = W$ and p as the propositional function "$x + 0 = x$." Clearly, whatever whole number is substituted for x in "$x + 0 = x$" will make the equation true. Hence the truth set for p is U. More generally, if a propositional function is always true, then its truth set is the universal set, and vice versa.

At the opposite extreme is the completely false propositional function. An instance, again taking $U = W$, is q, "$x + 1 = x$." Now, for every whole number it is certainly *false* that adding 1 to it yields a sum equal to the original number. So there are no replacements which make q true — i.e., its truth set is $\varnothing$. Which *general* statement can be made about propositional functions which are always false?

EXERCISES 3.3

1. Supply arithmetic equations similar to those used at the beginning of this section to illustrate the third and fourth lines of the truth table for $p \wedge q$ (see Table 3.2). Do the F values in the third column seem reasonable?

2. Consider the following propositions:

$$p\colon\ 9 + 4 = 14 \qquad q\colon\ 7 \in W \qquad r\colon\ \{1\} \subset \{3, 1, 2\}$$

$$s\colon\ A \cap \varnothing = A$$

   Write each of the following as simply as possible in sentence form:

   (a) $p \wedge q$ (b) $p \wedge s$ (c) $\sim p$
   (d) $(\sim p) \wedge r$ (e) $r \wedge q$ (f) $s \wedge q$
   (g) $s \wedge (\sim q)$ (h) $(\sim r)$ (i) $\sim(\sim r)$
   (j) $(\sim p) \wedge (\sim r)$ (k) $\sim(p \wedge r)$

3. Each of the statements, p, q, r, s, of Problem 2 has an obvious truth value. Using these, determine the truth value of each of the compound statements in parts (a) through (k).

4. Let $U = W$ and consider the following propositional functions:

$$p\colon\ x + 8 = 11 \qquad q\colon\ x + 11 = 8 \qquad r\colon\ x + 8 = 8 + x$$

   (a) Find and describe the truth sets of p, q, and r.
   (b) What is the truth set of a propositional function which
   (i) is always true? (ii) is always false?

5. Figure 3.2 shows C and F in U, with the four regions into which C and F divide U numbered for easy reference.

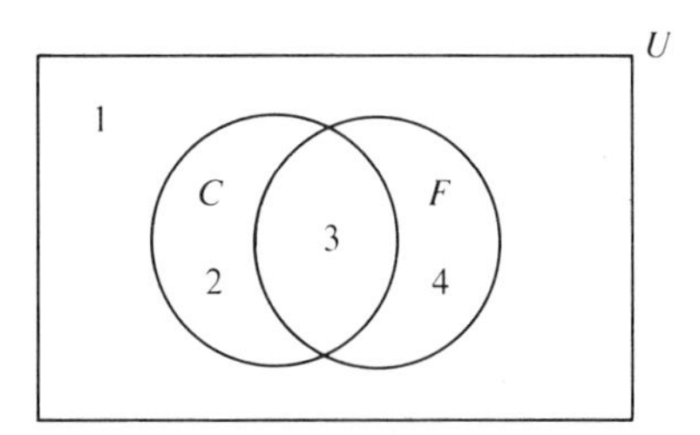

Figure 3.2

(a) In the text we already ascertained that the truth set for $p \wedge q$ is $C \cap F$, where p is "x is a commuting student" and q is "x is a freshman." Which numbered region(s) is (are) contained in the set $C \cap F$?

(b) The first line of the truth table for $p \wedge q$ (Table 3.2) describes the case where p and q are both true. The region of Figure 3.2 which shows both p and q true is #3. Is this region contained in the truth set for $p \wedge q$ as determined in part (a)?

(c) The second line of the truth table describes the case where p is true, but q is false. Now, p is true in regions #2 and #3, and q is false in #1 and #2; so #2 is the only region in which p and q have the values given in the second line. Hence #2 depicts the first two columns of the second line. Is #2 contained in the truth set for $p \wedge q$ as determined by part (a)? Does your answer justify the entry F in the third column of the second line?

(d) In a fashion similar to that of part (c), identify the region which depicts the first two columns of the third line, and see whether it is contained in the truth set for $p \wedge q$. Does your answer justify the entry in the third column of the third line?

(e) Work out part (d) for the fourth line of the table.

6. (a) Let p be the proposition "It is raining" and q be "I forgot my umbrella." Write the compound proposition $p \wedge q$; then write $q \wedge p$ and compare their meanings intuitively. Do they mean the same thing?

(b) If two logical expressions have the same truth table (i.e., if the entries match line by line), they are said to be *logically equivalent*. Set up a truth table for $q \wedge p$ and compare it with that for $p \wedge q$. (In order to compare on the same basis, the first two columns of the tables should be identical; but you should combine p and q in reverse order, considering q first.) Are $p \wedge q$ and $q \wedge p$ logically equivalent? Compare your conclusion with that for part (a).

(c) Since in part (b) p and q may refer to *any* propositions (or propositional functions), not merely the examples given in the text, we may immediately generalize the results of part (b). Accordingly, is the binary operation $\wedge$ on propositions commutative?

(d) Compare the results of part (c) with a similar property relating to the truth *sets* of p and q.

7. Recall that negating a proposition reverses all the truth values, as the truth table for $\sim p$ (Table 3.1) shows. The same outcome applies in negating compound propositions: every T in the table is changed to F, and vice versa. As an example we negate the *compound* proposition $p \wedge q$ to get the truth table for $\sim(p \wedge q)$. Copy and complete the following truth table:

p	q	$p \wedge q$	$\sim(p \wedge q)$
T	T	T	F (T of third column changed to F)
T	F	F	T
F	T	F	
F	F		

8. Instead of combining p and q to get $p \wedge q$ and then negating the compound proposition to get $\sim(p \wedge q)$, as we did in the preceding exercise, we could *first* negate the simple propositions individually, obtaining $\sim p$ and $\sim q$, and *then* combine $\sim p$ and $\sim q$ to obtain $(\sim p) \wedge (\sim q)$. Would the result be logically equivalent to that of Problem 7? Answer this question by the following two procedures:

(a) Let p be "It is raining" and q be "I forgot my umbrella." Write out the statements for $\sim(p \wedge q)$ and $(\sim p) \wedge (\sim q)$, and try to decide intuitively whether they have the same meaning.

(b) If part (a) was confusing, the method of truth tables may be easier. Complete the following table, then compare its last column with that of Problem 7:

p	q	$\sim p$	$\sim q$	$(\sim p) \wedge (\sim q)$
T	T	F	F	F (Combine $\sim p$ and $\sim q$ by using Definition 3.2)
T	F	F	T	
F	T	T		
F	F			

What answer does this table give to the question raised in the second sentence of this exercise?

(c) Do the results of this problem parallel those of set theory? Explain.

9. (a) If three propositional functions are to be combined — say, p, q, and r — how many different combinations of truth values are possible? Try to list them in orderly fashion, using columns for p, q, and r.

(b) One method of combining three functions with *and* is first to combine p and q, obtaining $(p \wedge q)$; then to combine this result with r, obtaining $(p \wedge q) \wedge r$. Copy and complete the following truth table, whose last column gives the truth values of $(p \wedge q) \wedge r$ for all possible cases:

p	q	r	$p \wedge q$	$(p \wedge q) \wedge r$
T	T	T	T	T [obtained as the conjunction
T	T	F	T	F of $(p \wedge q)$ and r]
T	F	T	F	
T	F	F		
F	T	T		

etc.

(c) In a similar fashion set up the truth table for $p \wedge (q \wedge r)$, where the simple functions appear in the same order, p–q–r, but are grouped differently.

(d) Compare the truth tables of parts (b) and (c) to see whether they are logically equivalent. Which property of the binary operation of conjunction does this comparison illustrate?

(e) Is there a parallelism between part (d) and a certain property of sets? Explain.

*10. Let p, q, r represent any three propositional functions.

(a) Set up truth tables for $p \wedge (q \wedge r)$ and for $(p \wedge q) \wedge (p \wedge r)$, and compare results.

(b) Relate the results of part (a) to parallel results for sets mentioned in *Section 2.6.

Section 3.4 Disjunction: Another Binary Operation on Propositions

The parallelism we are building between sets and logical operations is becoming more and more evident. With regard to operations, we have so far linked negation of propositions with complementation of sets and conjunction of propositions with intersection of sets. It is natural to seek an operation on propositions which corresponds to union of sets. This operation is called *disjunction*, and it uses the connective *or*. As an example, let r be the proposition "It is raining" and s, "The sun is shining." Using *or* we obtain the compound statement, "It is raining or the sun is shining." This statement is abbreviated "$r \vee s$," where $\vee$ is the symbol for *or*.

Here it is necessary to raise a question of language. Does the proposition $r \vee s$ mean that *either* it is raining, *or else* the sun is shining; or does it include the possibility that both events are happening simultaneously, as they do when we see a rainbow? Dictionaries appear to favor the first meaning; but the second also occurs commonly in conversation. Frequently *or* is used *exclusively*, in the sense of *or else*, as in the sentence "Today is Wednesday or Thursday; I forget which." Here the possibility that today could be simultaneously Wednesday and Thursday does not make sense. The connective *or* can, on the other hand, be used *inclusively*, permitting the possibility of both alternatives at once. An example is the sentence "This family medical insurance policy covers all members of the family who are under 19 or are college students." Such policies presumably cover 18-year-old college students, who satisfy both alternatives. In fact, in legal documents, when the inclusive *or* is meant, it is often written in the form *and/or*.

Since mathematics cannot tolerate ambiguity, it is necessary to choose between the two meanings. In formal logic we agree to take *or* in the *inclusive* sense, as *and/or*.[7] Among the reasons for this choice is that it corresponds to the concept of union in set theory. Under this agreement we proceed to develop the truth table for $p \vee q$ by means of examples and intuition, leading to a formal definition and the expected link with sets.

We may as well use the illustrations of Section 3.3, with p as "$2 + 3 = 5$" and q as "$7 - 6 = 1$." Then $p \vee q$ represents "$2 + 3 = 5$ or $7 - 6 = 1$." Now, since p and q are both true and *or* is inclusive, it is reasonable to call $p \vee q$ true, thus obtaining the first line of Table 3.3. For the second line we take p as before, but change q to $7 - 6 = 2$. Is the compound statement "$2 + 3 = 5$ or $7 - 6 = 2$" true or false? Since *or requires* only one alternative to be true — though it permits both — we conclude that, because p is true, so is $p \vee q$. You are invited to supply examples to complete the table.

TABLE 3.3

p	q	$p \vee q$
T	T	T
T	F	T
F	T	T
F	F	F

The above intuitive considerations impel us to make the following:

Definition 3.3 Disjunction If p and q are any two propositions, the *disjunction* of p and q, "p or q" (in symbols, $p \vee q$), is a proposition which is false when p and q are both false, and true otherwise.

We now investigate the parallelism with set theory. Using again the two propositional functions drawn from Ivy College,[8] we let p be "x is a commuting

---

[7]See Problem *10 for a treatment of the exclusive *or*.

[8]See Section 2.4.

student" and q be "x is a freshman." As before, the corresponding truth sets are C, the set of commuters, and F, the set of freshmen. What, then, is the truth set for $p \vee q$? According to Definition 3.3 or Table 3.3, $p \vee q$ is true whenever either or both of p, q are true. This means that all replacements for x which are elements of C or of F or of both will render $p \vee q$ true. This means that its truth set consists of all elements which are in C and/or in F, namely, the set $C \cup F$. A comparison of the definitions of set union (Definition 2.7) and proposition disjunction (Definition 3.3) reveals that both use the inclusive interpretation of *or*. Indeed, with this agreement concerning the meaning of *or*, the predicate of Definition 2.7 may be shortened to "the set of elements which are in A or B."

EXERCISES 3.4

1. (a) Provide a sentence of your own to illustrate the exclusive use of *or* and another for the inclusive use.

   (b) Which one of the two meanings does the symbol $\vee$ have?

2. Supply arithmetic equations similar to those used in this section to illustrate the third and fourth lines of the truth table for $p \vee q$ (Table 3.3). Do the values in the third column seem reasonable?

3. Consider the following propositions:

   p: $9 + 4 = 14$ $\quad$ q: $7 \in W$ $\quad$ r: $\{1\} \subset \{3,1,2\}$

   s: $A \cap \varnothing = A$

   Write each of the following in sentence form as simply as possible:

   (a) $p \vee q$ (b) $p \vee s$ (c) $(\sim p) \vee r$

   (d) $r \vee q$ (e) $s \vee q$ (f) $s \vee (\sim q)$

   (g) $(\sim p) \vee (\sim r)$ (h) $\sim(p \vee r)$ (i) $p \wedge (q \vee r)$

   (j) $p \vee (q \wedge r)$ (k) $(p \vee q) \wedge (p \vee r)$

4. Each of the statements p, q, r, s of Problem 3 has an obvious truth value. Using these, determine the truth value of each of the compound statements in parts (a) through (k).

5. For convenience we copy Figure 3.2, which shows the four numbered regions into which C and F divide U.

   (a) In the text we ascertained that the truth set for $p \vee q$ is $C \cup F$, where p is "x is a commuting student" and q is "x is a freshman." Which numbered region(s) is (are) contained in the set $C \cup F$?

   (b) The first line of the truth table for $p \vee q$ (Table 3.3) describes the case where p and q are both true. The region of the diagram which

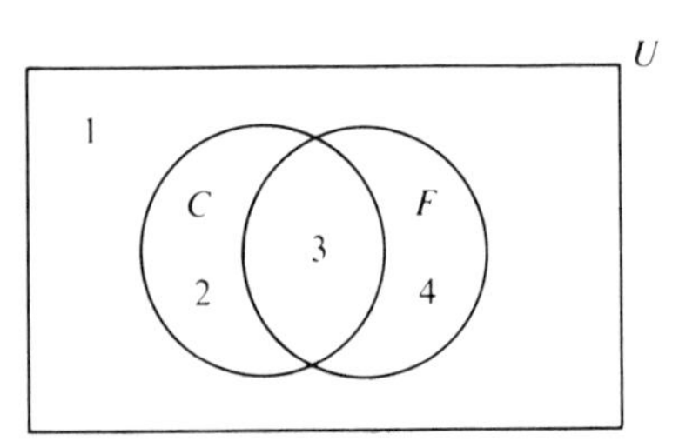

Figure 3.2

shows both p and q true is #3. Is this region contained in the truth set for $p \vee q$ as determined in part (a)?

(c) Region #2 of Figure 3.2 parallels the entries in the first two columns of the second line of the table for $p \vee q$. Is #2 contained in the truth set for $p \vee q$ as determined in part (a)? Does your answer justify the entry T in the third column of the second line?

(d) Similarly discuss the third and fourth lines of the truth table for $p \vee q$, and justify the entries in the third column for these lines.

6. (a) Let p be the proposition "It is raining" and q be "I forgot my umbrella." Write the compound proposition $p \vee q$; then write $q \vee p$ and compare their meanings intuitively. Do they mean the same thing?

   (b) Set up truth tables for $p \vee q$ and $q \vee p$ and determine from them whether these two expressions are logically equivalent. Compare your conclusion to that of part (a).

7. (a) Make a study of the effects of negation on disjunction, as follows:

      (i) Let p and q be as in Problem 6. Write out the statements for $\sim(p \vee q)$ and $(\sim p) \vee (\sim q)$ and try to decide intuitively whether they have the same meaning.

      (ii) Work out truth tables for $\sim(p \vee q)$ and for $(\sim p) \vee (\sim q)$ and compare them.

   (b) Now make a "grand comparison" among the *four* truth tables obtained in Exercises 3.3, Problems 7 and 8, and in part (a)(ii) of this problem. Are any of the four expressions logically equivalent to each other? If so, which ones?

   (c) Do the above results parallel those for set theory? Explain.

8. Discuss the logical expressions $(p \vee q) \vee r$ and $p \vee (q \vee r)$ in the same manner as that requested in Exercises 3.3, Problem 9.

9. Now that we have introduced both conjunction and disjunction, we should ask whether either operation distributes over the other.

   (a) Use truth tables to show that $p \wedge (q \vee r)$ and $(p \wedge q) \vee (p \wedge r)$

are logically equivalent; hence conjunction *does* distribute over disjunction.

(b) Similarly investigate whether disjunction distributes over conjunction.

*10. The symbol $\veebar$ is used for the *exclusive or*. For example, if p is the statement "Today is Wednesday" and q is the statement "Today is Thursday," then $p \veebar q$ is the compound statement "Today is Wednesday or else it is Thursday."

(a) Write a general definition for the exclusive *or*, $p \veebar q$.

(b) Construct the truth table for $p \veebar q$ on the basis of part (a).

(c) Let P and Q be the truth sets for p and q, respectively, and consider Figure 3.3. Which of the numbered regions depict(s) the truth set of $p \veebar q$?

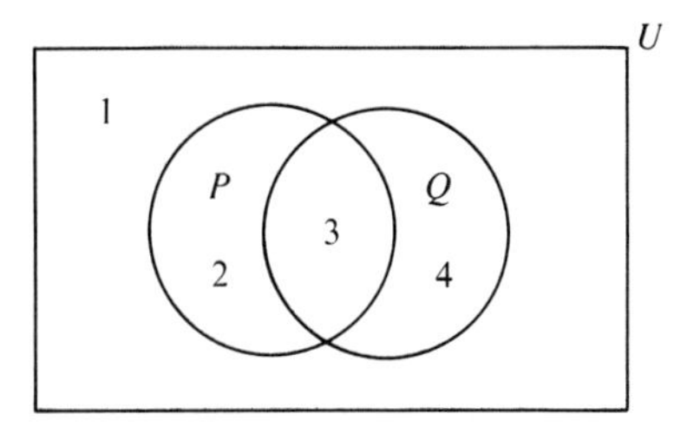

Figure 3.3

(d) Explain why the truth set of $p \veebar q$ is not equal to $P \cup Q$ or $P \cap Q$.

(e) The truth set of $p \veebar q$ is called the *symmetric difference* of the sets P and Q; and it is symbolized by $P \Delta Q$. It may be defined by the equation $P \Delta Q = (P \cap Q') \cup (Q \cap P')$, or by the equation $P \Delta Q = (P - Q) \cup (Q - P)$.[9] Show that $P \Delta Q$ yields exactly the region(s) of part (c).

(f) Consider the logical expressions $(\sim p) \veebar (\sim q)$ and $\sim(p \veebar q)$.

(i) Try to discover the relationship of these two expressions intuitively by studying the example

p : The line ℓ is horizontal

q : The line ℓ is vertical

where ℓ is a fixed line in the plane.

(ii) Set up the truth tables for the two expressions and compare them with each other and with the results of part (i).

[9]See the second Suggestion for Investigation at the end of Chapter 2.

*11. In this section we have developed the parallelism between disjunction of propositions and union of sets, in addition to mentioning, at the beginning of this section, two other parallelisms. We have also noted that some of the properties of the operations are the same: e.g., intersection of sets and conjunction of propositions are commutative and associative.

(a) Similarly compare properties of union of sets and disjunction of propositions.

(b) Make comparisons involving complementation with union and intersection in set theory and the analogous relations among propositions.

(c) In set theory there are two distributive laws relating union and intersection. Is there an analogous situation for propositions?

(d) Investigate distributive properties (if any) between conjunction and *exclusive* disjunction, using the definition of Problem *10(a).

Section 3.5 The Conditional

In mathematics and in everyday life, one of the most useful compound statements is that of the form "If p, then q." In symbols, this compound statement is written $p \rightarrow q$, where the arrow represents a new binary operation on propositions called *the conditional*. The following are examples of the conditional:

1. If I have stayed up late, then I am sleepy the next day.
2. If a whole number ends with the digit 6, then it is even.
3. If the noon whistle sounds, then the hands of my watch point to 12.
4. If $x + 3 = 7$, then $x = 2$.
5. If George Washington was born in 1432, then gold is heavier than silver.

The meaning of at least the first four of these compound statements is clear. The first three, moreover, seem intuitively acceptable, while the fourth is plainly wrong. Though we might wish to dismiss the fifth as irrelevant or meaningless, we must agree that each of its parts is a simple statement and that they have been formally combined by means of the "if . . . , then. . . ." connectives in the same way as the others.

These examples indicate that the conditional is not a completely routine operation. It turns out, in fact, to be a rather tricky concept which, if explored carefully, involves some subtle distinctions. Without noting all of these distinctions, we shall describe and illustrate the conditional at some length. We begin by alerting you to three pitfalls.

1. Though the first compound statement might suggest that a conditional involves causation, this is not generally the case. Thus in the third conditional it is absurd to claim that the blowing of the noon whistle *causes* the hands of my watch to be in a certain position. Indeed, there need be no apparent relation at all between the *if clause*, or *hypothesis*, and the *then clause*, or *conclusion*, as the last example shows. What is important is that the *form* of all five statements is the same: "If p, then q," where p and q are propositions.

2. A brief experiment illuminates the second pitfall. Suppose we interchange the hypothesis and the conclusion of a conditional, thereby obtaining the statement, "If q, then p." From the first conditional example we thus obtain "If I am sleepy the next day, then I have stayed up late." Does this new statement mean the same thing as the original one? Certainly not, for although we would probably accept the original as true, the other is false, since sleepiness might as well be associated with poor health or boredom as with late hours. The point is even more obvious with respect to the second conditional example: for if a whole number is even, one cannot conclude that it ends with a 6: 12 and 34 provide counterexamples. Since a conditional apparently changes its meaning when the order within its clauses is reversed, it must be handled carefully.

3. The third pitfall concerns how to determine the truth value of a conditional. While intuition seems to be a fair guide in suggesting that the first three examples are true and the fourth false, it does not indicate exactly why this is so. Intuition, in addition, provides little guidance concerning the truth value of the fifth conditional. These examples do not lead directly to a truth table for $p \rightarrow q$; yet we certainly must assign *some* truth value to each line if $p \rightarrow q$ is to be a proposition.

As the arrow correctly suggests, the conditional is a unidirectional concept: $p \rightarrow q$ is not the same as $q \rightarrow p$. In each case the *hypothesis* — the clause which is verbally introduced by *if* — is the proposition at the tail of the arrow; and the *conclusion* — verbally introduced by *then* — is the one at its head. Thus in the conditional $p \rightarrow q$, p is the hypothesis and q the conclusion, whereas in the conditional $q \rightarrow p$, the roles of p and q are reversed.

We proceed now to develop the truth table for $p \rightarrow q$. As before, we shall let intuition carry us as far as possible, leading us to an arbitrary definition which, in turn, has an interesting and perhaps surprising relation to set theory.

To give intuition something to work on, let us suppose that an investment expert has just predicted in his syndicated column: "If corporation taxes are increased, then the stock market will fall." Let us consider under which circumstances we would recognize his prediction as correct, and under which we would proclaim it as a failure.

To begin with, the quoted statement is a conditional, $p \rightarrow q$, with p as "corporation taxes are increased" and q as "the stock market will fall." In the first line of Table 3.4 both p and q are true. This means that, in fact, corporation taxes are raised and that the stock market actually falls. Under these circumstances, we would certainly credit the expert with a successful prediction. So we mark T in the third column of the first line.

TABLE 3.4

p	q	$p \to q$
T	T	T
T	F	F
F	T	T
F	F	T

For the second line we consider the situation in which corporation taxes are increased, but the stock market does not fall (i.e., q is false). Clearly, the expert's prediction has failed; so we mark an F in the third column.

The third line refers to the situation in which corporation taxes are not raised, but the stock market falls. This case is more difficult: we might be tempted to say that the prediction does not apply. But we realize that the third column must be filled by either T or F if $p \to q$ is to be a proposition. Now, it would be unfair to accuse the expert of failure, since he offered no judgment as to what would happen if corporation taxes were not raised. For this reason, we decide to mark a T in the third column. By a similar argument, which you are invited to reconstruct, we also place a T in the last line of the third column. If these last two decisions seem a bit arbitrary, the reader should recall that a definition, by its nature, is arbitrary. It is surprising, however, how well these decisions harmonize with set theory.

The truth table for the conditional is perhaps a bit hard to remember. We note that $p \to q$ is true, *except* when the hypothesis is true *and* the conclusion is false. (This case is the only one in which the expert could be accused fairly of failure in his prediction.) Accordingly, we formalize the truth table for $p \to q$ by means of the following:

Definition 3.4 Conditional If p and q are any two propositions, the *conditional* "If p, then q" (in symbols, $p \to q$) is a proposition which is false if p is true and q is false; otherwise it is true.

A related definition, already given and used informally, is

Definition 3.5 Hypothesis, Conclusion Proposition p is called the *hypothesis* of the conditional $p \to q$, and q is called its *conclusion.*

We noted earlier that, for a given p and q, $p \to q$ may have a different meaning from $q \to p$. In fact, failure to recognize this difference constitutes one of the most common and stubborn *fallacies*, or errors, in reasoning. The celebrated "man in the street" is very apt to make this error; and even the logician has his unguarded moments. "Use Superlustre toothpaste and your teeth will gleam," proclaims an advertisement. The meaning apparently is, "If you use Superlustre toothpaste, then your teeth will gleam." Perhaps so, but even if we accept this statement, it does not follow that "If your teeth gleam, then you have used Superlustre toothpaste": for other dentifrices might be equally, or more, effective. Yet part of the ad writer's aim is to persuade the reader to use *only* Superlustre. A more serious example is the following: "If a person is a Communist, then he opposes U. S. foreign policy." Perhaps this conditional seems reasonable; but again, it is not logical to conclude from it that "If a person opposes U. S. foreign policy, then he is a Communist." You are advised to look for more examples of the fallacy of "assuming the converse": you will find them in advertisements, editorials, letters to the editor, and radio and television programs, as well as in your own thoughts!

To see that $p \rightarrow q$ is not logically equivalent to its *converse*, $q \rightarrow p$, we have only to construct their truth tables (see Table 3.5). In working out the truth table for $q \rightarrow p$, we note that, since q is the hypothesis and p the conclusion, the conditional $q \rightarrow p$ has the value F when q is true and p is false. Since this combination occurs only in the third line of the table, only the third line of the column for $q \rightarrow p$ has the value F. A comparison of the last two columns shows that they differ in truth value in the two middle lines.

TABLE 3.5

p	q	$p \rightarrow q$	$q \rightarrow p$
T	T	T	T
T	F	F	T
F	T	T	F
F	F	T	T

Since converses occur frequently, we shall make the following:

Definition 3.6 Converse Given a conditional $p \rightarrow q$, the conditional formed from it by interchanging the hypothesis and the conclusion (in symbols, the conditional $q \rightarrow p$) is called the *converse* of the given conditional.

Most of our examples so far have been of acceptable conditionals whose converses are questionable, to say the least. This need not be the case; for sometimes a conditional *and* its converse are both acceptable. An instance is, "If a triangle has three equal sides, then it has three equal angles." Memories of high school geometry or a few quick sketches will convince you of the correctness of this original conditional. Its converse, which you should write down and check, is also correct. We must conclude, unfortunately, that acceptance of a conditional — or rejection of it, for that matter — provides no clue as to the acceptability of its converse. This is the real significance of the truth table demonstration that a conditional and its converse are not logically equivalent. Stated in another way, the operation symbolized by $\rightarrow$ is *not* commutative.

EXERCISES 3.5

1. Use the conditional "If a person is a native of Brazil, then he speaks Portuguese" to justify the truth table for $p \rightarrow q$ intuitively, as the text did by means of the stock-market example.

2. Identify the hypothesis and the conclusion of each of the following conditionals, first changing its form if necessary:

   (a) If a person is a resident of Pennsylvania, then he is a resident of the United States.

   (b) If $10 + 2 = 12$, then $6 - 1 = 4$.

   (c) It will rain if the barometer falls rapidly. (*Hint:* Does this sentence mean the same as "If the barometer falls rapidly, then it will rain"?)

   (d) If the groundhog sees his shadow on February 2, then there will be six more weeks of winter.

   (e) I'll go swimming if the sun comes out.

   (f) If 2/3 is a whole number, then 17 is a fraction.

   (g) Every square is a rectangle. (*Hint:* Does this sentence mean the same as "If a figure is a square, then it is a rectangle"?)

   (h) If $x + 3 = 7$, then $x = 4$.

   (i) All rats are rodents.

   (j) If two lines in a plane are perpendicular, then they meet at a point.

   (k) If 4 is greater than 5, then a triangle has three sides.

   (l) Rattlesnakes are poisonous.

   (m) If a snake is poisonous, then it is a rattlesnake.

3. According to Definition 3.4, a conditional is true *except* when the hypothesis is true and the conclusion is false. For some of the conditionals in Problem 2 this exception does not occur. In Problem 2(a), for example, the facts of geography eliminate this exception; so this conditional must be accepted as true. Which of the remaining conditionals of Problem 2 must be accepted as true on the basis of generally available knowledge? For the other conditionals, explain why they are not acceptable.

4. (a) Write the converse of each conditional in Problem 2.

   (b) Which of the converses must be accepted as true on the basis of generally available knowledge?

(c) Does the list in part (b) differ from that in Problem 3? Is your answer consistent with the truth tables of this section? Explain.

5. (a) Make up a true conditional of your own whose converse is not true.

   (b) Make up a true conditional of your own whose converse is true.

   (c) Make up a conditional of your own which is not true, but whose converse is true.

6. The following definition is useful:

Definition 3.7 Inverse of a Conditional If $p \rightarrow q$ is a conditional, its *inverse* is the conditional $(\sim p) \rightarrow (\sim q)$.

Note that the inverse is formed by negating both parts of the given conditional. Thus the inverse of "If a whole number ends with the digit 6, then it is even" is "If a whole number does not end with the digit 6, then it is not even."

(a) Write the inverse of each of the parts of Problem 2.

(b) Which of the inverses of part (a) are true conditionals? Compare your answers with those of Problem 3.

(c) On the basis of the results of part (b), decide whether or not the inverse of a conditional is logically equivalent to it.

(d) Verify your answer to part (c) by setting up a truth table for $(\sim p) \rightarrow (\sim q)$ and comparing it with that for $p \rightarrow q$. This may be done in the following way:

p	q	$\sim p$	$\sim q$	$(\sim p) \rightarrow (\sim q)$
T	T	F	F	T
T	F	F	T	

etc.

(e) Compare the truth table for $(\sim p) \rightarrow (\sim q)$ with that for $q \rightarrow p$ (Table 3.5). What do you conclude?

(f) Illustrate your conclusion in part (e) by comparing your answers to part (b) with those of Problem 4(b).

7. A further definition completes the cycle. It is

Definition 3.8 Contrapositive If $p \rightarrow q$ is a conditional, its *contrapositive* is the conditional $(\sim q) \rightarrow (\sim p)$.

Note that the contrapositive is formed by simultaneously negating and interchanging both parts of the original conditional. Thus the contrapositive of "If a whole number ends with the digit 6, then it is even" is "If a whole number is not even, then it does not end with the digit 6."

(a) Write the contrapositive of each of the parts of Problem 2.

(b) For the contrapositive, answer questions similar to those of Problem 6(b), (c), and (d).

(c) Summarize the relations among $p \rightarrow q$, $(\sim p) \rightarrow (\sim q)$, $q \rightarrow p$, and $(\sim q) \rightarrow (\sim p)$, stating which are logically equivalent to which.

8. (a) Do you think the statements "Satisfaction guaranteed, or your money back" and "If you are not satisfied, then you get your money back" mean the same thing?

   (b) Construct truth tables for $p \rightarrow q$ and $(\sim p) \vee q$, and compare.

   (c) What relationship is there between parts (a) and (b). Do the results of part (b) support your answer to part (a)?

9. Construct a truth table for each of the following logical expressions:

   (a) $(p \wedge q) \rightarrow p$ (b) $(p \vee q) \rightarrow q$ (c) $p \rightarrow (p \vee q)$

   (d) $[p \wedge (p \rightarrow q)] \rightarrow q$ (e) $[q \wedge (p \rightarrow q)] \rightarrow p$

   (f) $[(\sim q) \wedge (p \rightarrow q)] \rightarrow (\sim p)$ (g) $[p \wedge (\sim q)] \rightarrow [\sim(p \rightarrow q)]$

   (h) $[p \vee (q \rightarrow (\sim p))] \rightarrow [q \wedge (\sim q)]$ (i) $[(p \rightarrow q) \wedge (q \rightarrow r)] \rightarrow (p \rightarrow r)$

10. A logical expression which has the value T for all lines of the truth table is called a *tautology*; one which always has the value F is called a *contradiction*. List the expressions in Problem 9 which are tautologies; list separately those which are contradictions.

*11. Collect one or more examples from your own reading or experience of the fallacy of "assuming the converse" and of the equivalent fallacy of "assuming the inverse."

*12. (a) Construct a truth table for $[(p \wedge q) \rightarrow r] \rightarrow [p \rightarrow (q \rightarrow r)]$.

   (b) Construct a truth table for $[p \rightarrow (q \rightarrow r)] \rightarrow [(p \wedge q) \rightarrow r]$, and compare it with that for part (a).

   (c) Note that the expressions of parts (a) and (b) have the overall form of conditionals; and that, from this point of view, they are converses of each other. Does the comparison in part (b) disprove our earlier warnings against "assuming the converse"? Explain.

Section 3.6 The Implication

In the preceding section we defined the conditional $p \rightarrow q$ as a compound statement of the form "if . . . , then. . . ." having a certain truth table with one F and three T entries. Initially we were concerned with the form of the statement as a whole more than with the specific contents of p and of q.

Later, however, we called attention to certain conditionals which seemed "reasonable" or "acceptable" or "true." Let us examine more carefully what is meant by a true conditional.

Intuitively, by a true statement of any kind we mean a statement which is correct under *all* circumstances: it admits of no exceptions. Thus in accepting as true statements like "$2 + 3 = 5$" and "If a triangle has two equal sides, then it has two equal angles," we are proclaiming that 2 and 3 always have a sum of 5 and that no triangle exists which has two equal sides but all its angles unequal. To claim, in particular, that a *conditional* is true means to assert the impossibility of its conclusion being false when its hypothesis is true. Since we are thus forbidden a certain combination of truth values in the conditional, we are really establishing a certain relationship between its hypothesis and conclusion. The verb *imply* is often used to describe this relationship. We may, for example, replace the quoted conditional by the stronger statement, "In a triangle, the equality of two of its sides *implies* the equality of two of its angles." More generally, we are led to the following:

Definition 3.9 Implication A statement p *implies* a statement q, or the conditional $p \to q$ is an *implication* (in symbols, $p \Rightarrow q$) iff the conditional $p \to q$ is never false.

By Definition 3.9, then, an implication, as distinct from a conditional, cannot have a true hypothesis and a false conclusion. For this reason, we may say, when $p \Rightarrow q$, that the conclusion, q, *follows* (logically) *from* the hypothesis, p.

In at least two circumstances it is possible to demonstrate that a conditional is in fact an implication. The first is illustrated by those parts of Exercises 3.5, Problem 9 which are tautologies. As a simple example, let us consider Problem 9(a): $(p \wedge q) \to p$. It is intuitively clear already that, from the hypothesis of $(p \wedge q)$, p alone — or q alone, for that matter — surely follows. Table 3.6 confirms this.

TABLE 3.6

p	q	$p \wedge q$	$(p \wedge q) \to p$
T	T	T	T
T	F	F	T
F	T	F	T
F	F	F	T

Note that in none of the four lines does the combination of a true hypothesis and a false conclusion occur: so the conditional is an implication and the expression is a tautology. Moreover, $(p \wedge q) \Rightarrow p$, whatever the content of p and q; so this implication form applies in mathematics and in other fields.

If you protest that this is a trivial example, you should review the more complicated tautologies of Problem 9 or consider the imposing list in Stoll.[10]

A second circumstance is of a subtler and more limited nature, since it depends strongly on the content of p and q — i.e., on the system to which the logic is being applied. As one example, let us use the conditional "If a whole number ends with the digit 5, then it is divisible[11] by 5." Here p is the statement, "a whole number ends with the digit 5," and q is "it (the whole number) is divisible by 5." This conditional will be an implication if we can somehow show that the T–F combination of the second line of the truth table is impossible for this particular choice of p and q.

Now, this example relates to whole numbers, written in Hindu-Arabic notation, which obey certain laws or restrictions that can be proved logically, though we have not yet done so. One consequence of these restrictions is the fact that no whole number exists which ends in 5 (so that p is true) and fails to be divisible by 5 (so that q is false). Thus, for this particular condition, the second line of Table 3.6 is eliminated: it is an implication. What about the other lines of the table? Undoubtedly you can think of whole numbers which illustrate them.

A second mathematical example of this kind of implication is, "If a quadrilateral is a square, then its diagonals are of equal length." Lines 1, 3, and 4 of the truth table are depicted, in order, in Figure 3.4. Since it is impossible to have a square whose diagonals are of unequal length, line 2 of the truth table for the corresponding conditional does not occur.

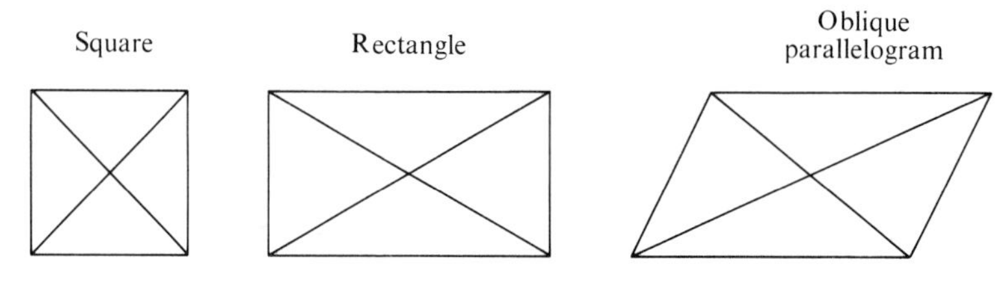

Figure 3.4

As a third example, a nonmathematical one, consider the following: "If a person is serving legally as a Senator in the United States Congress, then he is at least 30 years old." You should check to see that lines 1, 3, and 4 of the truth table for this conditional are possible, and note that line 2 is impossible, according to the United States Constitution.

Let us complete the picture by citing three examples of conditionals which are not implications, following them by one strange example which is an implication:

1. If a quadrilateral has four right angles, then it is a square.

---

[10]See the Stoll reference, Theorem 2.4, p. 73.

[11]A whole number is *divisible by n* (where n is a *nonzero* whole number) iff, when the number is divided by n, the remainder is zero.

2. If two whole numbers are odd, then their sum is odd.
3. If it is now raining in Saskatchewan, then today is Monday.
4. If $2 + 3 = 7$, then today is Monday.

Since some rectangles are not squares, it is logically possible for the hypothesis of the first conditional to be true and the conclusion false; so it is not an implication. Which lines of the truth table for $p \rightarrow q$ can occur in this example?

The second conditional is patently false, for not only is the second line of the truth table for $p \rightarrow q$ possible, but the first line cannot occur.

The third illustration may seem bizarre; but it accords with the definition of a conditional, since it has the required structure. The possibility that rain may fall in Saskatchewan on Tuesdays shows, however, that it is not an implication. Indeed, we note that, since the truth values of the hypothesis and conclusion are quite unrelated, all four lines of the truth table for $p \rightarrow q$ are possible.

By contrast, the fourth example, despite the lack of relation between its two clauses, is an implication: for the fact that the hypothesis, $2 + 3 = 7$, cannot be true, whatever the day of the week, rules out the second line of the truth table for $p \rightarrow q$. Which other line(s) is (are) also logically impossible?

Among all the examples of this section, only the implications are valuable for deductive reasoning. Their significance and power consist in the fact that, when the hypothesis is true, the conclusion *must* be true. For this reason, much effort in mathematics and in other fields where deductive reasoning is applied goes into proving that specific conditionals are implications.

This is what the high school geometry course (and larger and larger portions of other mathematics courses) is attempting to do by establishing theorems. Although theorems are not always written in the "if . . . , then. . . ." form, they can always be rewritten that way. An important theorem in high school geometry, for instance, is "The base angles of an isosceles triangle are equal." Rewritten as a conditional, it becomes, "If a triangle has two equal sides, then the angles opposite these sides are equal." Once this conditional has been proved to be an implication it can be applied to *any* triangle with two equal sides, in full confidence that the angles opposite these two sides must be equal. More generally, an implication can be considered as a prototype of deductive reasoning, which, when applied to an extensive system, accepts as true a complex set of hypotheses from which, by the application of laws of logic, a set of conclusions or theorems is derived. Moreover, if the truth of the hypothesis is assumed, then the truth of the conclusion is guaranteed. This matter will be discussed more fully in Section 3.11.

Since previous logical connectives, such as $\wedge$ and $\vee$, have been easy to associate with specific operations on sets, it is natural to search for a set-theoretic link with implication, $p \Rightarrow q$. To pursue this, let us construct a Venn diagram to depict the implication "If a whole number ends with the digit 6, then it is even." (You should check that this statement *is* an implication, according to Definition 3.9.) Clearly $U = W$, and the truth set of p (the hypothesis) is $S = \{6, 16, 26, 36, \ldots\}$. The truth set of q (the conclusion), moreover, is $E = \{0, 2, 4, 6, 8, \ldots\}$. Now it is evident that $S \subseteq E$: the subset concept parallels implication for propositions. Specifically, the conditional $p \rightarrow q$ is an implication, $p \Rightarrow q$, iff the truth set of p (here, S) is a subset of the truth set of q (here, E).

We have not previously encountered the Venn diagram for the subset relation. It uses the fact that, of the four regions into which two sets such as S and E divide U, region #2 is empty, and so must be shaded. Thus Figure 3.5 shows the Venn diagram for $S \subseteq E$. This diagram also reveals that there are no elements of W for which p is T and q is F, so that the second line of the truth table does not occur, and $p \Rightarrow q$ is indeed an implication. Similarly, the third and fourth lines are consistent with $S \subseteq E$. Thus the truth table for $p \rightarrow q$, as we somewhat arbitrarily defined it, is consistent with the subset relation. Had we made other choices for the third or fourth lines, this would not have been so. Thus we have a belated confirmation of the appropriateness of the conditional, as defined in Definition 3.4.

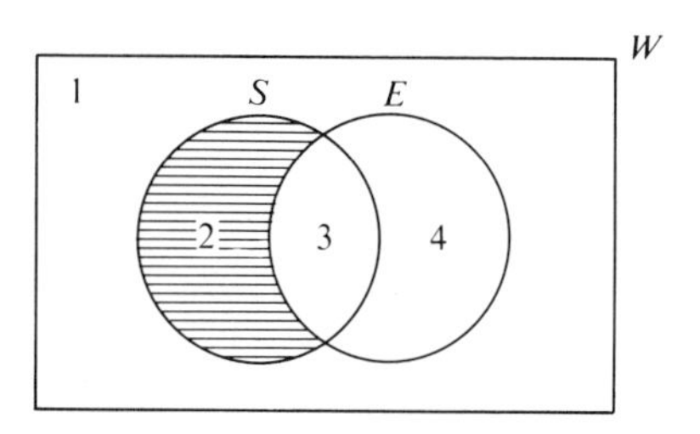

Figure 3.5

EXERCISES 3.6

1. (a) Which of the conditionals of Exercises 3.5, Problem 2 are implications?

   (b) How do your answers to part (a) relate to Exercises 3.5, Problem 3?

2. For each of the following, give an example of your own, if possible; if not possible, explain why no such example exists:

   (a) A conditional that is an implication.

   (b) A conditional that is not an implication.

   (c) An implication that is not a conditional.

   (d) An implication whose converse is an implication.

   (e) An implication whose converse is not an implication.

   (f) An implication whose contrapositive is an implication.

   (g) An implication whose contrapositive is not an implication.

3. The discussion in the text shows that $(p \wedge q) \Rightarrow p$ is correct — i.e., that

$(p \wedge q) \rightarrow p$ is an implication. Which of the following are correct? In each case, explain or justify your answer.

(a) $(p \wedge q) \Rightarrow q$ (b) $p \Rightarrow (p \wedge q)$

(c) $p \Rightarrow (p \vee q)$ (d) $[p \wedge (p \rightarrow q)] \Rightarrow q$

(e) $[q \wedge (p \rightarrow q)] \Rightarrow p$ (f) $[(p \rightarrow q) \wedge (q \rightarrow r)] \Rightarrow (p \rightarrow r)$

4. Let U be the set of living creatures and consider the conditional $p \rightarrow q$: "If it is a butterfly, then it is an insect."

   (a) Let B and I be the truth sets of p and q, respectively. Is either B or I a subset of the other? Explain.

   (b) Draw a Venn diagram showing the relations among U, B, and I, shading any empty regions of U.

   (c) Does this conditional have the same meaning as "All butterflies are insects"? Explain by reference to part (a).

   (d) Is this conditional an implication? Justify your answer by reference to the subset relation as explored in the preceding parts.

*5. Solve the following puzzle, noting as or after you do so the role of unstated implications in your reasoning (e.g., "If two persons are twins, then they are of the same age").

   Mr. Scott, his sister, his son, and his daughter are tennis players. The following facts refer to the people mentioned:

   1. The best player's twin and the worst player are of opposite sex.
   2. The best player and the worst player are the same age.

   Which one of the four is the best player?[12]

Section 3.7 The Biconditional and Logical Equivalence

Let U be the set of all persons who are living in the United States, and consider the conditional $p \rightarrow q$: "If a person is 18 years old, then he is entitled to vote in federal elections." It is evident that this is not an implication, because an 18-year-old alien, for example, is not entitled to vote. The converse, $q \rightarrow p$, is the conditional, "If a person is entitled to vote in federal elections, then he is 18 years old." This conditional is also not an implication. Why not? Sometimes, however, we are interested in the *conjunction* of a conditional and its converse, namely, $(p \rightarrow q) \wedge (q \rightarrow p)$. You can easily illustrate this expression simply by joining the two quoted sentences with an *and*.

It is instructive to work out the truth table for $(p \rightarrow q) \wedge (q \rightarrow p)$, as in Table 3.7. An examination of the last column shows that $(p \rightarrow q) \wedge (q \rightarrow p)$

---

[12]George J. Summers, *New Puzzles in Logical Deduction* (New York: Dover Publications, Inc., 1968), p. 3. Reprinted through permission of the publisher.

TABLE 3.7

p	q	$p \to q$	$q \to p$	$(p \to q) \wedge (q \to p)$
T	T	T	T	T
T	F	F	T	F
F	T	T	F	F
F	F	T	T	T

is true if and only if p and q, separately, have the same truth values, as they do in the first and last lines. You may note that the middle two lines are exactly the cases which prevent the quoted conditional and its converse from being implications.

Since $(p \to q) \wedge (q \to p)$ is rather cumbersome to write, it is usually abbreviated to $p \leftrightarrow q$. The symbol $\leftrightarrow$ is actually quite ingenious, for it combines the arrow in $p \to q$ with the arrow in $q \to p$. This new symbol, moreover, has an appropriate name: it is called the *biconditional*. Leaving out the intermediate columns, its truth table reduces to Table 3.8. In view of the symmetric roles

TABLE 3.8

p	q	$p \leftrightarrow q$
T	T	T
T	F	F
F	T	F
F	F	T

which p and q play in the biconditional, we may write $p \leftrightarrow q$ and $q \leftrightarrow p$ interchangeably (i.e., $\leftrightarrow$ is commutative). For a reason which we shall shortly investigate, the symbol $p \leftrightarrow q$ (or $q \leftrightarrow p$) may be read: "p if and only if q" or "p iff q." It is time, however, for the formal

Definition 3.10 Biconditional If p and q are any two propositions, the *biconditional* "p if and only if q" (in symbols, $p \leftrightarrow q$) is a proposition which is true when p and q have the same truth values, and is false otherwise.

Thus the phrase "if and only if" is made explicit by Definition 3.10. We make it seem intuitively reasonable by means of Table 3.9. Obviously the first two statements have the same meaning: for the *contents* of the hypothesis and conclusion are not disturbed, although the order of writing them is reversed. We would, in addition, accept the specific illustrations for Statements 1 and 2 as true for integers. Statement 3, by contrast, is the converse of the first two in a new form: it is important to realize that "p only if q" has the same meaning as "if q, then p." Thus the specific illustration says that $x^2 = 9$ only when $x = 3$, so that, if $x^2 = 9$, x must then have the value 3. Of course, these statements are not true in this illustration, because $x^2 = 9$ *also* when

$x = -3$, an integer. Since Statement 3 of the illustration is not true, Statement 4 cannot be true.

TABLE 3.9

General Form	*Specific Illustration*
1. If p, then q. $(p \rightarrow q)$	If $x = 3$, then $x^2 = 9$.
2. q, if p. $(p \rightarrow q)$	$x^2 = 9$, if $x = 3$.
3. q only if p. $(q \rightarrow p)$	$x^2 = 9$ only if $x = 3$.
4. q if and only if p. $(q \leftrightarrow p$ or $p \leftrightarrow q)$	$x^2 = 9$ iff $x = 3$.

Sometimes, however, *both* conditionals, Statements 1 (or 2) and 3, are true, and so the biconditional, Statement 4, is true also. As a specific illustration we consider:

1.′ If $x = 3$, then $2x = 6$.

2.′ $2x = 6$ if $x = 3$.

3.′ $2x = 6$ only if $x = 3$.

4.′ $2x = 6$ iff $x = 3$.

In this case we should like to assert that the two *simple* propositions, "$2x = 6$" and "$x = 3$," mean the same thing, or are *logically equivalent*. Before making a formal definition, however, let us recall what we did in Section 3.6. There we singled out true conditionals and called them "implications"; we introduced the symbol $\Rightarrow$ and read $p \Rightarrow q$ as "p implies q." Similarly, we single out true biconditionals and write $p \Leftrightarrow q$ to indicate that "p implies q" and "q implies p" are both true, or that p and q, as simple propositions, have the same meaning, or truth value. So for the immediately preceding illustration, only the first and fourth lines of the truth table for $p \leftrightarrow q$ apply: either $2x = 6$ and $x = 3$ are both true, or they are both false. To put the matter another way, the second and third lines, for which $p \leftrightarrow q$ is false, *cannot* occur. For the first illustration, on the contrary, $x^2 = 9$ can be true even though $x = 3$ is false (i.e., if $x = -3$); in this instance, the biconditional has the value F, so those two equations do not have the same truth value. Thus we announce the following:

Definition 3.11 Logical Equivalence A proposition p is *logically equivalent* to a proposition q, or the biconditional $p \leftrightarrow q$ is a *logical equivalence* (in symbols, $p \Leftrightarrow q$), iff the biconditional $p \leftrightarrow q$ is never false.

You will note that we have been using the phrase "logically equivalent" ever since its introduction in Exercises 3.3, Problem 6; but we have delayed the formal definition and symbolism until the related concept of implication was developed. We now show, by an example, that the earlier use of the phrase "logical equivalence" accords perfectly with Definition 3.11.

Let r and s be any propositions. We wish to demonstrate one of the DeMorgan laws of logic — say, that $\sim(r \wedge s)$ is logically equivalent to $(\sim r) \vee (\sim s)$. It is an easy exercise to obtain the columns to the left of the double vertical line in Table 3.10. Indeed, Exercises 3.4, Problem 7(b) and (c), re-

TABLE 3.10

r	s	$\sim(r \wedge s)$	$(\sim r) \vee (\sim s)$	$[\sim(r \wedge s)] \leftrightarrow [(\sim r) \vee (\sim s)]$
T	T	F	F	T
T	F	T	T	T
F	T	T	T	T
F	F	T	T	T

quested the above truth tables and others. Since the third and fourth columns are identical, the expressions $\sim(r \wedge s)$ and $(\sim r) \vee (\sim s)$ obviously have the same truth values in all four cases. However, by Definition 3.11, this means that the biconditional is a logical equivalence, as we expected. As a practical matter, of course, a visual comparison of the third and fourth columns already demonstrates logical equivalence informally, so that the fifth column is unnecessary work. But this last column confirms that our use of the phrase "logically equivalent" has been consistent throughout this chapter.

Definitions are important examples of logical equivalence. For instance, a whole number is defined to be *even* iff it is divisible by 2. Thus the statement p: "x is even" is logically equivalent to q: "x is divisible by 2." The purpose of the definition is to replace the more or less complicated expression, "divisible by 2," by the single word "even." A more dramatic simplification is achieved in the definition of a circle given in Exercises 2.1, Problem 6.

In Section 3.6, a parallelism between the concepts of implication $(p \Rightarrow q)$ and subset $(P \subseteq Q)$ was uncovered. It is correspondingly easy to discover a set-theoretic association with the concept of logical equivalence $(p \Leftrightarrow q)$. We use as an example the theorem from high school geometry, "A triangle has three equal angles iff it has three equal sides." To aid our analysis we may rewrite this theorem as: "If a triangle has three equal sides, then it has three equal angles; and if a triangle has three equal angles, then it has three equal sides." Here U is the set T of all triangles, S is the truth set containing all triangles with three equal sides, and A is the truth set of all triangles with three equal angles. Associated with $p \Rightarrow q$, as we have seen, is the relation $S \subseteq A$; and with $q \Rightarrow p$ is the relation $A \subseteq S$. But since $S \subseteq A$ and $A \subseteq S$ both hold, by Definition 2.2 we have $S = A$. Thus the logical equivalence of two propositions is mirrored in the equality of their truth sets. The Venn diagram for these equal sets, shown in Figure 3.6, indicates by its shading that regions #2 and #4 are both empty. Thus it demonstrates that either p and q are both true (region #3) or both false (region #1), so that $p \Leftrightarrow q$.

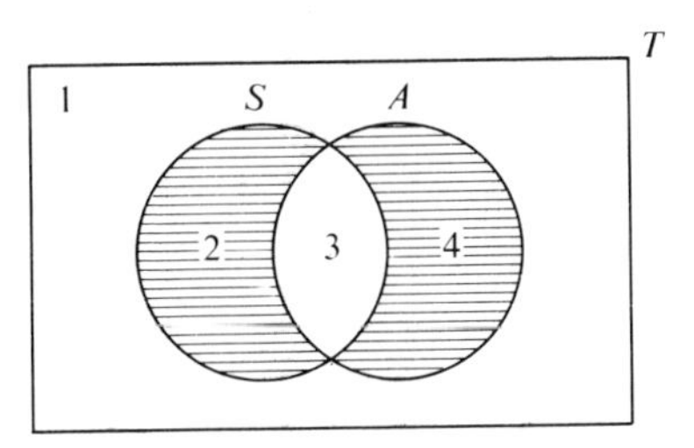

Figure 3.6

Although the logical symbols and their rules of combination have some significance in themselves and in relation to the theory of sets, for most mathematicians and laymen, logic is important chiefly as a basis for reasoning deductively about mathematics and other sciences. More specifically, parallel to the tautologies developed by truth tables are *rules of inference*, which guide the reasoning processes used in constructing proofs, developing models, and using the "scientific method."

As an example of a rule of inference, we choose probably the most frequently employed one, the *Rule of Detachment*, sometimes known by its Latin name, *modus ponens*. In Exercises 3.5, Problem 9(d), the expression $[p \wedge (p \rightarrow q)] \rightarrow q$ is shown to be a tautology, so that $[p \wedge (p \rightarrow q)] \Rightarrow q$, as given in Exercises 3.6, Problem 3(d), is correct. Now, suppose p and q are statements in some system, and that p is given as true, and also that $p \Rightarrow q$ has been established in the system. The Rule of Detachment then states that q must be true in the system. Schematically, this rule may be diagrammed as

$$\begin{array}{c} p \\ \underline{p \rightarrow q} \\ \therefore^{13}\ q \end{array}$$

You should notice that this rule corresponds to the first line of the truth table for $[p \wedge (p \rightarrow q)] \rightarrow q$, since p and $p \rightarrow q$ are both given as true. The first two propositions, p and $p \rightarrow q$, are called *premises* of the argument; q is called its *conclusion*.

As an example we cite the following argument:

p:	Today is Saturday.
$p \rightarrow q$:	If today is Saturday, then tomorrow will be Sunday.
$\therefore\ q$:	Tomorrow will be Sunday.

[13]The symbol $\therefore$ means "therefore."

In the Gregorian calendar system we use, $p \rightarrow q$ is indeed an implication. So if, in fact, today *is* Saturday, the Rule of Detachment guarantees that tomorrow will be Sunday.[14]

Another example is the following:

p:	$x + 3 = 11.$
$p \rightarrow q$:	If $x + 3 = 11$, then $x = 8$.
$\therefore q$:	$x = 8.$

In elementary algebra, the above argument is frequently abbreviated: "The solution of the equation $x + 3 = 11$ is $x = 8$." The logical form, however, emphasizes that the equation $x + 3 = 11$ must be given, and that the conditional, $(x + 3 = 11) \rightarrow (x = 8)$, must be an implication, as it is in our system of algebra.

What happens if the conditional is not an implication is illustrated by the following example:

p:	The wind is from the south.
$p \rightarrow q$:	If the wind is from the south, then the temperature will rise.
$\therefore q$:	The temperature will rise.

While this conditional is rather probable in the United States, weather records would doubtless show instances where it is false. For predictions in the southern hemisphere, on the other hand, the conditional might frequently be false. In any case, the possibility that $p \rightarrow q$ is false allows the second line of its truth table, permitting q to be false, and so denying the conclusion.

Other instances of *modus ponens*, other rules of inference, and several tempting fallacies appear in the exercises.

EXERCISES 3.7

1. For each of the following conditionals, $p \rightarrow q$, write its converse; then write $(p \rightarrow q) \wedge (q \rightarrow p)$; finally, write your previous answer in abbreviated form, $p \leftrightarrow q$.
   (a) If $x - 5 = 1$, then $x = 6$.
   (b) If two lines in a plane are perpendicular, then they meet at a point.
   (c) If a set contains no elements, then it is a null set.

---

[14]It is worth remarking that in the proposed World Calendar system, the conditional is *not* an implication: for the last day of December, which is always a Saturday, is not followed by Sunday. So the above argument does not succeed *in that system!*

(d) It will rain if the barometer falls rapidly.

(e) If $2 + 3 = 7$, then $8 - 2 = 5$.

2. Which of the biconditionals of Problem 1 are logical equivalences? Explain.

3. (a) Construct a truth table for $[(\sim p) \rightarrow q] \leftrightarrow (p \vee q)$.

   (b) On the basis of your answer to part (a), is it correct to write $[(\sim p) \rightarrow q] \Leftrightarrow (p \vee q)$?

   (c) Relate your answer to part (b) to Exercises 3.5, Problem 8.

4. Use truth tables to determine which of the following are correct:

   (a) $[\sim(\sim p)] \Leftrightarrow p$
   (b) $(p \leftrightarrow q) \Leftrightarrow (q \leftrightarrow p)$

   (c) $(p \rightarrow q) \Leftrightarrow [(\sim p) \rightarrow (\sim q)]$
   (d) $(p \rightarrow q) \Leftrightarrow [(\sim q) \rightarrow (\sim p)]$

   (e) $[\sim(p \rightarrow q)] \Leftrightarrow [p \wedge (\sim q)]$

   (f) $[(p \rightarrow q) \wedge (r \rightarrow q)] \Leftrightarrow [(p \vee r) \rightarrow q]$

5. Consider the implication "If a quadrilateral has its opposite sides parallel, then it is a parallelogram."

   (a) Write the converse of the implication.

   (b) Is your answer to part (a) also an implication? Note that the quoted statement is the *definition* of a parallelogram.

   (c) Taking U to be the set Q of all quadrilaterals, make a Venn diagram to show the relations among the sets Q, S (quadrilaterals with opposite sides parallel), and P (parallelograms), using correct shading.

   (d) How does this example illustrate the parallelism between the concept of logical equivalence in the algebra of logic and a certain relation (Which one?) between two sets in the algebra of sets.

6. An exclusive club has a private dining room in which all members, but no one else, may eat. Let p be "x is a member of the club" and q be "x may eat in the private dining room."

   (a) Use p and q to illustrate the four conditional forms which appear in Table 3.9.

   (b) Which of the conditionals of part (a) are implications? Is the biconditional a logical equivalence? Explain.

   (c) If a sign reading "Members Only" were placed at the entrance to the dining room, would it express completely the logical meaning of the first sentence of this exercise? Discuss.

7. Consider the statement "$\angle 1$ and $\angle 2$ are right angles only if they are equal." Choose r as "$\angle 1$ and $\angle 2$ are right angles" and s as "$\angle 1$ and $\angle 2$ are equal angles." Now answer about r and s questions similar to those about p and q in Problem 6(a) and (b).

8. Consider the conditional "If the weather is sunny and if the temperature is above 80° F., then I go swimming." Since the hypothesis is itself a compound statement, this conditional is of the form $(p \wedge q) \rightarrow r$.

   (a) Write the *complete* converse, $r \rightarrow (p \wedge q)$. Does it seem to have the same meaning as the original conditional?

   (b) Check your answer to part (a) by using truth tables to determine whether $[(p \wedge q) \rightarrow r] \Leftrightarrow [r \rightarrow (p \wedge q)]$ is correct.

   (c) In addition to the complete converse, $(p \wedge q) \rightarrow r$ has two *partial* converses, obtained by interchanging a single clause in the hypothesis with one in the conclusion: they are $(p \wedge r) \rightarrow q$ and $(r \wedge q) \rightarrow p$.

      (i) Write the two partial converses for this example.

      (ii) Answer part (b) for the partial converses, comparing them to each other and to the original $[(p \wedge q) \rightarrow r]$.

   (d) (i) Write the complete contrapositive of the original conditional, namely, $(\sim r) \rightarrow [\sim(p \wedge q)]$. Does it seem to have the same meaning as the original?

      (ii) Check your answer to part (i) by means of truth tables.

   (e) In addition to the complete contrapositive, $(p \wedge q) \rightarrow r$ has two partial ones, formed by interchanging the negation of a single clause in the hypothesis with the negation of one in the conclusion: they are $[p \wedge (\sim r)] \rightarrow (\sim q)$ and $[(\sim r) \wedge q)] \rightarrow (\sim p)$. Answer part (d) for the partial contrapositives.

9. (a) To which of the following simple arguments does the Rule of Detachment apply? In each case, assume that the given conditional is an implication in the system under consideration, and that the first premise is true.

      (i) Lines ℓ and m are perpendicular to line n. If two lines are perpendicular to a third line, then they are parallel.

      Therefore lines ℓ and m are parallel.

      (ii) This automobile is cheap.
      If an automobile is cheap, then it is economical to run.

      Therefore this automobile is economical to run.

      (iii) This automobile is economical to run.
      If an automobile is cheap, then it is economical to run.

      Therefore this automobile is cheap.

      (iv) This automobile is not cheap.
      If an automobile is cheap, then it is economical to run.

      Therefore this automobile is not economical to run.

(v) This automobile is not economical to run.
If an automobile is cheap, then it is economical to run.

Therefore this automobile is not cheap.

(vi) $2x - 5 = 1$.
If $2x - 5 = 1$, then $x = 3$.

Therefore $x = 3$.

(vii) $x = 3$.
If $2x - 5 = 1$, then $x = 3$.

Therefore $2x - 5 = 1$.

(b) For each of the exercises in part (a) that does *not* illustrate the Rule of Detachment, try to determine intuitively whether acceptance of the premises guarantees the truth of the conclusion.

*10. (a) Show by truth tables that $[(\sim q) \wedge (p \rightarrow q)] \rightarrow (\sim p)$ is a tautology.

(b) Associated with the tautology of part (a) is a second rule of inference, whose diagram is

$$\begin{array}{c} \sim q \\ p \rightarrow q \\ \hline \therefore \sim p \end{array}$$

To which arguments of Problem 9(a) does this rule of inference apply?

(c) To which of the following arguments does the rule of inference of part (b) apply?

(i) Triangle A and triangle B do not have the same area.
If two triangles are congruent, then they have the same area.

Therefore triangle A and triangle B are not congruent.

(ii) This boiler does not burst.
If the pressure inside this boiler exceeds 100 psi, then this boiler bursts.

Therefore the pressure inside this boiler does not exceed 100 psi.

(iii) The pressure inside this boiler does not exceed 100 psi.
If the pressure inside this boiler exceeds 100 psi, then this boiler bursts.

Therefore this boiler does not burst.

(iv) This boiler bursts.
If the pressure inside this boiler exceeds 100 psi, then this boiler bursts.

Therefore the pressure inside this boiler exceeds 100 psi.

(v) $x^2 - x \neq 0$.
If $x = 1$, then $x^2 - x = 0$.

Therefore $x \neq 1$.

(vi) $x \neq 1$.
If $x = 1$, then $x^2 - x = 0$.

Therefore $x^2 - x \neq 0$.

*11. Quite similar in appearance to the two rules of inference of Problems 9 and *10 are the following patterns:

$$\begin{array}{c} q \\ p \rightarrow q \\ \hline \therefore\ p \end{array} \qquad \text{and} \qquad \begin{array}{c} \sim p \\ p \rightarrow q \\ \hline \therefore\ \sim q \end{array}$$

(a) Construct the truth table for $[q \wedge (p \rightarrow q)] \rightarrow p$, and show that it is not a tautology.

(b) After examining the answer to part (a), explain why the truth of q, together with that of $p \rightarrow q$, do not imply the truth of p. Which line(s) of the truth table show(s) this?

(c) The answer to part (b) shows that the first pattern is not a rule of inference; on the contrary, it is a *fallacy*, or incorrect pattern of reasoning. By a procedure similar to that of parts (a) and (b), show that the second pattern is also a fallacy.

(d) Which parts of Problems 9 and *10 illustrate the two fallacies of this exercise? For each argument, explain why the truth of the first two premises does not imply the truth of the conclusion.

*12. For conditionals with four or more clauses distributed among hypotheses and conclusions, more possibilities for partial converses and contrapositives arise. Two examples from plane geometry are rather neat:

(a) Consider the conditional "If a line passes through the center of a circle and is perpendicular to a chord, then it bisects the chord and the arc of the chord." This is of the form $(p \wedge q) \rightarrow (r \wedge s)$. Form all the converses which have two clauses in the hypothesis, e.g., $(p \wedge r) \rightarrow (q \wedge s)$. It can be proved in geometry that all these converses — as well as the original conditional — are implications. Thus it is possible to state and remember six theorems with no more effort than for one.

(b) Consider the conditional "If a line passes through the center of a circle, and if a perpendicular is drawn to it at its point of intersection with the circle, then the perpendicular is tangent to the circle." This conditional, of the form $(p \wedge q \wedge r) \rightarrow s$, is usually abbreviated: "A line perpendicular to a radius at its outer extremity is tangent to the circle." Here, again, it can be proved that the original and its partial converses of the same form are implications. Write out these partial converses, preferably in abbreviated form.

*Section 3.8 A Formal Comparison of Sets and Logic

In the preceding sections, as we took up a new logical symbol, we showed that it paralleled a concept from set theory. This we did by considering the truth sets of the propositional functions p and q, aiding our intuition by Venn diagrams. Table 3.11 summarizes these parallelisms, where we take A and B to be the respective truth sets of p and q.

TABLE 3.11

Set Theory	*Logic*
Sets: A, B	Propositional functions: p, q
Complementation: A'	Negation: $\sim p$
Intersection: $A \cap B$	Conjunction: $p \wedge q$
Union: $A \cup B$	Disjunction: $p \vee q$
Subset: $A \subseteq B$	Implication: $p \Rightarrow q$
Set equality: $A = B$	Logical equivalence: $p \Leftrightarrow q$
Universal set: U	Universally true function: t
Null set: $\varnothing$	Universally false function: f

We do not have any useful analog in logic for the cartesian product of two sets. Although the truth value of a propositional function usually depends on the specific element of U which is substituted into the function, there are propositions which are true for every element $x \in U$, and also others which are false for every element $x \in U$. Examples of both were given at the end of Section 3.3, where they were shown to have U and $\varnothing$, respectively, as truth sets. We denote a universally true function by the letter t; similarly, f represents a universally false function.

It is now possible to exhibit, in detail, an analogy between sets and logic, using the information provided in Table 2.1. In order to leave something to the reader, we copy only the first two columns of that table, replacing the third by one containing the logical expressions analogous to the second column. Table 3.12 is the result.

TABLE 3.12

Property	*Set Theory*	*Logic*
1. Closure	$A \cap B$ is a unique set.	$p \wedge q$ is a unique propositional function.
2. Commutativity	$A \cap B = B \cap A$	$(p \wedge q) \Leftrightarrow (q \wedge p)$
3. Associativity	$(A \cap B) \cap C = A \cap (B \cap C)$	$(p \wedge q) \wedge r \Leftrightarrow p \wedge (q \wedge r)$
4. Identity	$A \cap U = A$	$(p \wedge t) \Leftrightarrow p$
5. Distributivity	$A \cap (B \cup C) = (A \cap B) \cup (A \cap C)$	$p \wedge (q \vee r) \Leftrightarrow (p \wedge q) \vee (p \wedge r)$
6. Self-Distributivity	$A \cap (B \cap C) = (A \cap B) \cap (A \cap C)$	$p \wedge (q \wedge r) \Leftrightarrow (p \wedge q) \wedge (p \wedge r)$
7.	$A \cap \varnothing = \varnothing$	$(p \wedge f) \Leftrightarrow f$
8. Idempotency	$A \cap A = A$	$(p \wedge p) \Leftrightarrow p$
9. Complementation	$A \cap A' = \varnothing$	$p \wedge (\sim p) \Leftrightarrow f$
10. DeMorgan law	$(A \cap B)' = A' \cup B'$	$[\sim (p \wedge q)] \Leftrightarrow [(\sim p) \vee (\sim q)]$
11.	$A \cap B \subseteq A$	$(p \wedge q) \Rightarrow p$
12.	$A \subseteq B$ iff $A \cap B = A$	$p \Rightarrow q$ iff $(p \wedge q) \Leftrightarrow p$

While the logical expressions in the third column can be justified by means of Venn diagrams, using the set parallelism, they may also be confirmed by use of truth tables. Property 4, for example, is done as follows:

p	t	$p \wedge t$	$(p \wedge t) \leftrightarrow p$
T	T	T	T
F	T	F	T

By definition, t has only the truth value T: this eliminates two cases from the truth table. We note that the first and third columns are identical, indicating logical equivalence; the fourth column confirms this, showing that the biconditional is a tautology and so can be written as $(p \wedge t) \Leftrightarrow p$.

As another example, the truth table analysis of Property 11 is as follows:

p	q	$p \wedge q$	$(p \wedge q) \rightarrow p$
T	T	T	T
T	F	F	T
F	T	F	T
F	F	F	T

Thus $(p \wedge q) \rightarrow p$ is a tautology: $(p \wedge q) \Rightarrow p$.

*EXERCISES 3.8

*1. Record the other half of the analogy exhibited above by pairing the third column (Union) of Table 2.1 with a new column of corresponding logical expressions.

*2. Supply truth tables to justify some of the logical expressions (especially Properties 7–12) occurring in Table 3.12 and its extension in Problem *1.

Section 3.9 Inductive Reasoning

In Section 1.3 we met an example of inductive reasoning. You were asked to add certain odd numbers and discover a pattern in their sums. The first step was to observe a number of particular sums,

$$1 = 1^2 \quad 1 + 3 = 2^2 \quad 1 + 3 + 5 = 3^2 \quad \text{etc.}$$

and to note that they are successive perfect squares. The next step was to create a general statement (in this instance, an equation) which would include all the particular observations as special cases. This procedure is typical of inductive reasoning: first collect a set of individual facts or statements; on the basis of these, try to frame a generalization which includes the particular facts but extends to cover all facts of the same kind. Drawing general conclusions from a set of particular facts is a creative and valuable process which is responsible for many advances in mathematics, science, and other fields. It can also, however, lead to serious errors: for it is possible that later facts will fail to fit the generalization; the generalization must then be rejected and the search renewed for a more comprehensive one. Fortunately, the more different facts which can be adduced to support the generalization, the more confidence we can have in it, though we can never be 100% sure.

We discuss two other instances of inductive reasoning to illustrate the caution necessary in reaching conclusions by this method. To begin with, suppose a public official wishes to learn his constituents' views on an important issue. He may call up two of his friends to get their opinions; and if they are the same, he may conclude that all his constituents have similar views. He has reasoned inductively, but dangerously so, since his two friends may have been biased and, in any case, are a negligible fraction of his constituents. Stated otherwise, his conclusion has a probability of being correct that is hardly better than guessing or flipping a coin, and may even be worse. The official may, on the other hand, hire an expert pollster to interview a carefully chosen sample of 500 of his constituents. The expert may then be able to make a generalization from these data which has a rather high probability of being correct; and he may even be able to estimate this probability.

As another example, consider the sun. In all recorded history, and presumably before that, the sun has risen every day. On the basis of billions of observations by millions of people across the centuries, we may conclude that the sun rises every day; but even this generalization is not certain, though it has an extremely high probability of being correct. Astronomers, to be sure, may be able to furnish more powerful reasons on which to base such a conclusion. But on the basis of simple repeated observations by mankind through the millenia, the occurrence of sunrise tomorrow is not an assured, though highly probable, event.

Thus we may characterize inductive reasoning as a process of drawing general conclusions from a set of particular facts, with the realization that such conclusions are never more than probable.

EXERCISES 3.9

1. A nurse who has responsibility for training diabetics to give themselves insulin injections has trained 250 persons during the past year. She found, in most cases, that the first self-administered injection was much harder for the patient than subsequent ones. Explain how this description illustrates inductive reasoning. Can the nurse be sure that the next patient she helps will have less difficulty with the second shot than with the first? Why or why not?

2. Give several examples of inductive reasoning, being careful to identify the particular facts and to state clearly the resulting generalization. If possible, give an estimate of your degree of confidence in the conclusion.

*3. A triangle has three sides and no diagonals (A *diagonal* of a polygon is a segment joining two nonadjacent vertices); a quadrilateral has four sides and two diagonals, indicated by dotted segments in Figure 3.7. A pentagon has five sides. How many diagonals does it have? By actual trial it is possible to construct exactly five diagonals in a pentagon. Thus we have the start of a table:

Diagonals of a Polygon

Number of Sides	*Number of Diagonals*
3	0
4	2
5	5

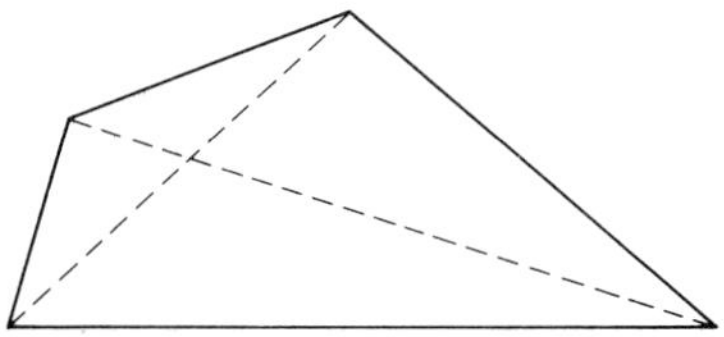

Figure 3.7

We could continue the table by drawing figures and constructing and counting diagonals as far as we like: but there ought to be a more efficient way. Is it possible to discover by inductive reasoning a general pattern or formula by which to determine, without continuing to draw figures, the number of diagonals in a polygon of n sides?

(a) As an aid in discovering and checking such a formula, extend the table a few lines.

(b) Describe or explain any pattern or formula you find.

(c) If possible, present an argument that guarantees the correctness of your formula.

(d) Would the argument requested in part (c) be an instance of inductive reasoning? Explain.

Section 3.10 Deductive Reasoning: Syllogisms

Is it true that the sum of two odd numbers is an even number? We may cite as examples: $1 + 1 = 2$, $7 + 9 = 16$, $3 + 111 = 114$, and $15 + 19 = 34$. One student in a college general mathematics course did just that and then concluded, "Therefore, two odd numbers, when added together, *must* always equal an even number." When the class's opinions were asked, more than 40% accepted this argument as a proof.[15] In subsequent experiments with his general mathematics classes, the writer obtained similar results, varying from 41% to 59% at the beginning of the course to 8% (still!) near the end.

The difficulty, of course, is that we are relying on *inductive* reasoning, which, though a powerful suggester, leads only to a *probable* conclusion. We cannot establish a statement beyond doubt by examining individual cases unless we examine *every* case. This we cannot do here, because the set of odd numbers is infinite. A proof[16] can be made only by *deductive* reasoning, based on certain assumptions about odd and even numbers.

Perhaps you are tempted to dismiss the statement about the sum of two odd numbers as being so obvious that it does not need proof. If so, you are invited to consider the following problem, presented by Leo Moser:

> Two or more points are placed anywhere on the circumference of a circle. Every pair is joined by a straight line. Given n points, what is the maximum number of regions into which the circle can be divided?[17]

Figure 3.8 shows that for $n = 2$ we obtain two regions, and that introducing a third point yields two more regions or a total of four. Experiments with

---

[15]H. Sitomer, "Motivating Deduction," *Mathematics Teacher*, LXIII (Dec., 1970), 661.

[16]See Exercises 4.10, Problem *6(b).

[17]This problem is discussed in M. Gardner's "Mathematical Games," *Scientific American*, 221:2 (Aug., 1969), 121, first paragraph. See also the Gibbs reference.

$n = 4$ and $n = 5$ lead to an obvious and simple formula. For $n = 6$, however, the formula *fails!* Here the "obvious" is wrong.

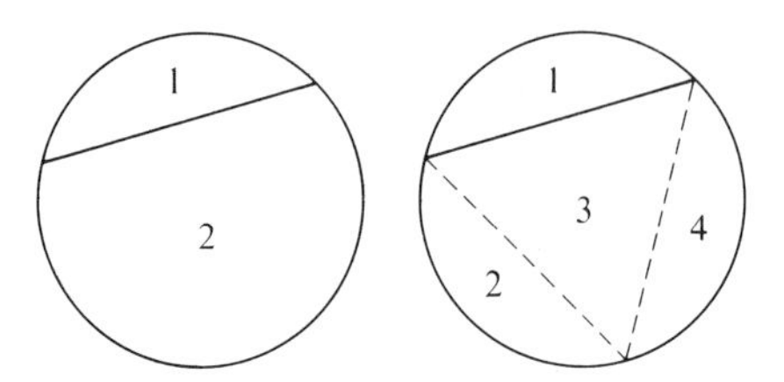

Figure 3.8

Many more examples could be adduced to show the danger of relying solely on inductive reasoning, unsupported by deductive argument. But what exactly is the nature of deductive reasoning, and why is it so helpful? In this chapter, we have already had examples of it, though they were not so labeled. Such were the proofs by truth tables of the equivalence of various logical expressions, and the applications of the Rule of Detachment and other rules of inference. In all cases, acceptance of certain given statements led us to a guaranteed result. Indeed, the *certainty* of the conclusions reached, once a set of given statements is accepted, is the hallmark of deductive reasoning. We now extend our discussion of deduction by introducing a simple kind of logical exercise, the *syllogism*, which has been studied by philosophers and logicians since the time of Aristotle.

A syllogism consists typically of three propositions describing relations among three sets. The first two propositions are called the *hypotheses*, or *premises:* they are to be accepted without question. From them it may be possible, by the application of the laws of logic, to derive the third proposition, called the *conclusion*. If this is so — if the conclusion follows inescapably from the hypotheses in the sense that accepting the latter makes the conclusion absolutely certain — then we have an instance of deductive reasoning.

A simple syllogism is:

Example 1

Hypotheses: All sparrows are birds.
All birds are living creatures.

Conclusion: All sparrows are living creatures.

Most people would agree intuitively that this syllogism exhibits sound, if somewhat trivial, reasoning. Here the "laws of logic" are quite transparent, since the hypotheses state, in effect, that sparrows constitute a subset of birds, which, in turn, is a subset of living creatures. From Exercises 2.2, Problem 8, for which a formal proof could have been given, we already know that if $A \subseteq B$ and $B \subseteq C$, then $A \subseteq C$. To apply this theorem to Example 1, we let A, B, and C represent the sets *sparrows*, *birds*, and *living creatures*, respectively. The certainty of the conclusion, it must be admitted, has been bought with a price: we have to begin by accepting the hypotheses, or we will have nothing to reason about. So the certainty of the conclusion is relative to our willingness to accept the hypotheses; but once we do, the conclusion follows automatically.

But we have been side-stepping a difficulty. How do we know that the reasoning was actually conducted according to the laws of logic? In brief, how do we know that the argument is strong or *valid?*[18] In Example 1, fortunately, we recognize the subset concept at work; but we need a more general method of testing the validity of an argument. One way is to use Venn diagrams.[19] Let us represent by circles the three sets mentioned in the syllogism: the sparrows (S), birds (B), and living creatures (L). (See Figure 3.9.) The procedure is to depict on the diagram the information given by the hypotheses,

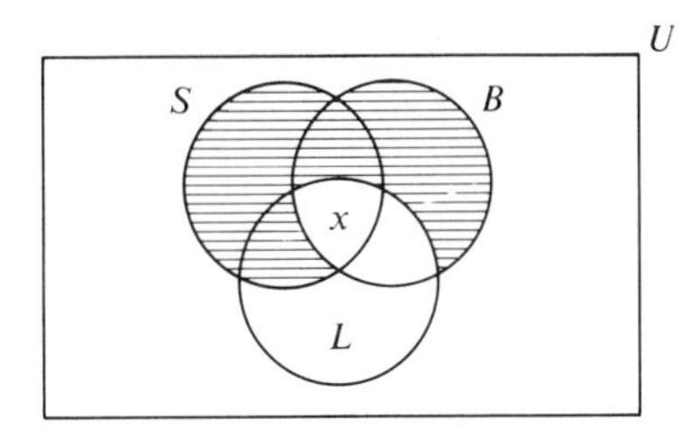

Figure 3.9

and then to determine whether this information *requires* that the diagram be consistent with the conclusion. We see that the first hypothesis, "All sparrows are birds," insures that the part of S outside B is empty, since there are no sparrows which are not birds. We shade this part to indicate that it is empty. Similarly, the second hypothesis requires us to shade the portion of B which is not in L. The fact that, in this case, the shaded regions intersect need not trouble us since the intersection of two null sets is certainly empty. Now the conclusion gives a relation between the sets S and L. From the shading that we have done, it is clear that the portion of S which is outside L is empty. Since this shading was required by the hypotheses, we are forced to admit that whatever elements S contains (e.g., x) are in the *unshaded* portion of the "S" circle, which is *inside* L; so they are living creatures. This means that "All sparrows are living creatures," which is the stated conclusion. Since the conclusion is inescapable, the reasoning is valid, and Example 1 is a case of deductive reasoning.

Let us turn to another syllogism.

---

[18]This word is derived from *validus*, which in Latin means "strong."

[19]Venn diagrams yield quite satisfactory tests of validity and emphasize that, in valid reasoning, the conclusion is inescapable. They are nevertheless "tests" rather than proofs: for a proof cannot rely on visual examination of a geometric figure. To give proofs of the validity of syllogisms would require an abstract development of logic and set theory, which we have not done. For a historical discussion and further examples of Venn's methods, consult Chapter 2 of the Gardner reference.

Example 2

Hypotheses: All skunks are fragrant flowers.
All fragrant flowers are made of seaweed.

Conclusion: All skunks are made of seaweed.

Is this a valid argument? We recall that the hypotheses are to be accepted without question: whether they are true or not is beside the point. The matter to determine here is whether, once having accepted the hypotheses, we are then forced by logic to accept the conclusion. We begin by identifying the three sets as: skunks (S), fragrant flowers (F), and things made of seaweed (T). Our task is to depict, in diagram form, the information contained in the hypotheses, and then check whether the conclusion is inescapable. Using this procedure, we obtain Figure 3.10, which *forces* us to conclude that "All skunks are made of seaweed." So in Example 2 the reasoning is valid. This is so, even though we do not wish to consider the conclusion as true.

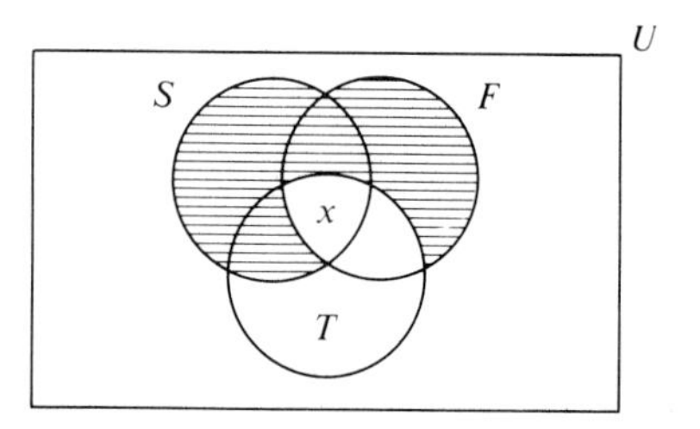

Figure 3.10

The alert reader, of course, will already have noticed that Example 2 is almost identical to Example 1. The diagrams are the same, except for labeling; S in Example 1 corresponds to S in Example 2, B corresponds to F, and L corresponds to T. Evidently these syllogisms are valid because they are both instances of a valid pattern, or *structure*. This particular structure is revealed by the abstract syllogism:

Hypotheses: All A is B.
All B is C.

Conclusion: All A is C.

Here, A, B, and C may be *any* sets. You are now invited to draw the Venn diagram for this abstract syllogism, thus justifying simultaneously the validities of Examples 1 and 2. Since, moreover, validity is evidently a matter of structure, the meaning or content of the statements in a syllogism does not affect the correctness of the reasoning. More generally, whether a chain of reasoning is logically correct (i.e., deductive) is a question of the structural relations among the propositions which compose it: it is a matter of abstract form rather than of content. Though it might seem that we are ignoring the substance of mathematics in making this statement, it is precisely this property of deductive reasoning that gives mathematics the flexibility and generality to permit its application to many fields of human endeavor. It would be helpful to reread Section 1.2 (especially the last three "characteristics" of mathematics) in the light of the above comments.

It is time to consider a different pattern. We illustrate it by

Example 3

Hypotheses:	All state-certified teachers are college graduates.
	All dentists are college graduates.
Conclusion:	All state-certified teachers are dentists.

Here we are dealing with the sets: state-certified teachers (T), dentists (D), and college graduates (C). The first hypothesis assures us that the set of state-certified teachers who are not college graduates is empty; so in Figure 3.11 we shade the portion of T which is outside C. The second hypothesis similarly requires us to shade the part of D which is outside C. The question now is: Does the resulting diagram force us to conclude that: "All state-certified teachers are dentists"?

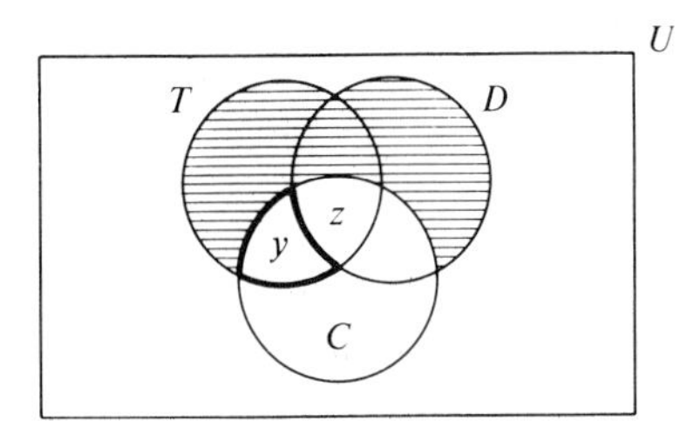

Figure 3.11

Consider the region which is heavily outlined in Figure 3.11. Since this region is not shaded, the hypotheses do not *require* that it be empty. Thus it is quite possible that this region contains an element such as y. What does y represent? Since y is in T but not in D, it represents a state-certified teacher who is not a dentist. Hence the existence of y contradicts the stated conclusion of the syllogism. In this way we have escaped the conclusion, even though we accepted the hypotheses and shaded the diagram in accordance with them. This means that the reasoning is weak or *invalid*— in fact, it does not deserve to be called deductive reasoning. So the abstract structure illustrated by Example 3 is not valid. You should compare this structure with that of Examples 1 and 2 to see how the two patterns differ.

We are not quite finished with Example 3, for we wish to anticipate two possible objections to the preceding discussion. One is, might not the heavily outlined region of Figure 3.11 be empty? Of course it might be; but the hypotheses do not *force* it to be empty, as they do in Example 1, A second is, might there not be an element, z, in both T and D which represents a teacher who is a dentist? Again, this is a possibility, but the hypotheses do not *force* it. Actually, both of these questions are quite irrelevant, for validity demands that the conclusion be inescapable, not merely possible; and the presence of y demonstrates a way of escaping our conclusion. Deductive reasoning is uncompromising in its insistence that the conclusion must follow with absolute certainty from the hypotheses: this is its chief virtue.

A fourth example deals with a different one of the four special types of propositions listed in Exercises 3.2, Problem 4. It is

Example 4

Hypotheses: All asteroids revolve around the sun.
Some comets are asteroids.

Conclusion: Some comets revolve around the sun.

Here, as before, our task is to test the validity of the argument, not to judge the truth of the individual statements. So you are invited first to decide intuitively whether the reasoning is valid; then to check the decision by means of a Venn diagram.

The Venn diagram for Example 4 again involves three sets: asteroids (A), objects which revolve around the sun (R), and comets (C). To make this example clearer, the first hypothesis should be reworded: "All asteroids are objects which revolve around the sun," and the conclusion should be similarly reworded. This is necessary in constructing the diagram, since sets (represented by circles) are *nouns* rather than verbal expressions. The first (reworded) hypothesis requires the shading indicated in Figure 3.12. The second hypothesis, stating that *at least one* comet is an asteroid, guarantees that the set $C \cap A$ contains at least one element, say x. Where is x located? Since part of $C \cap A$ is empty, as shown by the shading from the first hypothesis, x *must* be in the other part, which is heavily outlined in the diagram. But then x is also in R, thus *forcing* us to the conclusion that at least one comet revolves around the sun. Since there is no escape from this conclusion, the reasoning is valid.

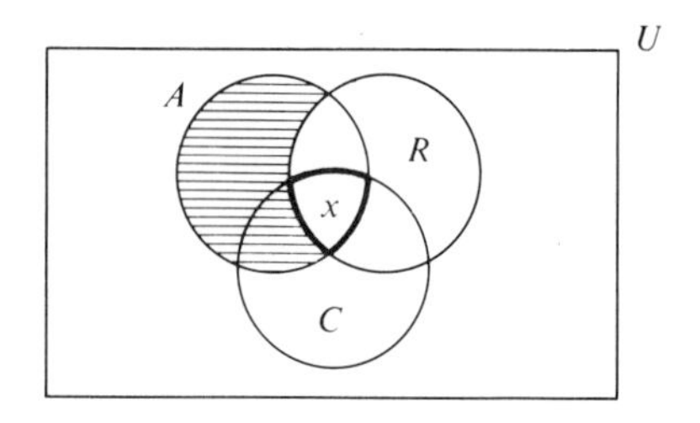

Figure 3.12

We consider a fifth example, illustrating the two other special types of propositions. It is

Example 5

Hypotheses: All triangles have three sides.
Some pentagons do not have three sides.

Conclusion: No pentagon is a triangle.

You are again invited to begin by deciding intuitively whether the syllogism is valid. Next comes the Venn diagram involving triangles (T), three-sided figures (S), and pentagons (P). (See Figure 3.13.) The first hypothesis is diagrammed as usual. The second says that at least one element of P is not in S: call it x. Because of the shading, x must be located in P as shown. Note that we have now used all the information contained in the hypotheses and have no right to assume any more knowledge. Turning to the conclusion, we see that it states that $P \cap T$ is empty. Our diagram, however, has left the heavily outlined region, which is part of $P \cap T$, unshaded. This means that the hypotheses do not insist that this region be empty; so it might very well contain an element, say y, which contradicts the conclusion. Since in this way the conclusion can be avoided, the reasoning is not valid.

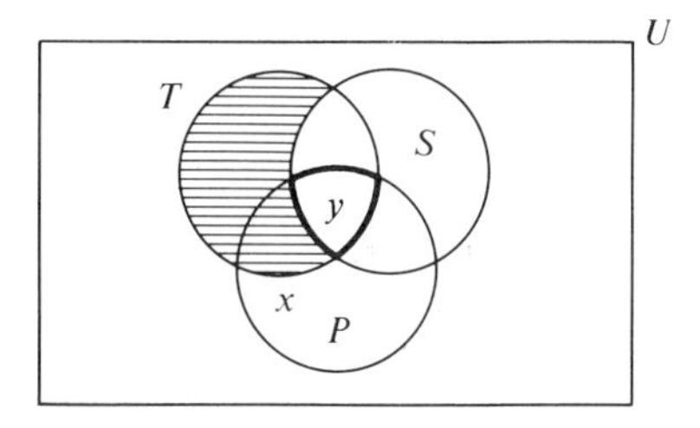

Figure 3.13

We end with a sixth example, given as an abstract pattern:

Example 6

Hypotheses: Some A is B.
Some B is C.

Conclusion: Some A is C.

Again we draw three interlocking circles. (See Figure 3.14.) This time, however, we do no shading, as neither hypothesis states that a region is empty. Indeed, the first states that $A \cap B$ is *not* empty; for at least one element of A is in B. Where is such an element? It may be in the position of x or that of y, or there may be an element in each position. Similarly, the second hypothesis guarantees the existence of an element y or z or both. The above sentences, however, do not *require* that all of x, y, and z exist. Turning our attention to the conclusion, we ask whether we are forced to accept that $A \cap C$ is nonempty. In other words, *must* $A \cap C$ contain an element such as y? The answer is

No, since the two hypotheses are satisfied by the elements x and z, respectively, so that y is not required to exist. Thus we conclude that this pattern is invalid.

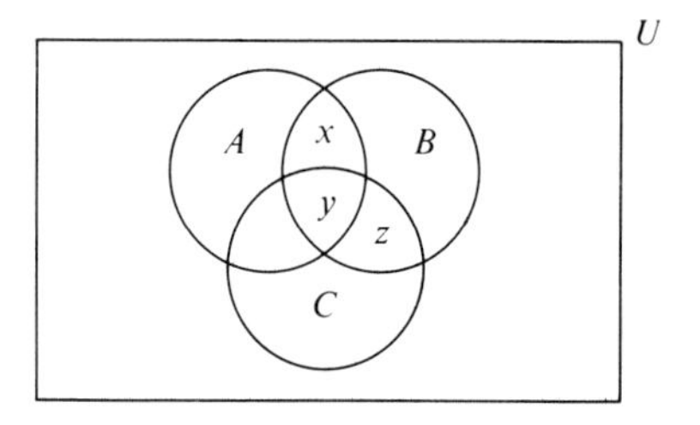

Figure 3.14

Let us pause to examine the relationships (if any) between the truth values of the individual propositions of a syllogism and the validity of the reasoning. Since validity is a question of structure rather than content, one would expect little or no connection between it and the truth values. In Example 3 the reasoning is invalid. We would doubtless accept the hypotheses as true, but the conclusion is certainly false. The same situation occurs in Example 5, except that this time the conclusion is true. The exercises contain instances of invalid reasoning with other combinations, such as false hypotheses and true or false conclusions.

When we turn to examples of valid reasoning, we find cases where the hypotheses and the conclusion are all true (Example 1) or all false (Example 2), or where one of the hypotheses is false and the conclusion is true (Example 4). Other combinations occur in the exercises. Significantly, however, we do not find any syllogism with true hypotheses and a false conclusion. Let us see why this is so. Let p_1 and p_2 be the two hypotheses and q the conclusion of the syllogism. Therefore, we have:

Hypotheses: p_1

p_2

Conclusion: q

Since, moreover, the reasoning is valid, q follows inescapably from $p_1 \wedge p_2$. This means that the conditional $(p_1 \wedge p_2) \rightarrow q$ is an implication. So the syllogism has the logical structure

$$\begin{array}{c} p_1 \wedge p_2 \\ \underline{(p_1 \wedge p_2) \Rightarrow q} \\ \therefore q \end{array}$$

This is immediately recognizable as the schematic form of the Rule of Detachment, with a compound hypothesis consisting of the conjunction of two premises.

The power of a valid syllogism, then, resides in the fact that, once the premises are accepted, the conclusion must be accepted. In direct applications to the real world, "acceptable" premises consist of facts observed about physical objects: briefly, the premises must be "true in the real world." Then, since the syllogism is valid, its conclusion must also be objectively true, and theoretically we could save ourselves the trouble of verifying it. In practice, however, the simple syllogism with only two hypotheses is too trivial to reflect the complexities of the real world.

More generally, suppose deductive (i.e., valid) reasoning is applied to a complex set of many hypotheses and leads through a chain of arguments to a set of conclusions. *If* the objective truth of *all* the hypotheses can be established, then the conclusion *must* be objectively true and need not be verified by observation. We must point out, however, that at least three crucial assumptions give us pause. The first is that a person's senses, and the instruments created to extend them, enable him or her to determine with sufficient accuracy the facts needed. One of many practical difficulties here is that the very act of observing a physical object may change the behavior of the object. The second assumption relates to the specific model that is chosen to represent objective reality; any physical situation is so complicated that the investigator cannot possibly comprehend all the myriad facts, let alone deal completely with them. The only recourse is to concentrate on the small number of facts which appear to be relevant and capable of observation and mathematical treatment. Such a limitation means that the investigator is setting up a theory or model of the physical situation and is no longer studying it directly. But this means that the reasoning, although valid, is subject to the (usually doubtful) assumption that the model is correct and adequate. The topic of models is so important that Section 3.11 is devoted to it. The third and fundamental assumption is that events in the real world happen in accordance with the laws of logic which man has created — i.e., that the real world is rational and not chaotic. On this assumption, or article of faith, all model building and, indeed, all scientific inquiry are based.

EXERCISES 3.10

1. Test the validity of each of the following syllogisms by means of a Venn diagram:

   (a) Hypotheses: All terriers are dogs.
   All dogs are animals.

   Conclusion: All terriers are animals.

   (b) Hypotheses: All dogs bark.
   Fido is a dog.

   Conclusion: Fido barks.

   (*Hint:* The three sets are dogs (D), animals which bark (B), and

Fido (F). Even though F contains only one element, it is a set, and so can be depicted by a circle.)

2. Some writers — even some dictionaries — present inductive and deductive reasoning as opposites in the following way: the former proceeds from the particular to the general, and the latter from the general to the particular. The first part of this claim is correct, according to our description of inductive reasoning in Section 3.9.

   (a) Examine the syllogisms of Problem 1, both of which illustrate deductive reasoning (Why?). Does each satisfy the claim that deductive reasoning proceeds from the general to the particular? Explain.

   (b) What are the major characteristics of deductive reasoning emphasized in Section 3.10? Do they have any relation to the generality or particularity of the propositions on which the reasoning is done? Explain.

Test the validity of each of the syllogisms in Problems 3–13 by means of a Venn diagram. Also give your opinion, when possible, as to the truth in the real world of each proposition in the syllogisms. Do your observations agree with the text's discussion of the relationship between the validity of the reasoning and the truth of the hypotheses and conclusion of a syllogism?

3. Hypotheses: All freshmen study Greek.
   John Classicus studies Greek.

   Conclusion: John Classicus is a freshman.

4. Hypotheses: Some pears are apples.
   All apples are blue in color.

   Conclusion: Some pears are blue in color.

5. Hypotheses: No rose is a tulip.
   All tulips have thorns.

   Conclusion: No roses have thorns.

6. Hypotheses: No roses have thorns.
   All tulips have thorns.

   Conclusion: No rose is a tulip.

7. Hypotheses: Puerto Rico is a state of the United States.
   Some states of the United States have a coastline on the Caribbean Sea.

   Conclusion: Puerto Rico has a coastline on the Caribbean Sea.

8. Hypotheses: All odd numbers are divisible by 2.
   Some numbers divisible by 2 are not even numbers.

   Conclusion: No odd number is an even number.

9. Hypotheses: All elephants are lions.
   Some elephants are not carnivorous.

   Conclusion: Some lions are not carnivorous.

10. Hypotheses: Some A is not B.
    All C is A.

    Conclusion: No B is C.

11. Hypotheses: Some A is B.
    No A is C.

    Conclusion: Some B is not C.

12. Hypotheses: All A is B.
    All B is C.
    All C is D.

    Conclusion: All A is D.

13. Hypotheses: All A is B.
    No C is B.
    All D is C.

    Conclusion: No A is D.

The hypotheses of several syllogisms are given in Problems 14–21. If possible, derive a conclusion for each syllogism by means of valid reasoning. If not possible, explain why not, using a Venn diagram.

14. All residents of Quebec speak French.
    None who speak French are Canadians.

15. Some cats are animals.
    Some animals drink milk.

16. Some composers conduct orchestras.
    All composers write music.

17. No ostriches fly.
    No fish fly.

18. No whole numbers are negative.
    Some negative numbers are fractions.

*19. Babies are illogical.[20]
    Nobody is despised who can manage a crocodile.
    Illogical persons are despised.

*20. No terriers wander among the signs of the zodiac.[20]

---

[20]These exercises are due to Charles Dodgson, an English mathematician of the nineteenth century; they appear in "Sets of Concrete Propositions, proposed as Premises for Soriteses: Conclusions to be found" in his *Symbolic Logic*. The reader is already familiar with two of his works, *Alice in Wonderland* and *Through the Looking Glass.*

Nothing that does not wander among the signs of the zodiac is a comet.
Nothing but a terrier has a curly tail.

*21. No birds, except ostriches, are 9 feet high.[20]
There are no birds in this aviary that belong to any one but *me*.
No ostrich lives on mince pies.
I have no birds less than 9 feet high.

As we noted at the end of Section 3.6, the implication $p \Rightarrow q$ corresponds to the subset relation $P \subseteq Q$ in set theory. Indeed, the "A-type" simple proposition (see Exercises 3.2, Problem 4), of the form "All *a*'s are *b*'s," can be recast into the implication form, "If it is an *a*, then it is a *b*." In Exercises 3.5, Problem 2 you got a bit of practice in converting A-type propositions to implications. This conversion process can be extended to the valid syllogism pattern illustrated by Examples 1 and 2 of this section. That pattern is:

Hypotheses: All A is B.
All B is C.

Conclusion: All A is C.

In implication form the pattern becomes

$$\begin{array}{r} p \Rightarrow q \\ q \Rightarrow r \\ \hline \therefore\ p \Rightarrow r \end{array}$$

where p is the statement, "It is an element of A," q is "It is an element of B," and r is "It is an element of C." (That this is a valid form follows from an examination of the truth table for Exercises 3.5, Problem 9(i), which shows that $[(p \rightarrow q) \wedge (q \rightarrow r)] \rightarrow (p \rightarrow r)$ is a tautology.)

22. Recast Problem 1(a) of this section in implication form, beginning with, "If it is a terrier, then. . . ." Does it fit the preceding implication pattern? Is your answer consistent with the one you obtained for Problem 1(a)?

23. Similarly recast Examples 1 and 2 of this section and answer the questions of Problem 22 about them.

24. (a) Similarly recast Example 3 of this section.
(b) Does your answer to part (a) fit the preceding implication pattern? If not, what form does it fit?
(c) Discuss the validity or invalidity of the implication pattern for Example 3. Is this a pattern for correct reasoning?

25. Consider the implication pattern

$$p \Rightarrow r$$
$$\underline{q \Rightarrow r}$$
$$\therefore p \Rightarrow q$$

(a) Determine intuitively, using examples if helpful, whether this pattern is a law of correct reasoning.

(b) Check your answer to part (a) by constructing the truth table of an appropriate logical expression. Which line(s) of the table show that the reasoning is correct, or faulty?

*26. As we noted in the text, Examples 1 and 2 of this section are related to set theory. In fact, you can prove their validity by use of the set-inclusion theorem, "If $A \subseteq B$ and $B \subseteq C$, then $A \subseteq C$." You can also use set algebra to prove that syllogisms of certain other forms are valid. As an example, we take Problem 6:

Hypotheses: No roses (R) have thorns (H).[21]
All tulips (T) have thorns (H).

Conclusion: No rose (R) is a tulip (T).

Translated into set equations, this syllogism becomes

Hypotheses: $R \cap H = \varnothing$
$T \cap H = T$[22]

Conclusion: $R \cap T = \varnothing$

Justify the following chain of equalities, which uses the first two set equations to produce the third:

$$\begin{aligned} R \cap T &= R \cap (T \cap H) = R \cap (H \cap T) \\ &= (R \cap H) \cap T = \varnothing \cap T = \varnothing \end{aligned}$$

Equating the two ends, we obtain the conclusion.

*27. Use the method of Problem 26 to prove the validity of Problem 13.

Usually arguments presented in daily life are not in the form of syllogisms, but they can sometimes be recast into that form and then analyzed logically. Rewrite each of the arguments in Problems 28–31 in the form of a syllogism; then decide whether it is valid. If it is not valid, state carefully one (or more)

---

[21]Since the first hypothesis means, "No roses are plants with thorns," H stands for the set of "plants with thorns." Similarly for the second hypothesis.

[22]See Property 12 of Table 2.1.

additional hypothesis which, combined with the others, would render the argument valid.

28. Pompous people always talk a lot. My neighbor is very talkative. Therefore, he is pompous.

29. This steam boiler will withstand a pressure of 100 psi. Since it is not bursting, the pressure in it does not exceed 100 psi.

30. Helping the handicapped is always good. Therefore, leading a blind man across the street is always good.

*31. Persons who have been graduated from college have substantially higher average annual earnings than those who have only completed high school. The latter, in turn, earn more on the average than those who did not graduate from high school. Doctors and lawyers, as a group, are among the highest-paid professionals; and they have completed several years of graduate work. So, if you want to have a high income, you should continue your education as long as possible.

*32. Select a substantial argument appearing in an advertisement, editorial, or letter to the editor, for logical analysis. Clip it out if you can (otherwise, copy it), and append it to your discussion of the argument. Your discussion should treat fully at least the following:

    (a) What are the *stated* hypotheses of the argument?

    (b) What is the stated conclusion of the argument? (If there is more than one main conclusion, you have two arguments, and should discuss them separately.)

    (c) Are any of the stated hypotheses irrelevant or unnecessary to the argument? If so, which ones, and why? (Hopefully, none are actually inimical to the argument!)

    (d) Is the argument, as you have so far analyzed it, a valid one? Why or why not? (Venn diagrams may help here.)

    (e) If the argument is not valid, try to find one or more additional hypotheses which, if accepted, will render the argument valid. State them carefully. Explain why the argument is now valid. (Here again, Venn diagrams may help.)

    (f) What is your opinion as to the truth of the stated hypotheses? of the (tacit) hypotheses which you added in (e)? Give reasons.

    (g) What is your opinion as to the truth of the conclusion? Give reasons.

    (h) Is the argument in its original form (before you added hypotheses in (e)) an example of inductive, deductive, or some other type of reasoning? What about this *after* you added these hypotheses? Give reasons in both cases.

(i) Do you find that, after making this logical analysis, your opinion of the argument and of the writer has changed from your initial reaction? Explain.

Section 3.11 Abstract Deductive Systems

A valid syllogism constitutes a trivial example of a *deductive system*. It is an *abstract* system if it is devoid of specific content so that only the structure of the reasoning is revealed. This is the case with the abstract syllogism generalizing Examples 1 and 2 of Section 3.10, which we may rewrite in terms of elements rather than sets as follows:

Hypotheses: All *a*'s are *b*'s.
All *b*'s are *c*'s.

Conclusion: All *a*'s are *c*'s.

Here the *a*'s, *b*'s, and *c*'s are undefined elements, and the two propositions of which the hypotheses consist are assumptions about these elements. The conclusion, which follows logically from the hypotheses, can be considered a *proved*, as contrasted with an *assumed*, proposition, and is thus a theorem.

More generally, an abstract deductive system contains four kinds of components: undefined terms, axioms or postulates[23] (assumed statements), defined terms,[24] and theorems (proved statements). The latter two components are clearly desirable: defined terms are a convenience and a source of new concepts by which the system grows; and theorems are the principal results of the system — its main excuse for being. Why not, then, define every term and prove every statement as a theorem? The answer is disconcerting: such a goal is logically impossible to attain!

To explain the last statement and also to illustrate, by means of a very important example, the four components of an abstract deductive system, let us turn to high school plane geometry (Euclidean). Here we are not speaking of physical objects or streaks on paper but of abstract objects existing only in our minds. In geometry we meet many *defined terms*. One example is *triangle*, whose definition may be given as "the union of a set of three points not on the same line and the segments joining these points in pairs." This is a good definition, but we note that the term *triangle* is itself defined by means of other terms, such as *point*, *line*, *on*, *segment*, etc. To be logical, we should now define these other terms. If we try defining *point*, for example, it is necessary to use still other terms in that definition. Where will the process of definition stop? If we keep introducing new terms with each definition, the process will continue forever. If, on the other hand (inadvertently or by design), we reuse a term that has already appeared in the sequence of definitions, then we are

---

[23]In accordance with modern usage, we consider *axiom* and *postulate* synonyms.

[24]The system of the preceding paragraph was so trivial that no defined terms were introduced.

guilty of circular definition. For instance, if in the process of defining *point* we should reintroduce the term itself, we would be saying, in effect, "A point is a point." This is a true statement, but of no help in defining the term. Since neither circular definition nor an infinite sequence of definitions is an acceptable procedure, we are forced to another alternative: to begin with a (hopefully small) number of terms as *undefined*, and use them to define all other terms. In geometry we commonly take *point*, *line*, *on*, and *between* as undefined terms.

This situation in defining terms may be likened to an attempt by an American who does not know a word of French to learn what French words mean by consulting a French dictionary, entirely in French and without pictorial or other aids. This dictionary will, of course, define every French word. Thus it may define *pluie* as "*Eau qui tombe par gouttes*." Since, however, these five words have no meaning for her, our puzzled American is no wiser than before. She may now look up *eau*, but its definition, being entirely in French, will not help her either; *but*, if the American has at her command even a small basic vocabulary of French words whose English equivalents she knows, then eventually she can learn all other French words from the dictionary. This "basic vocabulary" in a mathematical system is the set of undefined terms for that system.

Now *terms*, whether defined or undefined, are single words or phrases, not complete sentences. A deductive system (whether a syllogism or a larger structure) must also contain propositions, which are entire sentences. These propositions are of two kinds: axioms (or postulates) and theorems. Let us see why axioms are necessary.

In deductive reasoning we have a wonderful means of *proving* a proposition: if we can construct a valid argument, of which this proposition is the conclusion, then it is inescapable. In high school geometry, for example, this very idea is used in proving theorems. Thus, the statement "The sum of the interior angles of any triangle is 180°" comes at the end of a rather long chain of valid arguments.

But in any valid argument, such as a syllogism, we must start with hypotheses. Where do they come from? Since they are also propositions, perhaps they, in turn, could be established as conclusions of previous valid arguments. But the previous arguments would also have to have hypotheses. Where would they come from? When would this process end? It never could, unless we illogically reasoned in a circle, consciously or unwittingly. To put the matter another way, suppose we sit down with a small pile of undefined terms and a paper and pencil to create a new deductive system. We would like to prove our *first* theorem. What shall we use for hypotheses? At this stage, we have no previously proved theorems nor any other statements to invoke. Thus, in order to start proving theorems at all, we must first assume a (hopefully small) number of propositions which we *cannot* prove. These assumed statements are the axioms or postulates of our system. Examples of statements usually taken as axioms in geometry are: "One and only one line can be drawn through two points" and "If equals are added to equals, the sums are equal."

If we have chosen wisely the undefined terms and axioms, all other terms which we need may be defined with the aid of the undefined terms; and all other propositions which are deducible in the system may be proved as theorems, based ultimately on the axioms.

The relations among the four components are represented schematically by Figure 3.15. The bases of all abstract deductive systems are the fundamental concepts and laws of language and logic, including rules of inference, which we do not analyze, but take as given. A specific system builds on these bases by starting with a particular set of undefined terms and a particular set of axioms describing their behavior. Defined terms are introduced as needed. The first theorem (T1) is based on some of the axioms, which contain undefined and/or defined terms. Subsequent theorems (T2, T3, . . .) may be based partly or wholly on previously proved theorems: but their proofs can be traced back ultimately to the axioms. Although Figure 3.15 is meant to suggest these interrelationships, it is only an approximation and should not be taken too literally.

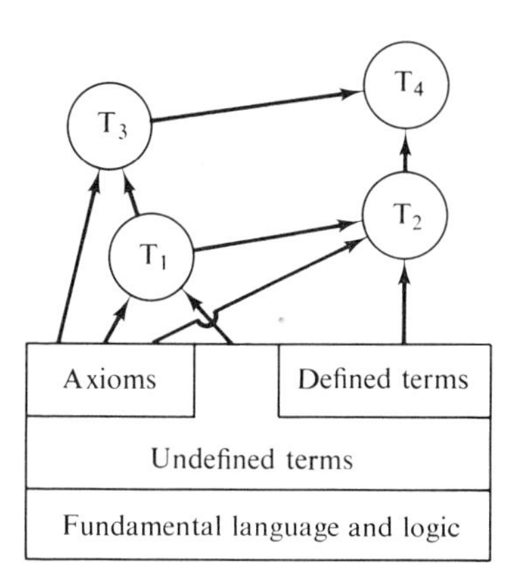

Figure 3.15

For a more specific example we turn to a well-known theorem of Euclidean geometry: "For every triangle, the sum of the measures of the interior angles is 180°." On which statements is the proof of this theorem based? A rather standard proof appeals to the following statements, as Figure 3.16 will remind you:

1. A theorem on construction of parallel lines.
2. The definition of alternate interior angles.
3. A theorem that alternate interior angles of parallel lines are congruent.
4. An axiom on adding angles.
5. An axiom on adjacent supplementary angles.

The defined terms, including those implicit in the proof, such as *triangle*, *angle*, and *parallel*, can be traced back ultimately to the undefined terms of the

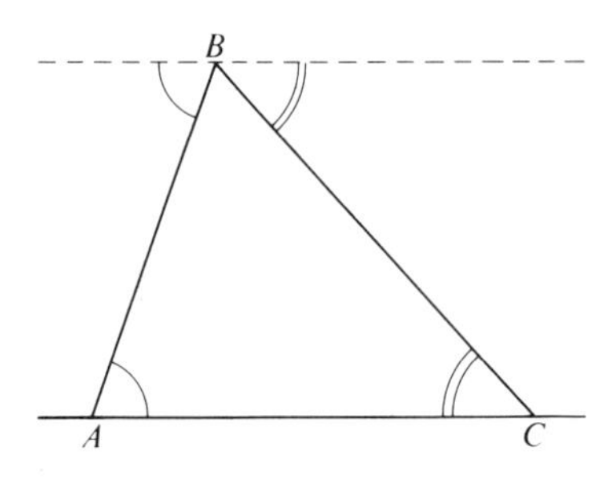

Figure 3.16

system; similarly, the two theorems can be traced back to its axioms. The second theorem, for example, depends partly on the parallel postulate and the theorem's converse, already proved. Though we decline to analyze the converse's proof, the point is that we *could* do so, and, with patience, would arrive at axioms. All the proofs, moreover, make use of rules of inference, some of which we have explicitly studied. Thus a diagram of the complete "genealogy" of this theorem would resemble that of Figure 3.15. The entire system of Euclidean geometry, however, is so incredibly rich that a comprehensive diagram of the relations among even the theorems which have so far been discovered could not possibly be drawn.

In the preceding paragraph we explored Euclidean plane geometry as an abstract deductive system. It is possible to develop set theory similarly, starting with certain undefined terms (set, element, etc.) and with a few fundamental relationships among sets as axioms; then we could define the remaining terms and prove as theorems the other relationships. In Chapter 2 we did not do this, because, for our purposes, this approach would have been too sophisticated and time-consuming.

The systems considered so far have been mathematical in content; but this need not be the case. It is possible to have abstract deductive systems whose content is theology: in fact, St. Thomas Aquinas (1225?–1274) developed just such a system of Catholic theology, as have other theologians and ecclesiastical bodies before and since. Darwin's and Newton's theories, considered abstractly, are examples of such systems in biology and physics, respectively; there have been several different "schools," or systems, less precisely worked out, in psychology; capitalism and Communism can be viewed as theoretical systems in the economic sphere. In all cases, these systems start with certain basic terms (God, species, particle, organism, commodity, etc.) and certain fundamental assumptions, and apply to them deductive reasoning, defining further terms or concepts as seem appropriate along the way.

Why are abstract deductive systems so important? First, since an abstract system is stripped of irrelevant content, the reasoning can proceed unimpeded by considerations of interpretation or emotion. Second, since the undefined terms are abstract, the system can be applied in a variety of contexts — sometimes even in different fields of knowledge — provided that the undefined terms, when interpreted concretely in the particular context, become true propositions.[25] Third, if the system has been appropriately applied to a given context, then its theorems, which have been validly derived from the axioms, become inescapable conclusions, and so furnish predictions, some of which can be verified in the concrete situation.

In this way, an abstract deductive system functions as a complex implication. Its axioms constitute the hypothesis (p) of the implication, and the theorems its conclusion (q). If the axioms are a true description of the concrete situation, then the first line of the truth table for $p \rightarrow q$ insures that the theorems are also true; indeed, *logically* there is no need to verify them experimentally. In practice, however, the truth of many sets of axioms is hard to establish: there is only a reasonable presumption that they are true. Considering this, it is essential to verify the truth of as many theorems as can be tested in the concrete application. Even if these theorems were all corroborated, one could not then conclude that the axioms were true: for though $q \rightarrow p$ is the converse of a true implication, it is not necessarily an implication. On the other hand, if one of the theorems, when tested experimentally in the application, is false, then the contrapositive, $(\sim q) \Rightarrow (\sim p)$, which is equivalent to the original implication, informs us that at least one of the axioms contained in the hypothesis (p) is false: so the abstract system is not appropriately applied. Thus, if we attempt to verify an application of an abstract system — or scientific *theory*, or *model*, as it is often called — by testing the theorems (q) of the system, no amount of successful testing will guarantee the truth of the axioms (p) of the model. To be sure, the more theorems which turn out to be true, the more confidence we can have *inductively* in the model. But if even one theorem is shown to be false — after due allowance for experimental error — we must logically reject the model.

Examples of the creation and rejection of models abound, especially in the history of science. An early model of the universe placed the earth at the center, with the planets, moon, sun, and stars all revolving in circles about it. One of the consequences (theorems) of the model was that the planets should move always in the same direction. Observations, however, showed that at times they reversed the direction of motion for relatively brief periods of time, though the stars did not. The original model, obviously wrong, was then changed by the addition of a new assumption: *epicycles* were created for the orbits of the planets to explain their retrograde motions. But although the new model "worked" for a time, in that it correctly predicted the motions of the planets, its success in doing so did not guarantee the existence of the epicycles. As further observations were made, this model also was rejected in favor of our familiar model of the solar system, with the earth and the other planets revolving around the sun. This model is not only a simpler one than the other — an advantage in itself — but so far it has predicted successfully the motions of the planets and has even helped in the discovery of two of them, Neptune and Pluto. But continued success, although strengthening our belief in the heliocentric theory, cannot prove that it is a correct description of planetary motion.

[25]See the fourth Suggestion for Investigation at the end of the chapter.

Finally, we investigate the so-called scientific method, which is now being applied in a variety of fields other than the traditional sciences. Does this method use deductive or inductive reasoning? You would do well to try to answer this question to your own satisfaction before proceeding. Many people picture the scientist's primary activities as collecting facts in the laboratory or in the field, evaluating them, and framing generalizations based on them; but this is only part of the story. For the scientist interested in discovering a tenable theory to explain specific observations, the next step after collecting relevant data is to choose abstract, undefined terms which can be interpreted or applied appropriately to the physical situation under investigation. Then the scientist creates a set of abstract axioms which describe the behavior of the undefined terms and whose interpretation in the physical situation seems to accord with the observed facts. The scientist may set up this model formally or informally, consciously or unconsciously: in any case, he or she will be creating it primarily by means of inductive reasoning, since the axioms will be abstract generalizations from the data.

The creation of a model, primarily by inductive reasoning, is only the beginning. Next the scientist uses the model to predict results for testing by further observation. In logical terms, the scientist applies deductive reasoning to the abstract system in order to prove theorems, interpreting them in the physical situation and checking by experimentation to see whether they are true. The collection of the experimental evidence and comparison with the theorems again involve inductive reasoning. This work may suggest new theorems in the model, whose proofs require more deductive reasoning. If experimental results cause rejection of the model, the cycle begins over again with a new model. Thus the scientist alternates between using deductive reasoning in the abstract model and inductive reasoning in relating it to the concrete application of the model. Both kinds of reasoning are essential to the task of explaining in a rational way man's observations about himself and the world in which he lives.

Just how significant abstract systems and their applications are to civilization — even to the "man in the street" who knows little and cares less about them — is shown by the model called the *theory of relativity*. In the late nineteenth century, the Newtonian model of the universe, widely accepted for two centuries, began to be shaken by such disturbing facts as the 1877 Michelson-Morley experiment on the speed of light. This and other scientific events led Einstein to construct a new model of relativity,[26] which he proposed in papers in 1905 and 1915. One of the deductions from this model was the possibility of nuclear energy. The crash technical development of this potentiality in the Manhattan Project of World War II, culminating in the detonation of atomic bombs in New Mexico and over Japan in 1945, ushered in the atomic age. Since then, not only has the destructive power of this new energy source been enormously increased by the development of thermonuclear fusion, but constructive uses also have been developed on a widening scale. We must

[26]See the first Gardner and first Lieber references for an explanation for laymen.

emphasize again that it was the creation of an abstract model by a man of genius, utilizing the results of many painstaking and dramatic experiments, that raised the curtain on the atomic age. What the remaining acts of this drama of civilization will be we do not know.

But of one thing we can be certain: this is not the end of the story. One reason is the impossibility of predicting all the applications that can or will be made of the theory. A deeper reason is implicit in the nature of a deductive system as a complex implication. The fact that the predictions (theorems) of Einstein's model have so far been borne out does not *prove* that it is a complete, or even a correct, description of the physical universe. That rival theories (models) have been constructed is indicated both by Problem 4 of the exercises of this section and by the more recent news item quoted below. You should examine the ways in which inductive and deductive procedures play their roles and should also note the considerable practical difficulties which often arise in testing a theory against physical reality.

New Tests Back Einstein Theory But Challenger to Persist in Rival Idea of Relativity [27]

The most serious recent challenge to Einstein's so-called "general" theory of relativity appears to be in serious trouble, but its co-inventor is not yet ready to give up.

A series of reports, based on observations of spacecraft, planets, distant stars and quasars beyond the sun, was presented yesterday at the California Institute of Technology, and all tended to support the Einstein formulation.

The rival theory was developed by Dr. Carl H. Brans of Loyola University in New Orleans and Dr. Robert H. Dicke of Princeton. Both the Einstein and Brans-Dicke theories predict that a strong gravitational field, like that of the sun, will slow and bend a passing beam of light or radio waves.

. . . .

However, the Jet Propulsion Laboratory reported that Einstein's theory "has been upheld by results thus far" obtained by exchanging radio signals with two spacecraft beyond the sun. They were Mariners 6 and 7, which had previously flown close to Mars and then continued in orbit around the sun.

Signals were sent to the spacecraft, 250 million miles away, and the spacecraft replied. The round-trip, twice passing close to the sun, took 43 minutes. As reported to the meeting by Dr. John D. Anderson of the Jet Propulsion Laboratory, the lag induced by the sun's gravity reached a maximum of 204 millionths of a second.

The Einstein theory predicted 200 millionths and the Brans-Dicke indicated that the delay would be closer to 186 millionths, according to the J.P.L. report.

---

[27] *New York Times*, November 13, 1970, p. 21, col. 6.

EXERCISES 3.11

1. Explain briefly, in your own words, why an abstract deductive system must contain: (a) undefined terms (b) axioms

2. In the text, the claim was made that certain terms must be left undefined. If this is so, how does a dictionary "get by" with defining *every* word? In answering this question, experiment by looking up the adjective *small* in a dictionary, copying down its definition, tracing the important words used by looking them up in turn, etc. (In your work, you may concentrate on the meaning(s) of *small* that have to do with physical size, ignoring other meanings.) What do you discover?

3. The following quotation is related to the discussion in this section. Read it carefully, then answer the questions.

> In setting up a scientific theory, one chooses a framework in which the phenomena to be discussed may be expressed. This framework is given some kind of structure, which is then compared with various experiments. If this structure is given in a precise manner, so that one may draw rigorous conclusions about it, it is called a "mathematical model."
>
> The investigation then takes two separate directions:
>
> (a) One studies the model itself, and draws various conclusions that are consequences of the chosen structure of the model.
>
> (b) One tests the relation between the model and experiment.[28]

(a) In the quotation, the words *axiom* and *theorem* do not appear at all, but these concepts must be part of the model. Where in the quotation are they described or referred to? Be specific in your answer.

(b) Which parts of the quotation refer primarily to inductive reasoning? Which parts refer primarily to deductive reasoning?

4. A news article in the *New York Times* in early 1968 was entitled, "Einstein's Theory on Gravity Gets a New Test." The subheading read, "Radar Bouncing Off Mercury is Increasing Precision in Measuring Sun's Pull." The article indicated that previous methods of observing Mercury were not precise enough to decide between Einstein's general theory of relativity and other recently developed theories. It was stated that a new method, developed by Dr. Irwin I. Shapiro of M.I.T., "should narrow uncertainty regarding the effect of solar gravity" from 25% "to 5% or less." The article went on to say that "if the result does not conform to the clear-cut prediction of Dr. Einstein's equations, one of the rival theories must be considered."[29]

---

[28]Hassler Whitney, "The Mathematics of Physical Quantities, Part II," *The American Mathematical Monthly*, 75:3 (March, 1968), 227.

[29]Walter Sullivan, *New York Times*, February 28, 1968, p. 20.

(a) Considering the Einstein general theory of relativity as an abstract deductive system or model of the universe, which one of the four components of the system is described by the quoted phrase "clear-cut prediction"? Explain.

(b) Explain in logical terms why the last sentence quoted is true.

(c) If the result *does* conform, within the limits of experimental error, to the "clear-cut prediction of Einstein's equations," what (if anything) should be asserted about the correctness of Einstein's model? Explain.

*5. Read carefully several times the following poem. Try to appreciate how it captures the essense and power of deductive reasoning. To the extent possible, without "analyzing it to death," try to relate some of its contents to specific concepts of logic discussed in this chapter.

Paradox[30]

Not truth, nor certainty. These I foreswore
In my novitiate, as young men called
To holy orders must abjure the world.
"If . . . , Then . . .", this only I assert;
Any my successes are but pretty chains
Linking twin doubts, for it is vain to ask
If what I postulate be justified,
Or what I prove possess the stamp of fact.
Yet bridges stand, and men no longer crawl
In two dimensions. And such triumphs stem
In no small measure from the power this game,
Played with thrice attenuated shades
Of things, has over their originals.
How frail the wand, but how profound the spell.

Clarence R. Wylie, Jr.

Suggestions for Investigation

1. Investigate the discovery of the planet Neptune in 1846 as a result of the work of Adams and Leverrier, and the discovery of the planet Pluto in 1930 following predictions by Lowell and Pickering. The model used in both cases was Newton's theory of gravitation; this model has since been rejected, owing to the work of Einstein and others.

2. For a philosophical discussion on whether the basic laws of science are simple or complex, read the third Gardner reference.

3. To see what is involved in confirming or rejecting a scientific theory, read the Salmon reference.

---

[30] C. R. Wylie, Jr. *Scientific Monthly*, 67 (July, 1948), 63. Reprinted by permission of author and publisher.

4. Try to find examples of the use of models in your major field or in another field with which you are familiar. Identify the various components of the abstract system. Ascertain whether the model is accepted as correct or useful by experts in the field, and state why or why not.

5. One advantage of abstract models is that sometimes they can be applied to several seemingly different situations. Thus a model may serve as an organizing principle and an efficient tool for studying several problems simultaneously. For a simple illustration, read the Usiskin reference.

6. Read the MacDonald reference, which describes an application of truth tables and Latin squares to the Mendelian laws of heredity.

7. The third Polya reference provides an easy example of setting up a model to solve a political question: namely, how small a fraction of the popular vote could elect a President of the United States. This article uses the apportionment of the House of Representatives in force in 1961. You may wish to update the calculations to take account of the new apportionment based on the 1970 census.

8. Related to the algebra of sets of Chapter 2 is a special kind of Boolean algebra which has only two elements, 0 and 1. Since this algebra is two-valued, it can be interpreted as a system of logic, with 0 and 1 corresponding to false and true, respectively. This system also has interesting applications to electrical circuitry, with 0 and 1 interpreted as "no current flowing" and "current flowing," respectively. For a discussion and examples, read the last chapter of the Allendoerfer and Oakley reference, especially Section 15.6.

9. Many of the articles in the Bell reference concern applications of mathematics, often involving the explicit construction of models. You may enjoy delving into some of them. The levels of difficulty and interest may vary considerably for you: if one article loses you or "turns you off" in the first few pages, skip it and try another.

10. For a closely written, yet understandable history and explication of contemporary problems in the foundations of logic and set theory, read the DeLong article, which spans the centuries from Pythagoras to Bertrand Russell, Godel, and Church. The Tarski article covers some of the same ground, focusing on paradoxes and proofs.

11. Most abstract deductive systems are far too large and complicated to understand at a glance or to work out completely. For these purposes, systems of bizarre "geometries" containing only a finite number of points and lines serve admirably. The Schaaf reference describes a number of such systems, any one of which will repay study. The Eves reference has a collection of them in Section 8.7. Examine the set of axioms and theorems of one or more of these finite geometries and construct models for them. As an excellent logical exercise, prove some of their theorems.

Bibliography

Allendoerfer, C. B. "Deductive Methods in Mathematics." *Twenty-third Yearbook.* Washington: National Council of Teachers of Mathematics, 1957. Chapter 4.

——— and C. O. Oakley. *Principles of Mathematics*, 3rd ed. New York: McGraw-Hill Book Co., 1969.

Bartley, W. W. "Lewis Carroll's Lost Book on Logic." *Scientific American*, 227:1 (Jul., 1972), 39–46.

Bell, M. S., ed. *Some Uses of Mathematics: A Source Book for Teachers and Students of School Mathematics*, SMSG *Studies in Mathematics*, XVI. Pasadena, Calif.: A. C. Vroman, Inc., 1967.

Bruyr, D. L. "Some Comments about Definitions." *Mathematics Magazine*, 43:2 (1970), 57–64.

Christian, R. R. *Introduction to Logic and Sets*, 2nd ed. Waltham, Mass.: Blaisdell Publishing Co., 1965.

Cohen, M. R. and E. E. Nagel. *Introduction to Logic and Scientific Method.* New York: Harcourt, Brace and Co., 1934.

DeLong, Howard. "Unsolved Problems in Arithmetic." *Scientific American*, 224:3 (Mar., 1971), 50–60.

Dinkines, Flora. *Introduction to Mathematical Logic.* New York: Appleton-Century-Crofts, 1964.

Doig, Peter. *A Concise History of Astronomy.* New York: Philosophical Library, 1951.

Einstein, Albert. *Out of My Later Years.* New York: Philosophical Library, 1950.

Eves, Howard. *A Survey of Geometry.* 2 vols. Boston: Allyn and Bacon, Inc., 1963. Volume 1.

Exner, R. M. and M. F. Rosskopf. *Logic in Elementary Mathematics.* New York: McGraw-Hill Book Co., 1959.

Gardner, Martin. *Relativity for the Million.* New York: The Macmillan Co., 1962.

———. *Logic Machines, Diagrams and Boolean Algebra.* New York: Dover Publications, Inc., 1968.

———. "Simplicity as a Scientific Concept: Does Nature Keep Her Counts on a Thumbnail?" *Scientific American*, 221:2 (Aug., 1969), 118–21.

———. "A New Miscellany of Problems." *Scientific American*, 228:5 (May, 1973), 102, 104.

Gibbs, R. A. "Euler, Pascal and the Missing Region." *Mathematics Teacher*, LXVI (1973), 27f.

Kemeny, G. G., J. L. Snell, and G. L. Thompson. *Introduction to Finite Mathematics*, 2nd ed. Englewood Cliffs, N.J.: Prentice-Hall, Inc., 1966. Chapter 1.

Ley, Willy. *Watchers of the Sky.* New York: Viking Press, 1963.

Lieber, L. R. *The Einstein Theory of Relativity*. New York: Rinehart & Co., Inc., 1945.

______. *Mits, Wits and Logic*, 3rd ed. New York: W. W. Norton & Co., Inc., 1960.

MacDonald, T. H. "Truth-Table Models of Mendelian Trait Segregation." *Mathematics Teacher*, LXIV (1971), 215–18.

Mathematics in the Modern World. Readings from the *Scientific American*. San Francisco: W. H. Freeman and Co., 1968.

Chapter 28. Quine, "Paradox."
Chapter 29. Pfeiffer, "Symbolic Logic."
Chapter 33. Dyson, "Mathematics in the Physical Sciences."
Chapter 34. Einstein, "On the Generalized Theory of Gravitation."
Chapter 36. Moore, "Mathematics in the Biological Sciences."
Chapter 37. Stone, "Mathematics in the Social Sciences."

Newman, J. R., ed. *The World of Mathematics*. 4 vols. New York: Simon and Schuster, 1956.

Part V. "Mathematics and the Physical World."
Part VI. "Mathematics and Social Science."
Part XI. "Mathematical Truth and the Structure of Mathematics."
Part XII. "The Mathematical Way of Thinking."
Part XIII. "Mathematics and Logic."
Part XXII. "Mathematics of the Good."

Poincare, Henri. "Intuition and Logic in Mathematics." *Mathematics Teacher*, LXII (1969), 205–12.

Polya, George. *Mathematics and Plausible Reasoning*. 2 vols. Princeton: Princeton University Press, 1954.

______. *How To Solve It*, 2nd ed. Garden City, N.Y.: Doubleday & Co., Inc., 1957.

______. "The Minimum Fraction of the Popular Vote That Can Elect the President of the U.S." *Mathematics Teacher*, LIV (1961), 130–33. This article is reprinted in the Bell reference, pp. 53–56.

Salmon, W. C. "Confirmation." *Scientific American*, 228:5 (May, 1973), 75–83.

Schaaf, W. L., ed. *Finite Geometry*, SMSG *Reprint Series*, 13. Pasadena, Calif.: A. C. Vroman, Inc., 1969.

Smith, D. E. *The Poetry of Mathematics and Other Essays*. New York: Yeshiva College, 1934.

Stabler, E. R. *An Introduction to Mathematical Thought*. Reading, Mass.: Addison-Wesley Publishing Co., 1953, Chapters 1–5.

Stoll, R. R. *Sets, Logic and Axiomatic Theories*. San Francisco: W. H. Freeman and Co., 1961.

Summers, G. J. *New Puzzles in Logical Deduction*. New York: Dover Publications, Inc., 1968.

Suppes, Patrick. *Introduction to Logic*. Princeton: D. Van Nostrand Co., Inc., 1957.

Tarski, Alfred. "Truth and Proof." *Scientific American*, 220:6 (Jun., 1969), 63–77.

Usiskin, Zalman. "Six Non-trivial Equivalent Problems." *Mathematics Teacher*, LXI (1968), 388–90.

Van Engen, Henry. "Strategies of Proof in Secondary Mathematics." *Mathematics Teacher*, LXIII (1970), 637–45.

White, L. G. and V. H. Baker. "Systems in Non-Mathematical Disciplines." *Mathematics Teacher*, LXII (1969), 171–77.

Whitrow, G. J., ed. *Einstein, the Man and His Achievement*. London: British Broadcasting Corp., 1967.

Wiscamb, Margaret. "Graphing True-False Statements." *Mathematics Teacher*, LXII (1969), 553–56.

Wylie, C. R., Jr. "Paradox." *Scientific Monthly*, 67 (1948), 63.

4

The System of Whole Numbers

What are the two *fundamental operations on whole numbers? How do the laws for whole numbers compare with those for algebra? How are proofs constructed for theorems on whole numbers? How many prime numbers are there?*

Section 4.1 What Are Whole Numbers?

This is the first chapter to deal extensively with the topic of numbers, which, as indicated in Section 1.2, pervade much of mathematics. The different kinds of numbers and their interrelations will be the major themes of the remaining chapters. Beginning with the simplest set, $W = \{0,1,2,3,\ldots\}$, we shall study their properties, note some inadequacies in their performance, and enlarge the set to include all the integers in an effort to correct the deficiencies. But since the system of integers lacks certain desirable theoretical properties and capabilities in applications, in Chapter 7 we shall briefly consider still larger number systems, such as the rational and real numbers.

The question "What are whole numbers?" is partly rhetorical, for you are already quite familiar with the whole numbers, both from long experience, beginning with learning to count, and from using them in examples of sets in Chapter 2. Since, however, the phrase "system of whole numbers" implies a formal deductive structure, it seems that we should lay the groundwork by listing the undefined elements and axioms of the system, and then proceed to make appropriate definitions and prove theorems. This approach has, indeed, often been used by mathematicians as well as by writers of texts like this one. Peano, for example, took two undefined terms and five axioms, from which he derived the system of whole numbers and its extensions through the complex numbers. The elegant and ambitious project of developing our common number systems from as narrow a base as possible is very time-consuming and tedious, except for those who are zealous for creating so much from so little.[1] A simpler and much more rapid development uses a dozen or so axioms, many of which Peano proved as theorems. In this, as in other matters, the greater the head start, the easier — and less challenging — the race.

The approach we shall choose, however, is neither of those described. In Chapter 2 we built up set theory rather fully, though informally, partly for its own sake. But we can also use it as a foundation for the whole numbers, without introducing any new undefined terms or axioms. This means that we trace the whole numbers back to the undefined terms of set theory, such as *set* and *element*, and to the axioms which describe their behavior. We did not list or study these axioms, since too much time and mathematical sophistication would have been required.

The immediate background for the system of whole numbers derives from Section 2.7, particularly Definitions 2.8 and 2.9 on equivalent sets and cardinal numbers, respectively. At this point we became interested in the existence and behavior of infinite sets and transfinite cardinals. From an unexpected characteristic of proper subsets, moreover, we obtained definitions of finite and infinite sets. In the present chapter we concentrate on finite sets and their associated cardinal numbers.

We recall that, according to Definition 2.9, the cardinal number of a set A is the class of sets containing all sets equivalent to A. As a special case, let us take $A = \varnothing$. As suggested in Exercises 2.7, Problem 5(d), A is certainly finite. Then $n(\varnothing)$, the cardinal number of $\varnothing$, is the class[2] of all sets equivalent to $\varnothing$. The only question is which symbol to use for $n(\varnothing)$. We accept the conventional symbol by defining $n(\varnothing) = 0$.

Next we suppose B to be a set containing a single element, say x, so that $B = \{x\}$; and we invent a symbol for $n(B)$ by defining $n(B) = 1$. Continuing, for a set $C = \{x,y\}$, we define $n(C) = 2$, etc. Thus we generate a set of cardinal numbers, 0, 1, 2, 3, . . . , each of which can be shown to be finite. That this set constitutes the whole numbers is the substance of

---

[1]The details of this development appear in the Landau reference, a book of 134 pages. The Spreckelmeyer-Mustain reference contains its beginnings in simplified form.

[2]Since these sets are indistinguishable from one another, the abstract class $n(\varnothing)$ really contains only the element $\varnothing$.

Definition 4.1 Whole Number A *whole number* is the cardinal number of a finite set.

We pause to remark that the *concepts* of cardinal number, null set, and set with a single element, are abstract, not tied to any particular language or symbols. The specific words or symbols have varied from one country or civilization or epoch of history to another.

While Definition 4.1 is unequivocal, it is well to remind you of some distinctions. Although the set W is infinite, each element of W is itself a finite cardinal number; aleph-null and the other transfinite cardinals are not whole numbers. Moreover, since $\varnothing$ is a finite set, W contains zero. There are occasions, such as in ordinary counting, when we wish to deal only with the nonzero whole numbers. In fact, the idea of a word or symbol to represent the absence of quantity is of relatively recent origin: few of the early systems for writing numbers had any symbol for zero, even as a placeholder, as we use it in 205. The subset N, obtained by deleting 0 from the set W, is called the set of *natural numbers;* thus $N = \{1,2,3, \ldots\}$. Formally, we have

Definition 4.2 Natural Number A *natural number* is any whole number except zero.

Sometimes natural numbers are called *counting numbers;* but the latter term is frequently used also to refer to W. Since the terminology is not standardized, you are cautioned, in reading other texts, to ascertain the author's meaning at the start.

Frequently we shall want to assert that two whole numbers are equal: thus $4 + 0 = 4$ and $2 + 2 = 4$. This assertion means that the expressions on the two sides of the equality are merely different symbols for the *same* cardinal number. So $4 + 0$, $2 + 2$, and 4 all refer to the same class of equivalent sets, namely, the one which contains the set $\{1,2,3,4\}$. Formally, we have

Definition 4.3 Equality Two whole numbers, a and b, are said to be *equal* (in symbols, $a = b$) iff they are the same cardinal number.

This definition may strike you as being absurdly simple. If $a = b$ means that a and b represent the same cardinal number, why bother with two different symbols? In particular, if $x = 144$, why use x at all? But suppose we wish to measure the height of a cliff by throwing a stone horizontally off the top and timing its fall to the bottom. Before performing the experiment, we know neither the height nor the time. If we represent the height in feet by x and the time in seconds by t, elementary physics tells us that these two quantities, neglecting air resistance, are related by the (approximate) equation $x = 16t^2$. Now, if we measure the time and find it to be 3 seconds, we can compute $x = 16(3)^2 = 144$ feet. But these equations, useful as they are, are merely statements that two or more expressions represent the same number,[3] just as Definition 4.1 states.

For completeness, we make another distinction — that between cardinal and ordinal numbers, though we shall not have occasion to discuss the latter again. This distinction is illustrated, albeit imperfectly, by comparing *one* and *first*, *two* and *second*, *three* and *third*, etc. The first word of each pair refers to a cardinal number, and the second to the corresponding ordinal. Note that the ordinal does not tell how many elements a set contains, but rather the position, or order, of a particular element in the set. The distinction is one of use rather than grammatical form. Consider the sentence "Chapter 2 of this text begins on page 13 and contains 44 pages." Clearly, 44 is a cardinal number for the number of the set of pages in Chapter 2. But 13, despite its form, is ordinal, because it refers to the thirteenth page in order from the beginning of the text. Which kind of number is the *2* of Chapter 2?

EXERCISES 4.1

1. In each of the following passages identify each number as a cardinal number or as an ordinal number:

   (a) The nineteenth book of the Old Testament, Psalms, has 150 chapters, more than any other book of the Bible. In the King James version, Psalm 23, verse 3, consists of 16 words, the longest of which (the twelfth one) contains 13 letters.

   (b) On July 20, 1969, two astronauts set foot on the surface of the moon; this was on the fifth day of the 9-day flight of the Apollo 11.

2. Let W, N, E, D represent the sets of whole numbers, natural numbers, even numbers, and odd numbers, respectively.

   (a) Which of the following is (are) true?

$$E \subseteq W,\ N \subseteq W,\ W \subseteq N,\ N \subset W,\ D \subseteq N,\ N \subseteq D,\ N \subset D,\ N \subseteq E,\ E \subseteq N$$

   (b) Taking $U = W$, explain the relationship between the sets N and $\{0\}'$, the complement of $\{0\}$.

3. (a) What is the cardinal number of the set of days in a leap year? Is your answer a whole number?

   (b) Answer the same questions about the set W. Justify your answers.

   (c) Answer the same questions about the set N. Justify your answers.

   (d) Is there a largest whole number? Explain.

---

[3]The fact that, in the actual performance of this experiment, the numbers are unlikely to be simple whole numbers, does not blunt this argument: instead, it indicates that the concept of equality will have to be extended to more complicated kinds of numbers.

(e) Does your answer to part (d) contradict Definition 4.1, which says that a whole number is the cardinal number of a finite set? Explain.

*4. Examine the first few pages of Landau's book (listed in the chapter bibliography), noting which are the undefined terms and what is the content of the five axioms. Give illustrations of the axioms if possible, using the set N. Then follow the building up of the theorems and definitions as far as your interest and understanding hold.

Section 4.2 Addition

Since at least the beginning of elementary school you have had much experience with addition of whole numbers. Intuitively, you may picture the process as combining two sets of objects — say, 2 apples and 3 more apples to get one set containing 5 apples. Indeed, as happens often in mathematics, experience in the physical world powerfully motivates the development of the abstract system — in this case the whole numbers. Logically, of course, this development cannot be based on concrete objects. But it will turn out, nevertheless, that our definition of addition, while formal and devoid of reference to physical things, recalls the joining of two sets of objects.

Before proceeding to the definition, however, we pause to explore another intuitive approach to adding whole numbers, one with a geometric flavor. The number line provides a graphic way of representing whole numbers. We choose any point on a (usually horizontal) line and let it represent the number 0. Then we take a convenient length and lay it off repeatedly on the line to the right of the point representing 0, marking each point so obtained. We label these points in order, from left to right: 0, 1, 2, 3, . . . (see Figure 4.1), in this way representing the whole numbers as equally spaced points on the line extending from 0 to the right. We can use this number line, not only to represent single numbers, but also to perform addition and other operations with pairs of numbers. For instance, we can add 2 and 3 geometrically by starting at the 0 point, mentally walking to the right 2 units, and then 3 units more, arriving at the point corresponding to the number 5, the required sum.

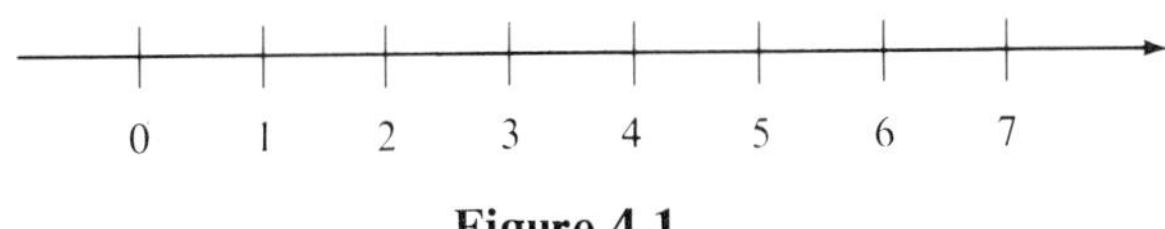

Figure 4.1

Let us now return to the first intuitive approach, that of combining sets, to see if we can abstract a formal definition of addition of whole numbers. A recent first-grade arithmetic workbook[4] introduces the word *set* through pictures (p. 4), uses *one-to-one matching* of sets (p. 6) and *nonequivalent sets* (p. 8),

[4]Gundlach, Welch, Buffie, *Sets, Numbers, Numerals*, 2nd ed. (River Forest, Ill.: Laidlaw Brothers Publishers, 1965).

brings in *subsets* (p. 21), and then develops the words and symbols for the first nine natural numbers. After these preliminaries, it deals with *joining sets* (p. 44). Figure 4.2 shows the nature of the first exercise. Exercises on the next page then ask the pupil to "draw a ring around each numeral for two" (*Choices:* 1, 2, 1 + 1) and to "draw a ring around the simplest numeral for two" (*Choices:* 1 + 1 and 2). What term do *we* use for the process of *joining sets?* Generalizing the procedure, the following statement apparently holds: "If A and B are any two sets, then $n(A) + n(B) = n(A \cup B)$." Let us, however, probe further.

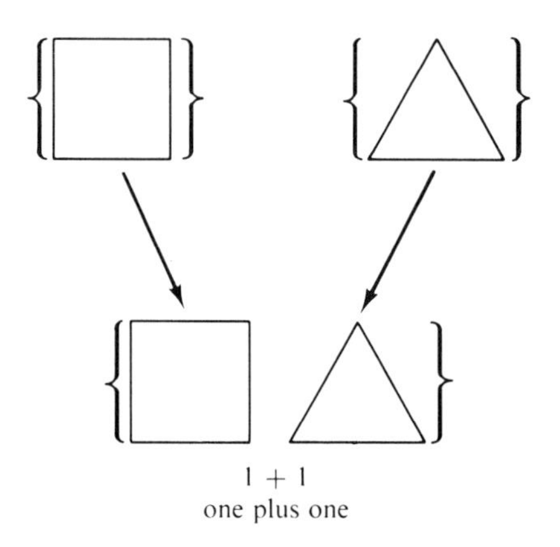

Figure 4.2

At a more complicated level, we consider the "Registrar's Problem." It may be illustrated by the figures for registration in a college which has three divisions. We are here dealing with three sets: Evening Division students (E), Junior College students (J), and Four-year Day Division students (D). Consider the following figures:

$$n(E) = 2{,}079 \qquad n(D) = 1{,}479 \qquad n(E \cup D) = 3{,}538$$

We try checking by adding 2,079 and 1,479; do we get 3,538? Is the registrar mistaken? No! The explanation is that some students are registered in both the day and evening divisions; and those students have been counted twice. How many such students are there? As a matter of fact, $n(E \cap D) = 20$. Since these students are counted twice in the sum 2,079 + 1,479, it is necessary to adjust this sum by subtracting their number once. Thus 2,079 + 1,479 − 20 *does* equal 3,538. Is there a general formula for $n(E \cup D)$? What about $n(E \cup D) = n(E) + n(D) - n(E \cap D)$? But there was no such problem in the first-grade workbook: there the quoted formula concerning sets A and B apparently worked. How can we explain the discrepancy?

As further practice the reader should find $n(E \cup J)$, given that $n(E) = 2{,}079$, $n(J) = 633$, $n(E \cap J) = 19$. (The answer is 2,693, not 2,712; why?)

On the other hand, where there is no double registration of students — i.e., where the intersection of the two sets is empty — no adjustment by subtraction is necessary. Thus if $n(D) = 1{,}479$ and $n(J) = 633$, as before, and if there are no students attending both the Four-year Day Division and the Junior College, then $D \cap J = \varnothing$; consequently, $n(D \cap J) = 0$, and we have $n(D) + n(J) = n(D \cup J) = 2{,}112$.

The foregoing examples suggest that we can add whole numbers by utilizing the union of *disjoint* sets (i.e., sets whose intersection is empty). To add 2 and 3, for example, we choose any two disjoint sets whose cardinal numbers are 2 and 3, respectively, perform their union, and take the cardinal number of the union as the sum of 2 and 3. We could select, for instance, $A = \{v,w\}$ and $B = \{x,y,z\}$, because $n(A) = 2$, $n(B) = 3$, and $A \cap B = \varnothing$. Then $A \cup B = \{v,w,x,y,z\}$, and $n(A \cup B) = 5$. Summarizing,

$$2 + 3 = n(A) + n(B) = n(A \cup B) = 5$$

so that

$$2 + 3 = 5$$

You should similarly calculate the sums $3 + 2$, $5 + 1$, $4 + 0$, and $0 + 0$. We are thus led to the following:

Definition 4.4 Addition If a and b are any two whole numbers and A and B are sets such that $n(A) = a$, $n(B) = b$, and $A \cap B = \varnothing$, then *the sum* $a + b = n(A \cup B)$.

Since definitions are arbitrary, it might seem safe to accept this one unquestioningly. In order to be useful, however, the definition of an operation should satisfy two criteria: the result of the operation must actually exist; and this result must be unique for any particular given numbers. In other words, we expect *an* answer and *only one* answer. Let us examine Definition 4.4 from this point of view, first asking whether, given a and b as whole numbers, the sum $a + b$ always exists as a whole number. We reason as follows: If a and b are whole numbers, which by Definition 4.1 are also cardinal numbers, we can certainly find[5] finite sets A and B which have a and b as cardinal numbers. Moreover, with care in the selection of the elements of A and B, we can arrange

[5]That even *one* set — e.g., the null set — exists can be taken as an axiom of set theory; then, with the aid of other axioms, it is possible to produce at least one set corresponding to every whole number — in fact, two disjoint sets, if necessary.

that $A \cap B = \varnothing$. Then, since the union of two sets always exists and is a unique set,[6] $A \cup B$ is a set, and so has a cardinal number, $n(A \cup B)$. But since A and B are finite, so also is $A \cup B$; hence $n(A \cup B)$, which we have defined as $a + b$, is a whole number, as we expected. Thus Definition 4.4 always produces a whole number.

But we also need to ask whether the number produced by Definition 4.4 is always the same for a given $a, b \in W$, regardless of the selection of disjoint sets A, B. This is the question of uniqueness. In particular, suppose, in computing $2 + 3$, we had chosen $A^* = \{\#, \%\}$ and $B^* = \{\&, \$, @\}$ to represent 2 and 3, respectively, with $A^* \cap B^* = \varnothing$. Would the sum still have come out 5? We shall omit a rigorous general argument; but we point out, for the example under discussion, that we must have, to begin with, $A \sim A^*$ and $B \sim B^*$, because both pairs of sets represent the same cardinals, respectively. Indeed, to show that $A \sim A^*$, we may pair v with # and w with %; similar pairing is possible for $B \sim B^*$. Now we may use the same correspondences to show that $(A \cup B) \sim (A^* \cup B^*)$; hence the cardinal numbers of the two union sets will necessarily be the same, in this case, 5.

In the preceding two paragraphs we have been discussing the *closure* property for addition of whole numbers. This is the same property that was shown to be true of intersection and of union of sets, but not of the cartesian product. Similarly, we could have shown that conjunction and disjunction are closed operations in the system of propositions. Although this argument for closure of addition in W is not rigorous, it could be made so. We shall therefore announce the result as our first theorem for W:

THEOREM 4.1 Closure of Addition W is closed under addition.

Let us see which other properties addition in W has. An obvious one is *commutativity*: even children recognize early that, since $2 + 3 = 5$, $3 + 2 = 5$ also. It is easy to supply further examples, but they prove nothing. What we need is a valid argument in support of the *general* statement that $a + b = b + a$ for any two whole numbers a and b. This general statement constitutes

THEOREM 4.2 Commutativity of Addition Addition in W is commutative.

In order to have general symbols for proving Theorem 4.2, we may rephrase it as follows: If $a, b \in W$, then $a + b = b + a$. But how do we prove it? There are no magical rules for creating a proof. We simply examine the hypothesis and conclusion, recall whatever previous information of the system appears

---

[6]See the paragraph following Definition 2.7.

relevant, and try to construct a chain of valid reasoning which leads from the hypothesis to the conclusion. Since Theorem 4.2 involves addition, which has barely been introduced, Theorem 4.1 and Definition 4.4 are the only tools at hand. Now, Definition 4.4 refers to the union of two sets; and we happily recall that set union is commutative.[7] Perhaps we could try justifying $a + b = b + a$ by "translating" it into a statement involving union of sets: $n(A \cup B) = n(B \cup A)$. This equation provides a link; it is true, since $A \cup B = B \cup A$; so $a + b = b + a$ must also be true, and we have a proof for the theorem. But this process of discovery is usually done mentally, perhaps with brief written notes to keep track of the argument. Though this process is by far the most important part of the proof, it lies submerged, like the lower 90 percent of the iceberg. The 10 percent which appears in print is the formal proof, an orderly presentation of the steps which lead to the conclusion, sometimes revealing little about how they were discovered or about the ideas which were tried and found wanting. The proof goes like this:

THEOREM 4.2 If $a, b \in W$, then $a + b = b + a$.

Proof: It is given that a and b are whole numbers. To compute $a + b$, Definition 4.4 requires two disjoint sets, A and B, such that $n(A) = a$ and $n(B) = b$. By Definition 4.1 such sets can always be found.[5] Now, Definition 4.4 asserts that $a + b = n(A \cup B)$. Similarly, we have $b + a = n(B \cup A)$. But $A \cup B = B \cup A$, since union of sets is commutative. Hence $n(A \cup B) = n(B \cup A)$, because equal sets are equivalent and so have the same cardinal number.[8] Summarizing the previous equations, the following expressions are all symbols for the same cardinal number: $a + b$, $n(A \cup B)$, $n(B \cup A)$, and $b + a$. So by Definition 4.3 for the equality of whole numbers, $a + b = b + a$, which is the conclusion of the theorem.

The preceding proof appears in paragraph form, as writers of mathematics usually present proofs, though frequently they suppress the more obvious details. High school geometry texts, on the other hand, often use two parallel columns with numbered steps. Although this format is rather mechanical, it is orderly and it emphasizes the necessity of justifying each step by an "acceptable" reason. An acceptable reason in a formal proof is one of the following: an hypothesis, an axiom, a definition, a basic law of logic, or a previously proved theorem. As an example, we rearrange the proof of Theorem 4.2 in parallel-column form:

---

[7]See Exercises 2.5, Problem 2(a)(i).

[8]Since we have not developed set theory and logic as deductive systems, though this could be done, we cannot formally cite true statements about these systems as axioms or theorems. To speed up our work, however, we shall freely use such statements hereafter without further comment.

THEOREM 4.2 If $a, b \in W$, then $a + b = b + a$.

Proof:

Statements	*Reasons*
1. $a, b \in W$	1. Hypothesis
2. Choose sets A and B, disjoint, such that $n(A) = a$, $n(B) = b$.	2. Definition of whole number (Definition 4.1)[5]
3. $a + b = n(A \cup B)$	3. Definition of addition in W (Definition 4.4)
4. $b + a = n(B \cup A)$	4. Definition of addition in W (Definition 4.4)
5. $A \cup B = B \cup A$	5. Commutativity of union of sets[8]
6. $n(A \cup B) = n(B \cup A)$	6. Equal sets have the same cardinal number.[8]
7. $a + b = b + a$	7. Definition of equality in W (Definition 4.3), applied to Steps 3,4,6

A third form of proof, inspired by flow charts used in computer programming, has recently been recommended.[9] Although it requires more space and the learning of a few diagrammatic conventions, it has the advantage of clarifying the relationships among the steps of the proof. The conventions include representing the Rule of Detachment,

$$\frac{\begin{array}{c} p \\ p \Rightarrow q \end{array}}{\therefore q}$$

by

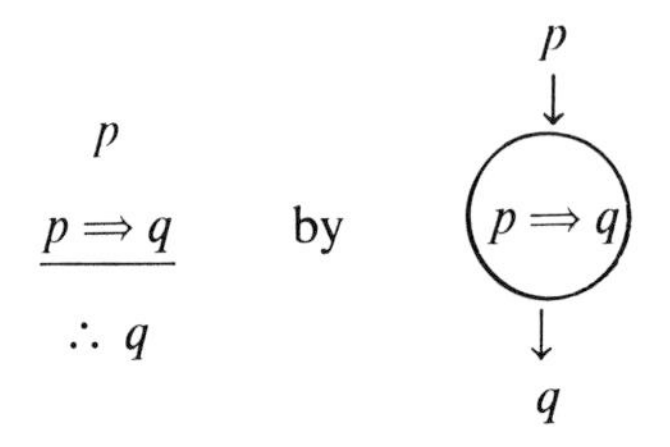

In most proofs, this diagram, consisting of an hypothesis (p), an implication ($p \Rightarrow q$), and a conclusion (q), is repeated again and again, forming a sequence of overlapping applications of the Rule of Detachment. In many proofs, the simple elements, p and q, are replaced by conjunctions or (occasionally) disjunctions: $p_1 \wedge p_2 \wedge \ldots$, instead of simply p, and similarly for the conclusion. The content of the circle, $p \Rightarrow q$, can be an axiom (A), a definition (D), or a theorem (T). The individual statements, $p, q, \ldots$, can be given statements (G), consequences of an implication (S), or conclusions of the theorem (C). By way of illustration, we apply these designations to Theorem 4.2, obtaining Figure 4.3 as a result.

[9]See the Hallerberg reference.

THEOREM 4.2 If $a, b \in W$, then $a + b = b + a$.

Statements

G_1: $a \in W$
G_2: $b \in W$
G_3: Set A
G_4: Set B
D_5: Definition 4.1[5]
S_6: $n(A) = a$
S_7: $n(B) = b$
S_8: $A \cap B = \varnothing$
D_9: Definition 4.4
S_{10}: $a + b = n(A \cup B)$
S_{11}: $b + a = n(B \cup A)$
T_{12}: $\cup$ is commutative.[8]
S_{13}: $A \cup B = B \cup A$
T_{14}: Equal sets have equal cardinal numbers.[8]
S_{15}: $n(A \cup B) = n(B \cup A)$
D_{16}: Definition 4.3
C_{17}: $a + b = b + a$

Diagram

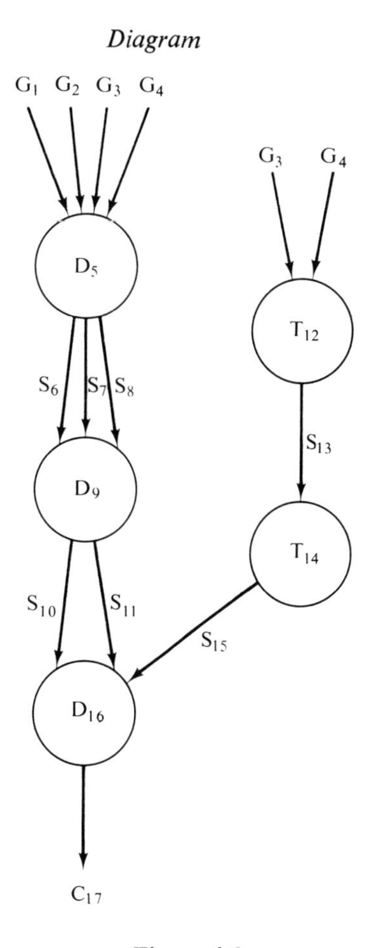

Figure 4.3

You should now reexamine this example to see how it confirms the given general description of the flow-diagram method.

The link between addition of whole numbers and union of sets with respect to commutativity tempts us to explore also the associative and identity properties. Indeed, we are not disappointed. It is easy to prove

THEOREM 4.3 Associativity of Addition Addition in W is associative.

In the exercises, you are invited to rephrase this theorem and to provide a proof. We pass on to consider a special property of zero — that zero is an[10] identity for addition. Thus we state

---

[10]It can be proved that 0 is the only additive identity; but our proof does not establish this.

THEOREM 4.4 Additive Identity If $a \in W$, then $a + 0 = 0 + a = a$.

Proof: Since $0 + a = a + 0$ follows from the commutative property (Theorem 4.2), we have only to prove that $a + 0 = a$. Again, by Definition 4.1, we select[5] a set A such that $n(A) = a$. By definition, 0 is the cardinal number of the null set: $n(\varnothing) = 0$. From set theory, $A \cap \varnothing = \varnothing$[8]; so A and $\varnothing$ are disjoint, as required by Definition 4.4. Now we may apply this definition to get $a + 0 = n(A \cup \varnothing)$. But from set theory again, $A \cup \varnothing = A$. Since equal sets have the same cardinal number, $n(A \cup \varnothing) = n(A)$; and so, by the definition of equality in W, $a + 0 = a$.

We close this section with a proof of the addition law for equality and its converse. Sometimes this law, in a more general form, is expressed as "If equals are added to equals, the sums are equal." It is much used in equations: for example, given that $x = 4$, with its aid one can conclude that $x + 3 = 4 + 3$, or $x + 3 = 7$. This theorem can be phrased as

THEOREM 4.5 Equality for Addition If $a, b, c \in W$ and $a = b$, then $a + c = b + c$ and $c + a = c + b$.

(You should try to discover a proof before looking at the one which follows.)

Proof: We select[5] sets A and C such that $n(A) = a$ and $n(C) = c$ and $A \cap C = \varnothing$. Since $a = b$ is given and $n(A) = a$, then by Definition 4.3, $n(A) = b$ also. Applying the definition of addition, $a + c = n(A \cup C)$; similarly, $b + c = n(A) + n(C) = n(A \cup C)$. But these two expressions, $a + c$ and $b + c$, are symbols for the same cardinal number, $n(A \cup C)$; hence $a + c = b + c$. By changing the order of c and C in the last two sentences, we can similarly prove $c + a = c + b$. (Thus we begin with $c + a = n(C \cup A)$, etc.)

In the exercises, you are asked to prove a generalization of Theorem 4.5.

THEOREM 4.6 Equality for Addition If $a, b, c, d \in W$ and if $a = b$ and $c = d$, then $a + c = b + d$.

It is this theorem which, expressed in words, begins, "If equals are added to equals. . . ."

For later use we shall need the converse of Theorem 4.5, which permits us to conclude from $a + c = b + c$ that $a = b$. We frequently use this converse in algebra for solving equations such as $x + 5 = 7$, which we may rewrite as $x + 5 = 2 + 5$ and then, "cancelling" the 5, conclude that $x = 2$. Formally we state

THEOREM 4.7 Cancellation for Addition If $a,b,c \in W$ and if $a + c = b + c$, then $a = b$.

Though the proof of this theorem could be based on set theory, it is more complicated than that of Theorem 4.5. For this reason we shall defer its proof until Section 4.7, when we can call the order properties in W to our aid.

In summary, in this section we have defined on W a binary operation called addition. We have seen, moreover, that addition is closed, commutative, and associative; that 0 is the additive identity; and that the addition law for equality holds. These properties are so fundamental and all-pervasive in mathematics that you are advised to commit them to memory in order to have them in readiness for use during the remainder of the text. Indeed, the fact that they have now been proved as theorems makes them available for use in proving more advanced theorems, and thus contributes to the building of the entire structure of our familiar number system.

EXERCISES 4.2

1. Draw a number line. Use it to add 3 and 4; 5 and 2; 6 and 0.

2. (a) For which of the following sets is it true that $n(A) + n(B) = n(A \cup B)$? In each case, write down the cardinal numbers involved.

   (i) $A = \{a,b,c,d\}$ $\quad B = \{e,f\}$

   (ii) $A = \{a,b,c,d\}$ $\quad B = \{d,e,f\}$

   (iii) $A = \{x \mid x \text{ is a month whose name begins with a } J\}$
   $B = \{x \mid x \text{ is a month in the first half of the year}\}$

   *(iv) $A = E$ (even numbers) $\quad B = D$ (odd numbers)

   (b) For which of the sets of part (a) is it true that $n(A) + n(B) - n(A \cap B) = n(A \cup B)$? Explain.

3. Apply Definition 4.4 to compute the following sums, providing in each case specific sets A and B:

   (a) $3 + 4$ (b) $4 + 3$ (c) $5 + 1$ (d) $6 + 0$

4. The following statements illustrate theorems concerning addition of whole numbers. In each case *name* the theorem: do not use numerical references, like "Theorem 4.1."

   (a) $7 + 5 = 5 + 7$ (b) $(7 + 5) + 1 = 7 + (5 + 1)$

   (c) $(7 + 5)$ is a unique whole number.

   (d) $3 + (2 + 4) = (3 + 2) + 4$

   (e) $3 + (2 + 4) = 3 + (4 + 2)$

   (f) $(2 + 4)$ is a unique whole number.

   (g) $3 + (2 + 4) = (2 + 4) + 3$

   (h) If $x = 5$, then $x + 1 = 5 + 1$.

   (i) If $6 = x + 3$, then $11 = x + 8$.

5. Using the commutative and/or associative theorems, write each of the following problems in a form which makes the mental computation of the sum easiest:

   (a) $(5 + 12) + 8 = ?$ (b) $22 + (22 + 7) = ?$

   (c) $64 + (18 + 36) = ?$ (d) $(29 + 6) + (4 + 1) = ?$

6. For each of the following sets, state whether it is closed under addition; then justify your answer:

   (a) E (even numbers) (b) D (odd numbers)

   (c) those whole numbers ending with the digits 0 or 5

   (d) $\{0,1\}$ (e) $\{0\}$ (f) N (natural numbers)

7. Prove the associative theorem for addition (Theorem 4.3), after first completing the rephrasing:

THEOREM 4.3 If $a, b, c \in W$, then. . . .

In your proof you may use either the paragraph or the parallel-columns form. (*Hint:* Obtain $(a + b) + c = n(A \cup B) + n(C) = n[(A \cup B) \cup C]$, and compare with similar expressions for $a + (b + c)$, recalling that union of sets is associative.)

8. Prove the generalized addition theorem for equality (Theorem 4.6).

*9. Make a flow-diagram proof for one or more of Theorems 4.3, 4.4, 4.5, and Problem 8.

*10. Suppose we were to define a weird kind of "addition" in W, symbolized by $\oplus$, by means of the equation $a \oplus b = n(A \cup B)$, where $n(A) = a$, $n(B) = b$, but A and B are not necessarily disjoint. Is the operation $\oplus$ closed in W? Try some examples, then justify your answer.

*11. Answer the preceding exercise after replacing the equation there by $a \boxplus b = n(A \times B)$, where $\times$ denotes the cartesian product.

*12. Returning to the registrar's problem of this section, but changing the data slightly, suppose we have

$$n(D) = 1{,}479 \qquad n(E) = 2{,}079 \qquad n(J) = 633 \qquad n(D \cap E) = 20$$

$$n(E \cap J) = 19 \qquad n(D \cap J) = 2 \qquad n(D \cap E \cap J) = 1$$

Venn diagrams may help sort out these data.

(a) Compute $n(E \cup J)$.

(b) Compute $n(D \cup E \cup J)$, the total enrollment of the three divisions of the college.

(c) Write down a general formula by which to compute $n(A \cup B \cup C)$, where A, B, C are any three sets. Also list the information required to use the formula in computing $n(A \cup B \cup C)$.

Section 4.3 Simple Proofs

In the previous section we established the operation of addition in W and proved some fundamental properties by means of set theory. With the aid of these theorems we can prove many more, some of them useful in themselves, and all of them good exercises in deductive reasoning. We are going to analyze a typical example in considerable detail and then request you to practice with several other theorems.

Let us consider a familiar method of checking the addition of a column of whole numbers. As a trivial example we use the sum

$$\begin{array}{c} 4 \\ 2 \\ 3 \\ \hline 9 \end{array}$$

Adding down, we get $(4 + 2) + 3 = 6 + 3 = 9$; adding up, we get $(3 + 2) + 4 = 5 + 4 = 9$. The check consists in ascertaining that the two sums, though differently obtained, are equal: $(4 + 2) + 3 = (3 + 2) + 4$. Now we examine this equation. Is it *merely* an instance of the associative law? To be sure, the grouping has changed; but so also has the order, from 4–2–3 to 3–2–4. So associativity is not a sufficient explanation of this equation, though it certainly is involved. Which other property (-ties) of the whole numbers is (are) involved?

Of course, we are not interested merely in a numerical example. We wish to prove that, in general, adding up and adding down produces the same answer. Is there a *general* equation (in terms of x, y, z) which states this fact for three addends? Certainly there is; and we incorporate it into the formal statement which follows:

THEOREM If $x, y, z \in W$, then $(x + y) + z = (z + y) + x$.

Before trying to prove this theorem, we shall establish a very obvious, but useful property of equality in W. We recognize intuitively that, if one number equals a second, and the second in turn equals a third, then the first number equals the third. We call this the *transitive*[11] property of equality in W. Stated formally, this becomes

---

[11]Several other important relations are transitive. In Exercises 2.2, Problem 8, we showed that the subset relations on sets are transitive, and in Exercises 3.6, Problem 3(f), that the corresponding implication relation on propositions is transitive.

THEOREM 4.8 Transitivity of Equality If $a, b, c \in W$ and if $a = b$ and $b = c$, then $a = c$.

The proof, based on the definition of equality in W, is immediate and is left as an exercise. This theorem can be extended readily to more than three whole numbers merely by applying it several times in succession. So, for instance, from $a = b$, $b = c$, $c = d$, we may conclude that $a = c$, and then that $a = d$, by two applications of Theorem 4.8. In such cases we shall cite Theorem 4.8 only once, assuming that it is applied as many times as necessary in the given situation.

Let us now devise a proof for the theorem $(x + y) + z = (z + y) + x$. It may be helpful to think of working out a proof in three stages. First, we examine the theorem to be sure we understand it: sometimes a good way to do this is to "try it out" in specific cases. This is what we did in discussing the example $(4 + 2) + 3 = (3 + 2) + 4$. Second, we try to discover informally some steps which lead from what is given to the conclusion. In the case of the theorem $(x + y) + z = (z + y) + x$, seeing that it involves associativity and commutativity, we might try applying these properties in some sequence or other to the left side of the equation, to see if we can change it to look more like the right side. For example, the application of associativity yields:

$$(x + y) + z = x + (y + z)$$

which is not yet the right side; but at least y and z are now grouped together. Now we need to change the order of these letters; how can we accomplish this? The end result of our efforts will be a chain of expressions leading from the expression on the left to that on the right, each step presumably accomplished by a legitimate mathematical maneuver. Here, such a chain might be $(x + y) + z = x + (y + z) = (y + z) + x = (z + y) + x$. The third stage is to write the theorem down formally and clearly, indicating *where* the proof begins and supplying it. Thus here,

THEOREM If $x, y, z \in W$, then $(x + y) + z = (z + y) + x$.

Proof:

1. $(x + y) + z = x + (y + z)$ because addition in W is associative.
2. . . . .
3. . . . .

Last. $(x + y) + z = (z + y) + x$ because. . . .

You should note that the equation to be proved appears as the last step. You are invited to write in the intermediate steps.

Although the chain of equations may guide the formal proof, the latter must also include a justification for every link of the chain. Sometimes this requires the insertion of additional steps if the proof is to be complete. To make this point clear, a complete proof of the theorem follows. You should compare it with your efforts to fill in the blanks. For clarity we use the parallel-column format. Note how the chain of equalities which we previously worked out guides the sequence of steps.

THEOREM If $x, y, z \in W$, then $(x + y) + z = (z + y) + x$.

Proof:

1. $(x + y) + z = x + (y + z)$	1. Assoc. ad. (Theorem 4.3)[12]
2. $(y + z)$ is a single (unique) whole number.[13]	2. Cl. ad. (Theorem 4.1)
3. $x + (y + z) = (y + z) + x$	3. Comm. ad. (Theorem 4.2)[13]
4. $y + z = z + y$	4. Comm. ad. (Theorem 4.2)[14]
5. $(z + y)$ is a single whole number.	5. Cl. ad. (Theorem 4.1)
6. $(y + z) + x = (z + y) + x$	6. Equal. ad. (Theorem 4.5)[15]
7. $(x + y)$, $(x + y) + z$, $x + (y + z)$, $(y + z) + x$, $(z + y) + x$ are single whole numbers.	7. Cl. ad. (Theorem 4.1)
8. $(x + y) + z = (z + y) + x$	8. Trans. equal. (Theorem 4.8)[16]

This proof seems unnecessarily complicated. In particular, we might wish to omit the three closure steps (2,5,7) as being obvious. For completeness, however, they should be included to call attention to the fact that, in some steps, expressions like $(y + z)$ and $(x + y) + z$ are treated as single whole numbers. Soon we shall relax somewhat these requirements for completeness, including the closure steps; but you should always be capable of supplying the omitted steps on demand.

In conclusion, we call attention to the contrast between the proofs of Section 4.2 and those of this section. In the former, basic properties of addition were proved by the use of Definition 4.4. Since this definition, given in terms of sets, was all we knew about addition in W at that point, the proofs had to be based on the properties of sets. In this section, however, the situation is different: having already established several extremely important theorems on addition, such as those on commutativity and on associativity, we can use them to make

---

[12]The theorem numbers are supplied only for the reader's convenience; henceforth they will be omitted whenever the theorem has a generally accepted name, as "associative theorem for addition."

[13]Commutativity applies to *two* numbers, not three. Step 2 permits considering $(y + z)$ as one number, to be commuted with x to produce Step 3.

[14]Here commutativity applies to a different pair of numbers, namely, y and z.

[15]In Theorem 4.5, let $a = (y + z)$, $b = (z + y)$, $c = x$.

[16]If we represent the last four expressions of Step 7 by the single letters a,b,c,d, respectively, then Steps 1 $(a = b)$, 3 $(b = c)$, and 6 $(c = d)$ imply Step 8 $(a = d)$.

other proofs, usually without referring back to Definition 4.4 or the properties of set theory. Only when we introduce the operation of multiplication, again in terms of sets, will it be necessary to utilize set properties for a time, in order to build up the important properties of multiplication in W.

EXERCISES 4.3

1. Prove Theorem 4.8. (*Hint:* What does the statement $a = b$ mean, according to the definition of equality for whole numbers?)

2. If a, b, c, d are any whole numbers, give a complete proof that
   (a) $a + (b + c)$ is a whole number.
   (b) $[a + (b + c)] + d$ is a whole number.
   (*Hint:* Do not make a big production of this: in part (a), only two steps beyond stating the hypothesis are required.)

3. Another method of proving that $(x + y) + z = (z + y) + x$ is to use the commutative property *first* and the associative later on.
   (a) Complete a chain of equalities leading from $(x + y) + z$ to the right side of the preceding equation, starting with $(x + y) + z = z + (x + y)$.
   (b) Write out a complete proof based on your answer to part (a).

4. In each of the following equations, all letters represent whole numbers. Consider each equation as a theorem to be illustrated, then proved, by the following procedure: first, provide (at least mentally) a numerical example; second, devise a chain of equalities leading from one side of the equation to the other; third, guided by this chain, write out a complete proof.
   (a) $(x + y) + z = (y + x) + z$ (b) $(x + y) + z = z + (x + y)$
   (c) $(x + y) + z = x + (z + y)$
   (d) $[(x + y) + z] + w = (x + y) + (z + w)$ (*Hint:* Treat $(x + y)$ as a single whole number.)

5. We have defined addition as a binary operation: so far, $a + b + c$, a triple sum, has no meaning because it has three addends. But we can assign a meaning by the following:

Definition 4.5 Triple Sum If $a, b, c \in W$, then $a + b + c = (a + b) + c$.

The right side already has meaning, since it is a succession of two binary sums. Use Definition 4.5 and other properties of whole numbers to prove the following:

THEOREM $a + b + c = a + (b + c)$.

6. (a) Extend Definition 4.5 to four addends, $a + b + c + d$, by using the expression $[(a + b) + c] + d$.

   (b) Now prove that

   (i) $a + b + c + d = (a + b) + (c + d)$

   (ii) $a + b + c + d = a + [(b + c) + d]$

   (iii) $a + b + c + d = a + (b + c) + d$

   (*Note:* From the above results it appears that parentheses can be inserted in $a + b + c + d$ to group the addends in all possible ways, and that the resulting expressions are all equal. Thus the associative property can be generalized to four addends and, similarly, to a larger number of addends.)

*7. The commutative property for addition was stated and proved for two addends (Theorem 4.2). By an analysis similar to that of Problem 6, investigate whether commutativity can be generalized to three or more addends: e.g., $a + b + c = b + a + c = a + c + b$, etc.

*8. Make a proof of the theorem $(x + y) + z = (z + y) + x$ by the flow-chart method.

Section 4.4 Multiplication

Addition is one of the two basic operations on whole numbers; the other is multiplication. Subtraction and division, as we shall see in a later section, are *inverse* operations, definable in terms of addition and multiplication, respectively. For multiplication, several approaches are possible. A common one is to consider it as repeated addition: thus $4 \times 3 = 3 + 3 + 3 + 3$. We shall later prove this.

The approach which we shall use now, however, goes back to the cartesian product of set theory. It was foreshadowed by Exercises 2.8, Problem 5(c), which hints that, for finite sets, the cardinal number of the cartesian product of two sets is equal to the arithmetic product of the cardinal numbers of the two sets. Exercises 2.8, Problem 5(b) provided an illustration. There we began with the sets $E = \{a,b\}$ and $F = \{d,e,f\}$, for which $n(E) = 2$ and $n(F) = 3$. We then found $E \times F$, which had as elements ordered pairs like (a,d) and (b,f)—six of them in all. Thus we obtained $n(E \times F) = 6$. For this exercise, then, we found that $n(E) \cdot n(F) = n(E \times F)$, an equation which was easy to generalize. This and other examples suggest

Definition 4.6 Multiplication If a and b are any two whole numbers and A and B are sets such that $n(A) = a$ and $n(B) = b$, then $ab = n(A \times B)$.

Concerning this definition, we note that A and B are not required to be disjoint, as they were in Definition 4.4 for addition. The product, moreover, is symbolized by ab, i.e., a and b juxtaposed without any operation sign, except where confusion might arise. An instance of the exception is the product of 2 and 3, which cannot be written as 23, since our numeration system interprets 23 as "2 tens plus 3 units." In such cases, an operation sign such as $\times$ or $\cdot$ (a *raised* dot) or a pair of parentheses is used to denote multiplication, as in the example $2 \times 3 = 2 \cdot 3 = (2)(3)$.

It is now our task to develop the properties of multiplication. A good point of departure is to check, using numerical examples, whether the properties of addition (closure, commutativity, associativity, identity, and equality) carry over to multiplication. You are invited to check these properties. We shall announce the results as theorems and discuss the proofs.

First, then, is multiplication closed in W? This question resolves itself into two parts: (1) If $a, b \in W$, is $ab \in W$ also? and (2) Is ab unique, once a and b are given? As for (1), the discussion on closure of addition indicated that, for any $a, b \in W$, it is always possible to find sets A and B, such that $n(A) = a$ and $n(B) = b$. From these two sets it is always possible to determine $A \times B$. Although $A \times B$ is not the same kind of set as its separate components, it *is* a set and contains elements of the form (x,y), unless A or B is empty; in the latter event, $A \times B$ is still a set, namely, $\varnothing$. Also, since A and B are finite, by definition of whole number, the paired elements (x,y) of $A \times B$ are also finite in number, hence $n(A \times B)$ is a whole number. In summary, ab always exists and is a whole number.

The question of the uniqueness of a product can easily be answered for the particular case, $2 \cdot 3$, which preceded Definition 4.6. For this product we chose $E = \{a,b\}$ and $F = \{d,e,f\}$. Suppose instead that we choose $E^* = \{a^*,b^*\}$ and $F^* = \{d^*,e^*,f^*\}$, where the starred elements and sets are different from the unstarred ones. But since $E^* \sim E$ and $F^* \sim F$, we also have $n(E^*) = n(E) = 2$ and $n(F^*) = n(F) = 3$. Then we can form $E^* \times F^*$, which has elements such as (a^*,d^*) and (b^*,f^*). Though all the ordered pairs in $E^* \times F^*$ are different from those in $E \times F$, the two sets can be placed in one-to-one correspondence in an obvious way, with $(a^*,d^*) \leftrightarrow (a,d)$, etc. Then clearly, $E^* \times F^* \sim E \times F$ and $n(E^* \times F^*) = n(E \times F) = 6$. The point is that choosing different sets to represent the cardinal numbers 2 and 3 has not changed the value of the product, $2 \cdot 3$, obtained by applying Definition 4.6.

To argue the question of uniqueness (2) for any product ab, we have only to generalize the preceding paragraph. Having obtained $ab = n(A \times B)$ (Definition 4.6) by choosing sets A and B so that $n(A) = a$ and $n(B) = b$, we now make a different choice of sets, A^* and B^*, still having $n(A^*) = a$ and $n(B^*) = b$. We note that $A^* \sim A$ and $B^* \sim B$, since the two pairs of sets have the same cardinal numbers, respectively. Then $A^* \times B^*$ will perforce contain ordered pairs of the form (x^*,y^*), which can be placed in one-to-one correspondence with the pairs (x,y) of $A \times B$ simply as $(x^*,y^*) \leftrightarrow (x,y)$. Hence $A^* \times B^* \sim A \times B$, so that $n(A^* \times B^*) = n(A \times B) = ab$, and the product is the same, regardless of the choice of sets to represent a and b.

Though the preceding two paragraphs do not quite constitute a formal proof, they could be modified to establish rigorously the following:

THEOREM 4.9 Closure of Multiplication W is closed under multiplication.

We now turn to commutativity. The key to proving that $ab = ba$ is to show that $A \times B \sim B \times A$. This question, raised in Exercises 2.8, Problem 5(e)(ii), can be answered in the affirmative, since only the *order* within the sets of ordered pairs differs between $A \times B$ and $B \times A$, not the *number* of pairs. Recalling that we are assuming the results of set theory,[8] we make the following brief proof:

THEOREM 4.10 Commutativity of Multiplication If $a, b \in W$, then $ab = ba$.

Proof: Given that $a, b \in W$, it is possible to find sets A, B, such that $n(A) = a$ and $n(B) = b$. By Definition 4.6, $ab = n(A \times B)$ and $ba = n(B \times A)$. But by properties of set theory,[8] $A \times B \sim B \times A$; hence $n(A \times B) = n(B \times A)$, so that by the definition of equality for whole numbers (Definition 4.3), applied to the last two equations, $ab = ba$.

The property of associativity for multiplication also follows rather easily. The only "sticking point" is working with the cartesian products

$$(A \times B) \times C \quad \text{and} \quad A \times (B \times C)$$

Since

$$A \times B = \{(x,y) \mid x \in A \text{ and } y \in B\}$$

we have

$$(A \times B) \times C = \{((x,y),z) \mid (x,y) \in A \times B \text{ and } z \in C\}$$

But $A \times (B \times C)$ yields elements of a different form: $(x,(y,z))$. Although we cannot claim that $(A \times B) \times C = A \times (B \times C)$, we can establish that $(A \times B) \times C \sim A \times (B \times C)$, which is enough to imply that $(ab)c = a(bc)$. The details are left as a starred exercise. Thus we arrive at

THEOREM 4.11 Associativity of Multiplication If $a, b, c \in W$, then $(ab)c = a(bc)$.

Theorem 4.4 stated that 0 is an additive identity in W. This suggests, by analogy, a theorem stating that 1 is a multiplicative identity in W. Thus we have

THEOREM 4.12 Multiplicative Identity If $a \in W$, then $a \cdot 1 = 1 \cdot a = a$. The proof of this theorem, based on Definition 4.6, is left to you.

A final multiplicative analog to the properties of addition is the following triple of theorems:

THEOREM 4.13 Equality for Multiplication If $a, b, c \in W$ and $a = b$, then $ac = bc$ and $ca = cb$.

A generalization is

THEOREM 4.14 Equality for Multiplication If $a, b, c, d \in W$ and if $a = b$ and $c = d$, then $ac = bd$.

Almost a converse of Theorem 4.13 is

THEOREM 4.15 Cancellation for Multiplication If $a, b, c \in W$, with $c \neq 0$, and if $ac = bc$, then $a = b$.

The proofs of the first two theorems are left as problems. The proof of Theorem 4.15 is left to Section 4.7, when the order properties are available.

It is instructive and helpful to the memory to summarize the corresponding properties which we have developed for addition and multiplication in W (see Table 4.1).

TABLE 4.1

Addition	*Multiplication*
Closure	Closure
Commutativity	Commutativity
Associativity	Associativity
Identity (0)	Identity (1)
Equality	Equality

So far, the two operations seem identical in their behavior, despite their different definitions, and yet we know that they are quite distinct. Before probing their differences, however, let us exploit some advantages of their likenesses. For example, it is easy to state the multiplicative analog of the equation $(x + y) + z = (z + y) + x$; it is

THEOREM If $x, y, z \in W$, then $(xy)z = (zy)x$.

How may we prove it? An examination of the proof of the addition theorem, $(x + y) + z = (z + y) + x$, proved in Section 4.3, reveals that the definition of addition is not explicitly used, but only the properties of closure, commutativity, and associativity, plus transitivity of equality, which involves neither addition nor multiplication. This means that to prove the corresponding multiplication theorem takes little trouble or even thought! We merely copy the proof of the addition theorem, changing the symbolism to multiplication, and for justification, substituting the theorem numbers which refer to multiplication.

Thus the first two steps convert to

1. $(xy)z = x(yz)$	1. Assoc. mult. (Theorem 4.11)
2. (yz) is a single whole number.	2. Cl. mult. (Theorem 4.9)

The remaining steps likewise require only trivial modifications. It is hardly worthwhile to write out the details.

But addition and multiplication *are* different operations. Where, then, does the difference lie? It begins to emerge in the final two theorems of this section and is fully revealed in the next section. The first theorem, called the *zero product theorem*, shows the result of *multiplying* a whole number by the *additive* identity, 0. This result, indeed, was foreshadowed by Exercises 2.8, Problem 5(d). It is

THEOREM 4.16 Zero Product Theorem If $a \in W$, then $a \cdot 0 = 0 \cdot a = 0$.

Proof: In view of the commutativity of multiplication, we need only show that $a \cdot 0 = 0$. We choose a set, A, such that $n(A) = a$; for 0, we perforce choose the empty set, since $n(\varnothing) = 0$. Then, by the definition of multiplication, $a \cdot 0 = n(A \times \varnothing)$. But by set theory, $A \times \varnothing = \varnothing$; hence $a \cdot 0 = n(A \times \varnothing) = n(\varnothing) = 0$. Since all four expressions are symbols for the same cardinal number, $a \cdot 0 = 0$.

The second theorem is actually a converse of the first. To see this, we first restate Theorem 4.16 as, "If $a, b \in W$ and $a = 0$ or $b = 0$, then $ab = 0$." Its converse is then

THEOREM 4.17 If $a, b \in W$ and $ab = 0$, then $a = 0$ or $b = 0$.

You may recall using this theorem in solving quadratic equations by factoring. For instance, given $x^2 - 5x + 4 = 0$, immediate factoring of the left side yields $(x - 1)(x - 4) = 0$. Considering that $(x - 1)$ and $(x - 4)$ are two numbers[17] whose product is zero, we can conclude at once that $x - 1 = 0$ or $x - 4 = 0$, from which we obtain 1 and 4 as solutions of the quadratic.

Since the proof of Theorem 4.17 involves many of the same ideas as that of Theorem 4.16, you are invited to fill in the details, including reasons, of the following outline:

1. Since $ab = 0$ is given, for appropriate sets A and B, such that $n(A) = a$ and $n(B) = b$, we must have $n(A \times B) =$ ______.
2. But 0 is the cardinal number of only one set; hence $A \times B =$ ______.

---

[17]In this example we assume that x, $x - 1$, and $x - 4$ represent whole numbers; so Theorem 4.17 applies. This theorem, moreover, can be easily extended to other number systems of algebra.

3. Now the cartesian product is a pairing operation involving elements of two sets. If $A \times B$ is empty, this implies that A or B is ______, so that $n(A) = a =$ ______ or $n(B) = b =$ ______.

Having established both Theorem 4.16 and Theorem 4.17 as implications, we can combine them into a single statement of logical equivalence, as

THEOREM 4.18 If $a, b \in W$, then $ab = 0$ iff $a = 0$ or $b = 0$.

It is interesting to note in passing that, while the familiar number systems obey Theorem 4.18, there are systems in which it does not hold. In Chapter 6 we shall examine some in which Theorem 4.16 holds, but Theorems 4.17 and 4.18 are false.

Enough has been said to show that zero is a special number which exhibits rather peculiar behavior. The consequences are far reaching and become apparent and unavoidable in connection with the operation of division.

EXERCISES 4.4

1. Apply Definition 4.6 to compute the following products, providing in each case specific sets, A and B:

   (a) 2×4 (b) 4×2 (c) 5×1 (d) 0×3

2. (a) If $a, b, c, d \in W$, prove that (i) $a(bc) \in W$ (ii) $[a(bc)]d \in W$

   (b) How closely do your proofs parallel those of Exercises 4.3, Problem 2?

   (c) If $a, b, c, d \in W$, prove that

   (i) $a + bc \in W$ (ii) $(a + b)(c + d) \in W$

3. Prove the multiplicative identity theorem (Theorem 4.12).

4. In each of the following equations all letters represent whole numbers. Consider each equation as a theorem to be illustrated, then proved, using the procedure of Exercises 4.3, Problem 4.

   (a) $(ab)c = (ba)c$ (b) $[(ab)c]d = (ab)(cd)$

   (c) $(ab)(cd) = (cd)(ab)$ (d) $ab + cd = dc + ba$

5. By analogy with addition, define the triple product abc, and prove that $abc = a(bc)$.

6. Prove the first two theorems for equality for multiplication, Theorem 4.13 and Theorem 4.14. (*Hint:* Refer to the corresponding addition proofs in Section 4.2.)

7. (a) Compare the two cancellation theorems, Theorem 4.7 and Theorem 4.15, and point out in which respect(s) the latter fails to be completely analogous to the former.

(b) Show by an example that, if the restriction $c \neq 0$ were omitted, Theorem 4.15 would be false.

8. (a) In Section 3.4 we agreed that *or* would be interpreted inclusively, unless otherwise specified. Is this, the correct interpretation in Theorems 4.17 and 4.18? Explain.

(b) Fill the blanks in the text's outline of the proof of Theorem 4.17.

*(c) Make a formal proof of Theorem 4.17. (*Hint:* In examining the third paragraph of the outline, consider what would happen if $A \neq \varnothing$ and $B \neq \varnothing$.)

*9. Prove the associative property for multiplication (Theorem 4.11).

*10. State the results for multiplication which correspond to those of Exercises 4.3, Problems 6 and 7. How can you justify these results without giving detailed proofs?

Section 4.5 The Distributive Property

In set theory there are two distributive laws: intersection distributes over union and vice versa. Let us examine the corresponding relations between multiplication and addition in W. The equation which states that multiplication distributes over addition is a familiar one: $a(b + c) = ab + ac$. Indeed, you may be inclined to dismiss it as merely an example of the algebraic technique of "multiplying out" by the factor a; or, by reading it from right to left, of "factoring out" the a. While these observations are correct, the emphasis is wrong: it puts the cart before the horse! The techniques of multiplying out and factoring derive their validity from the distributive law, not the other way about. It is *because of* the distributive property and other properties of our number systems that the algebraic techniques "work."

The importance of the distributive law in arithmetic is suggested by the following examples:

1. It can be used in computational short-cuts, as in the equation of Section 2.5: $29 \cdot 36 + 29 \cdot 64 = 29 \cdot (36 + 64) = 29 \cdot 100 = 2{,}900$.

2. It is used in multiplying a whole number by a mixed number. Examina-

$$
\begin{array}{r}
126 \\
\times\, 5\tfrac{1}{3} \\
\hline
630 \\
42 \\
\hline
672
\end{array}
\qquad\qquad
\begin{array}{r}
21 \\
\times\, 3 \\
\hline
3 \\
60 \\
\hline
63
\end{array}
$$

tion of the left-hand computation shows that we treat $126 \cdot 5\frac{1}{3}$ as $126(5 + \frac{1}{3}) = 126 \cdot 5 + 126 \cdot \frac{1}{3} = 630 + 42 = 672$.

3. In Chapter 5 we shall analyze multiplication problems like $3 \cdot 21$ as $3 \cdot 21 = 3(2 \cdot 10 + 1) = 3 \cdot (2 \cdot 10) + 3 \cdot 1 = 6 \cdot 10 + 3 = 63$. You should recognize that the right-hand computation is based on multiplying 3 by 1 and by 20, then adding the products.

Returning to theoretical considerations, we ask whether in W there is a second distributive law, with addition distributing over multiplication. You are requested to write down the equation which expresses this law. It is $a + bc = (a + b)(a + c)$; note that it is obtained from the equation $a(b + c) = ab + ac$ by interchanging $+$ with the juxtaposition which symbolizes multiplication. This equation certainly looks strange. Worse, it is *false*, as you may easily verify by substituting almost any three whole numbers for a, b, and c. Here, then, is the promised distinction between addition and multiplication in W: multiplication distributes over addition, but addition does *not* distribute over multiplication.

We shall henceforth refer to *the* distributive law in W, because there is only one. We state and prove it formally, using results from set theory, especially the distributive theorem in Section 2.8, which we introduced for just this purpose. By noting how the latter relates to the definitions of addition and multiplication in W, you can devise a proof of your own with which to compare the following:

THEOREM 4.19 Distributivity of Multiplication over Addition If $a, b, c \in W$, then $a(b + c) = ab + ac$.

Proof: Since $a, b, c \in W$, we may choose finite sets A, B, and C, such that $n(A) = a$, $n(B) = b$, $n(C) = c$, and $B \cap C = \varnothing$. By definition of addition, $b + c = n(B \cup C)$; then by definition of multiplication, $a(b + c) = n[A \times (B \cup C)]$. But from the theorem of Section 2.8, which asserts that cartesian product distributes over union, $A \times (B \cup C) = (A \times B) \cup (A \times C)$. Then since equal sets have the same cardinal number, $n[A \times (B \cup C)] = n[(A \times B) \cup (A \times C)]$ follows immediately. Now, working with the right side, by the definition of multiplication, $ab = n(A \times B)$ and $ac = n(A \times C)$. We note that $(A \times B) \cap (A \times C) = \varnothing$, since $B \cap C = \varnothing$, and so the second elements in the cartesian product couples must differ in $(A \times B)$ and $(A \times C)$. Hence we may now apply the definition of addition to get $ab + ac = n[(A \times B) \cup (A \times C)]$. But this is exactly the cardinal number we earlier found for $a(b + c)$; this proves that $a(b + c) = ab + ac$.

The distributive property is also indispensable in algebra. A simple example is the process of "combining terms," e.g., in the expression $3x + 2x$. How do we obtain $5x$, and why? Note that $3x + 2x = (3 + 2)x = 5x$, where the first equality is justified by the distributive law in the form of Theorem 4.21. To take another example, can we combine $3x + 2y$? No, because this expression is not in the form $ac + bc$ and so the distributive law does not apply. There need be no mystery about *like* and *unlike* terms: it is just that the distributive law can be used in the first case, with a resulting simplification; but it cannot be used in the second. More sophisticated algebraic manipulations involving multiplying out or factoring may use the distributive law, as well as other laws, several times. An example is

$$\begin{aligned} 2a^4 - 32b^4 &= 2(a^4 - 16b^4) = 2(a^2 + 4b^2)(a^2 - 4b^2) \\ &= 2(a^2 + 4b^2)(a + 2b)(a - 2b) \end{aligned}$$

As a further application in arithmetic we give an informal justification of the fact that multiplication in N (the set of natural numbers) can be interpreted as repeated addition. We say *informal* because a rigorous proof would require an appeal to the principle of mathematical induction, which is beyond the scope of this course. As preliminaries we note that $1 + 1 = 2$, $1 + 1 + 1 = (1 + 1) + 1 = 2 + 1 = 3$, $1 + 1 + 1 + 1 = 4$, etc. according to the definition of addition in W. More generally, by a method known as *mathematical induction* we could show that $1 + 1 + \cdots + 1 = n$, provided that the sum $1 + 1 + \cdots + 1$ contains n addends, each with the value one. We also need a generalization of the distributive property which states that

$$a(b_1 + b_2 + \cdots + b_n) = ab_1 + ab_2 + \cdots + ab_n$$

This generalization, which is made plausible in Problem *10, also requires mathematical induction in its proof. We now state the theorem and give an informal argument to make it seem plausible. It is

THEOREM 4.20 Repeated Addition If $a, b \in N$, then $ab = b + b + \cdots + b$, where the indicated sum contains a addends, all of them b's.

Informal Argument: If $a = 1$, then the multiplicative identity property (Theorem 4.12) gives immediately $1 \cdot b = b$ (one addend). If $a \in N$ but $a \neq 1$, then a can be written as $1 + 1 + \cdots + 1$, the sum being taken on a addends. Then $ab = (1 + 1 + \cdots + 1) \cdot b = 1 \cdot b + 1 \cdot b + \cdots + 1 \cdot b$ by the generalized distributive property, there being again a addends in the sum. But the multiplicative identity property, applied a times, permits simplification of the last expression: thus $1 \cdot b + 1 \cdot b + \cdots + 1 \cdot b = b + b + \cdots + b$, for a addends. By the transitive property for equality, the theorem follows.

The distributive property, as stated in Theorem 4.19, has the single factor, a, on the left, and so may be called the *left* distributive law. The *right* distributive law, $(a + b)c = ac + bc$, also holds. We could prove this by suitable modifications in the proof of Theorem 4.19; but there is an easier way, using properties of W rather than returning to set theory. You should pause to see if you can construct a chain of equalities in W, leading from $(a + b)c$ to $ac + bc$. An efficient chain is the following: $(a + b)c = c(a + b) = ca + cb = ac + bc$. The conversion of this chain to a complete proof, with reasons supplied, is left as an exercise. Thus we establish

THEOREM 4.21 (Right) Distributivity If $a, b, c \in W$, then $(a + b)c = ac + bc$.

Since there is little need to distinguish between Theorem 4.19 and Theorem 4.21 once we have proved them, we shall hereafter refer to both as *the* distributive law and use whichever one fits the context.

As a final illustration of the use of distributivity in algebra, we discuss the product of two binomials. You have probably learned an efficient way of calculating such products as $(x + 3)(x + 2)$. We simply multiply methodically the quantities joined by the numbered arcs, and then combine the resulting terms where possible: $(x + 3)(x + 2)$; ① ②

$$(x + 3)(x + 2) = x^2 + 2x + 3x + 3 \cdot 2 = x^2 + 5x + 6$$

① ② ③ ④

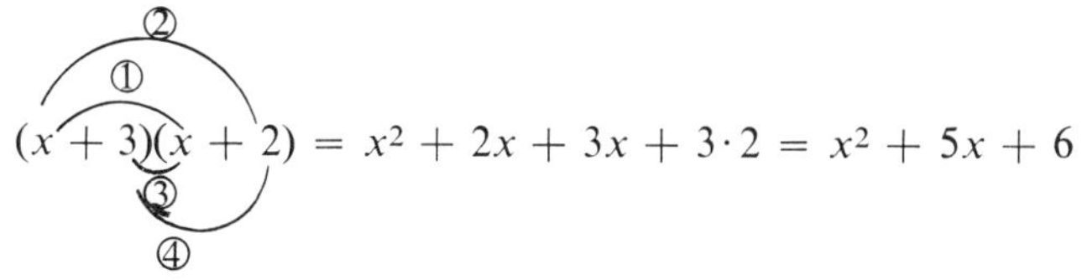

More generally, we get $(a + b)(c + d) = ac + ad + bc + bd$. The latter equation can be proved by using some of the properties of W listed in Table 4.1, together with the distributive law. At first glance it may seem hard to know how to obtain four terms at once from $(a + b)(c + d)$. The key is a standard ploy in mathematics: try to reduce an unknown problem to a known one. Here this may be done by treating $(c + d)$ as a single whole number, which we may designate, for convenience, by x. Then $(a + b)(c + d)$ becomes $(a + b)x$ which looks familiar: indeed, by the distributive law, it yields $ax + bx$ or $a(c + d) + b(c + d)$. Another application of the distributive law gives $(ac + ad) + (bc + bd)$, which is almost what we want. To complete the proof, we must appeal to the generalized associative law for addition, which says that three or *more* whole numbers may be grouped (or ungrouped) by addition in any meaningful manner. The particular application that is needed here is given by Exercises 4.3, Problem 6(b)(i), considering ac, ad, bc, and bd as single whole numbers. You should try to construct a formal proof and compare it with that of the following:

THEOREM 4.22 Binomial Expansion If $a, b, c, d \in W$, then

$$(a + b)(c + d) = ac + ad + bc + bd$$

Proof:

1. $a, b, c, d \in W$	1. Hypothesis
2. $(c + d)$ is a single whole number.[18]	2. Cl. ad.

[18]In the explanation before the theorem, $(c + d)$ was represented by x. You may do so here if it will clarify Step 3.

3. $(a + b)(c + d) = a(c + d) + b(c + d)$	3. Distr.
4. $a(c + d) + b(c + d) =$ $(ac + ad) + (bc + bd)$	4. Distr.
5. ac, ad, bc, bd, are single whole numbers.[19]	5. Cl. mult.
6. $(ac + ad) + (bc + bd) =$ $ac + ad + bc + bd$	6. Gen. assoc. ad. [Exercises 4.3, Problem 6(b)(i)]
7. All expressions on both sides of all the above equations are single whole numbers.[19]	7. Cl. ad. and mult.
8. Hence $(a + b)(c + d) =$ $ac + ad + bc + bd$	8. Trans. equal.

EXERCISES 4.5

1. (a) Complete the following numerical illustration of the distributive law in W: $4(2 + 3) =$ ______

   (b) Explain how the two dot diagrams illustrate part (a):

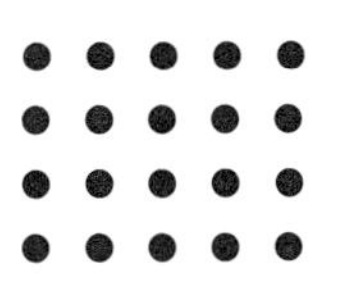

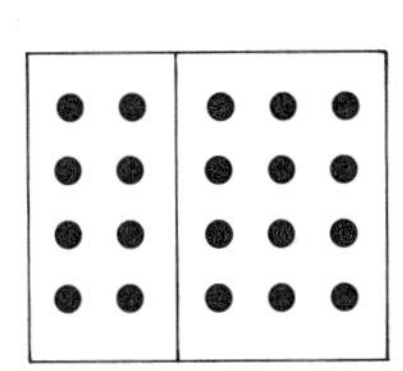

   (c) Complete and check the following numerical illustration of the right distributive law: $(2 + 3)4 =$ ______ .

   (d) Do the dot diagrams in part (b) confirm part (c) also? Explain.

2. Use the distributive property to simplify the following calculations so that they can be done easily in your head:

   (a) $7 \cdot 51 = 7(50 + 1) =$ ______ (b) $9 \cdot 42 =$ ______

   (c) $30 \cdot 21 =$ ______ (d) $33 \cdot 5 =$ ______ (e) $41 \cdot 61 =$ ______

3. (a) Explain how $4m + 5m$ can be simplified to $9m$, where $m \in W$.

   (b) Explain why $4m + 5p$, where $m, p \in W$ cannot, in general, be simplified.

   *(c) Are there any particular values of m and p for which $4m + 5p$ *can* be simplified? Explain your answer.

---

[19]Hereafter, obvious closure steps like 5 and 7 will be omitted. It is useful to retain Step 2, however, since it foreshadows the treatment of $(c + d)$ as a single number in the next step.

4. Many desk calculators multiply by addition: for example, to obtain the product 7×489, they add the number 489 to itself six times in rapid succession.

   (a) Write out this addition explicitly.

   (b) Upon which theorem of this section is this method of multiplying based?

5. Show by a numerical example that addition does not distribute over multiplication in W — i.e., that $a + bc = (a + b)(a + c)$ is false in W.

6. Name the theorem(s) illustrated by each of the following statements. Assume that letters represent whole numbers.

   (a) $5 + 4 = 4 + 5$ (*Answer:* Commutativity of addition)

   (b) $(7)(3) = (3)(7)$

   (c) $7(2 + 6) = (7)(2) + (7)(6)$

   (d) $6 + 8$ is a unique whole number.

   (e) $(3)(5 + x) = (5 + x)(3)$

   (f) $4 + 0 = 0 + 4 = 4$

   (g) $(2x)y = 2(xy)$

   (h) $(m + n)a = ma + na$

   (i) $5 \cdot 1 = 5$

   (j) $3 + (y + x) = (3 + y) + x$

   (k) $xy + z = z + yx$

   (l) $(x + 5)(x + 4) = x^2 + 9x + 20$

7. You purchase in a store two items priced at \$.34 and \$.86 plus a 7% sales tax.

   (a) Theoretically, the clerk should be able to compute the tax either on each item separately or on the sum of the two prices.

      (i) Show how both methods, if computed exactly, yield a tax of \$.0840, or \$.08 to the nearest cent.

      (ii) Explain how your computations illustrate the distributive law.

   (b) In practice, the clerk does not compute the tax, but reads it off from a table. Suppose the table shows the tax to be \$.03 on \$.34, \$.06 on \$.86, and \$.08 on \$1.20.

      (i) Figure the total tax according to the table by two methods, first taxing both items separately, then taxing the total purchases.

      (ii) Does the distributive law apply here? Explain.

   (c) We have proved Theorem 4.19, the distributive law, in the abstract system W. Does part (b) invalidate the proof, or does it reveal a concrete situation in which the system W does not apply? Discuss.

8. Write out a complete proof of Theorem 4.21 using the chain of equalities given in the text.[20]

---

[20]Obvious closure steps may be omitted, as suggested in note 19.

9. In each of the following equations, all letters represent whole numbers. Consider each equation as a theorem to be proved. Use the procedure recommended in Exercises 4.3, Problem 4.[20]

   (a) $x(y + z) = zx + yx$ (b) $(x + y)z = zy + xz$

   (c) $a(b + c + d) = ab + ac + ad$ (*Hint:* Use the definition of a triple sum in W.)

   (d) $(a + b)(c + d) = ac + bc + ad + bd$

*10. In Problem 9(c), the distributive law was extended to a sum of three terms.

   (a) Similarly state and prove its extension to a sum of four terms.[20] (*Hint:* See Exercises 4.3, Problem 6.)

   (b) Explain how the distributive law can be extended to more than four terms.

*11. In Problem 5 you showed that, for at least some values of $a, b, c \in W$, $a + bc = (a + b)(a + c)$ is false: hence, addition does not distribute over multiplication in W. Now find all the special values of $a, b, c \in W$ — if there are any — for which the equation is true. You may use elementary algebraic knowledge in addition to the laws of W we have already stated.

*Section 4.6 A Glimpse at Transfinite Arithmetic

By way of contrast to the finite arithmetic we have been developing, it is interesting to explore the arithmetic of transfinite cardinals. In particular, we ask to what extent the definitions and theorems of the preceding sections appear to carry over to transfinite numbers. We are, of course, severely limited in our investigation, since we know only one transfinite number, $\aleph_0$. We shall go as far as we can in investigating expressions like $\aleph_0 + \aleph_0$ and $2 \cdot \aleph_0$. For a more thorough treatment, you should consult the Lieber reference and that of Birkhoff and MacLane.

First we take up addition, starting with its definition in terms of sets. Let us consider the sum $1 + \aleph_0$. Invoking Definition 4.4 as an experiment — for it is stated only for whole numbers — we select sets $A = \{a\}$ and $B = W$ to represent 1 and $\aleph_0$, respectively. So far, the outlook is favorable: for $n(\{a\}) = 1$, $n(W) = \aleph_0$, and $\{a\} \cap W = \varnothing$. Now we form $\{a\} \cup W$, which yields

$$\{a\} \cup W = \{a,0,1,2, \ldots, k, \ldots\}$$

which we compare with

$$W = \{0,1,2,3, \ldots, k + 1, \ldots\}$$

Evidently $\{a\} \cup W$ can be put into one-to-one correspondence with W, using $a \leftrightarrow 0$ and $k \leftrightarrow k + 1$ for $k \in W$. So $n(\{a\} \cup W) = n(W) = \aleph_0$. Our work is summarized in the following chain of equations:

$$1 + \aleph_0 = n(\{a\}) + n(W) = n(\{a\} \cup W) = n(W) = \aleph_0$$

Equating the two ends reveals the startling equation, $1 + \aleph_0 = \aleph_0$. To see how unusual an equation this is, you should ask yourself for which values of $x \in W$ it is true that $1 + x = x$, in *finite* arithmetic!

So *if* we "stretch" Definition 4.4 to cover transfinite cardinals, then we get a logical, though strange result. Perhaps it would seem intuitively more acceptable if cast in story form. Suppose a manager is in charge of a hotel with an infinite number of rooms: $\aleph_0$ rooms, to be exact. Let each room be numbered: 0, 1, 2, ... Suppose further that one Saturday night every room is occupied by one person, so that there are $\aleph_0$ guests. Now, a weary traveler arrives and requests a room. Can the manager accommodate the new arrival? Yes! It is only necessary to move each person from the room he or she is occupying to the next one in succession, beginning with the guest in Room #0, who moves to Room #1. Room #0 is now empty and can be assigned to the new traveler. If you object that the manager will run out of rooms at the other end of the hotel, you have not understood the meaning of $\aleph_0$. Mathematically, the story is sound. Psychologically, we must admit that the problem of persuading $\aleph_0$ guests to move is indeed formidable!

But why stop with one new arrival? Let the imagination soar: consider two, three, perhaps even $\aleph_0$ late guests. Problems *1 and *3 ask for such sums as $2 + \aleph_0$.

After such success in the use of Definition 4.4, we should ask whether the properties of closure, commutativity, and associativity hold for addition involving $\aleph_0$. In addition to citing examples, a study of the proofs of Theorems 4.1–4.6 reveals that they can be readily adapted to the case where $a = \aleph_0$, once Definition 4.3 is broadened to include $\aleph_0$. We do not yet, however, have the proof of Theorem 4.7 on cancellation of addition; and it is precisely here that a difficulty emerges. Though $1 + \aleph_0 = 2 + \aleph_0$, as the exercises will show, an attempt to cancel the $\aleph_0$ leaves the absurd conclusion, $1 = 2$.

The ultimate sum with $\aleph_0$ is $\aleph_0 + \aleph_0$. How shall we select sets A and B so that $n(A) = n(B) = \aleph_0$ and $A \cap B = \varnothing$? An adaptation of the hotel story provides a clue. To accommodate $\aleph_0$ new arrivals in his already-full hotel, the manager may reassign each original guest to a room whose number is double that of the original room number. Then only the rooms numbered 0, 2, 4, . . . will be occupied, and the new guests can move into 1, 3, 5, ... That they will have sufficient accommodations is shown by the discussion of Section 2.7 and especially by Exercises 2.7, Problem 5(f).

Let us turn now to multiplication. Pressing into service, beyond its finite limitations, the definition of multiplication in W, we try the first nontrivial product, $2 \cdot \aleph_0$. A neat choice for the sets A and B of Definition 4.6 is $A = \{a,b\}$ and $B = E = \{0,2,4, \ldots\}$. Then we have

$$A \times B = \{(a,0), (b,0), (a,2), (b,2), (a,4), (b,4), \ldots\}$$

Comparing $A \times B$ with W, where

$$W = \{\ 1\ ,\ 2\ ,\ 3\ ,\ 4\ ,\ 5\ ,\ 6\ , \ldots\}$$

we see that $A \times B \sim W$ by a one-to-one correspondence which is easy to visualize and not too difficult to put into a formula. We have the chain of equalities

$$2 \cdot \aleph_0 = n(\{a,b\}) \cdot n(E) = n(\{a,b\} \times E) = n(W) = \aleph_0$$

so that

$$2 \cdot \aleph_0 = \aleph_0$$

It is of some interest to compare $2 \cdot \aleph_0$ with $\aleph_0 + \aleph_0$, and both expressions with Theorem 4.20. Further exploration of the properties of multiplication covered by Theorems 4.9–4.12 and the distributive theorem (Theorem 4.19) are left to the exercises, which also contain a counterexample to the cancellation theorem (Theorem 4.15).

Apparently, then, the sum or product of any nonzero whole number with $\aleph_0$ is always $\aleph_0$: for any $k \in N$, $k + \aleph_0 = k \cdot \aleph_0 = \aleph_0$. Moreover, even $\aleph_0 + \aleph_0$ and $\aleph_0 \cdot \aleph_0$ yield $\aleph_0$, though we have not examined the latter statement. Is there any operation we can perform with $\aleph_0$ which produces a cardinal number larger than $\aleph_0$? And what is meant by "larger than $\aleph_0$"? For definitive answers you must consult references on transfinite numbers. A partial answer to the last question, however, appeared in Definition 2.13 and Exercises 2.7, Problem 6.

For an answer to the first question of the preceding paragraph, we may consider the cardinal number of all the *subsets* of an infinite set, such as W, whose cardinal number is $\aleph_0$. Exercises 2.3, Problem *8 suggests that the number of subsets of U, when U is finite, is $2^{n(U)}$. It turns out that this formula also holds for infinite sets. This means that the number of subsets of W, for example, is given by $2^{\aleph_0}$. That this is a very large number is suggested by an attempt to list the subsets of W:

$$\varnothing;\ 0,\ 1,\ 2, \ldots;\ (0,1),\ (0,2),\ (0,3), \ldots\ (1,1),\ (1,2),\ (1,3), \ldots\ (2,2),\ (2,3), \ldots$$

So far, in this display we have only begun to indicate a few of the subsets which contain two elements; and there are even more combinations of elements producing subsets of 3, 4, 5, ... elements each. Although appearances are certainly not a reliable guide, particularly with regard to the infinite, you will hardly be startled to learn that $2^{\aleph_0} > \aleph_0$. In fact, it can be proved that $2^a > a$, where a is any cardinal number, whether finite or any kind of transfinite number.

*EXERCISES 4.6

*1. (a) Use Definition 4.4 to show that $2 + \aleph_0 = \aleph_0$, naming the sets and actually setting up the one-to-one correspondence between them,

(b) How would you modify your answer to part (a) to show that $3 + \aleph_0 = \aleph_0$ and that, in general, $k + \aleph_0 = \aleph_0$ for any $k \in W$, including $k = 0$.

*2. Is there any difficulty in using Definition 4.4 to show that $\aleph_0 + 1 = \aleph_0$? If so, explain how to resolve it; if not, verify that Definition 4.4 applies in the same way it does for the sum $1 + \aleph_0$.

*3. In the text, the hotel story is used to make plausible the equation $\aleph_0 + \aleph_0 = \aleph_0$. Justify this equation abstractly by means of Definition 4.4, naming appropriate sets and setting up the one-to-one correspondence between them.

*4. (a) Write down a verbal rule or formula for the one-to-one correspondence exhibited in the text's discussion of $2 \cdot \aleph_0 = \aleph_0$.

(b) Give at least an informal argument to justify that

(i) $1 \cdot \aleph_0 = \aleph_0$

(ii) $3 \cdot \aleph_0 = \aleph_0$

(iii) $k \cdot \aleph_0 = \aleph_0$, where $k \in N$.

*5. Use the results of Problem 4 to provide an example showing that Theorem 4.15 is false for transfinite cardinals.

*6. Consider the set $W \cup \{\aleph_0\}$, with addition and multiplication defined by Definitions 4.4 and 4.6, respectively.

(a) Explain why addition and multiplication are closed in the system. (*Hint:* You need consider only sums and products involving $\aleph_0$.)

(b) Study the proofs for commutativity of addition (Theorem 4.2) and multiplication (Theorem 4.10) in W. To what extent, if at all, do they need to be changed to establish commutativity in the enlarged system $W \cup \{\aleph_0\}$?

(c) Answer part (b) with respect to associativity.

(d) Answer part (b) with respect to distributivity.

*7. (a) Work Exercises 2.3, Problem *8 if you have not already done so.

(b) Illustrate the general theorem $2^a > a$, where a is a finite cardinal ($a \in W$).

(c) On the basis of the formula established in part (a), give at least an informal argument showing that $2^a > a$ for $a \in W$.

Section 4.7 Order and Subtraction

In listing the whole numbers, we generally use the *natural order:* 0, 1, 2, 3, 4, ... This is also the order in which we label the points of the number line, which gives a geometric picture of the whole numbers. The fact that the whole numbers can be *ordered* is a significant property which we shall now examine and link with the operation of subtraction, by linking both order and subtraction with addition. In Section 4.8 we shall examine an analogous relation, which we shall link with division and both with multiplication.

Returning to set theory, let us consider the null set and the set containing one element — e.g., $I = \{a\}$. By definition, we assign cardinal numbers to these two sets; they are $n(\varnothing) = 0$ and $n(I) = 1$. Now we can develop the other finite cardinals *in the usual order* by annexing one element at a time: thus we can take set $J = \{a,b\}$; then $n(J) = 2$; etc. We get a succession of cardinals: 0, 1, 2, 3, . . . ; by the annexation process it is clear that 2 is the *successor* to 1, 3 the successor to 2, etc. In effect, this gives us a portion of the addition table: $1 + 1 = 2$, $2 + 1 = 3$, $3 + 1 = 4$, ... We have to work a bit harder to develop the rest of the table: for example, to obtain $2 + 3 = 5$. To do so, it is necessary, according to Definition 4.4, to find two sets which represent 2 and 3, respectively. We may choose $J = \{a,b\}$ and $K = \{c,d,e\}$, noting that $J \cap K = \varnothing$, as required. Then $2 + 3 = n(J) + n(K) = n(J \cup K) = n(\{a,b,c,d,e\}) = 5$. By considering other sums, we may complete the addition table in W through $9 + 9$, or even further if we like.

Let us now turn the problem around: given one addend and the sum, can we find the other addend? For example, consider the equation $2 + ? = 5$: in modern elementary texts this is often written $2 + \square = 5$; in algebra it becomes $2 + x = 5$. In terms of sets, this calls for a *separation* of a set representing 5 (e.g., $L = \{a,b,c,d,e\}$) into two disjoint subsets, one of which is (or is equivalent to) set J and the other, set X, containing the remaining elements of set L, namely, $\{c,d,e\}$.

Since $2 + x = 5$ means $n(J) + n(X) = n(L)$, or $n(J \cup X) = n(L)$, $J \cup X$ must be equivalent (maybe equal) to L, or $(\{a,b\} \cup X) \sim \{a,b,c,d,e\}$; so that $X \sim \{c,d,e\}$, since X must be disjoint from J. Hence $n(X) = 3$. So the solution of $2 + x = 5$ is $x = 3$.

Consider now the equation $3 + x = 2$. We can represent 3 by $K = \{c,d,e\}$, as before, and let X be a set such that $n(X) = x$, with $K \cap X = \varnothing$. By Definition 4.4, we should have $n(K) + n(X) = n(K \cup X)$, with $n(K \cup X) = n(J) = 2$, by hypothesis. But $K \subseteq K \cup X$, for the union of two sets must contain the elements of each of its parts. So $K \cup X$ must contain at least the elements $\{c,d,e\}$ of K, which fail to correspond one-to-one with the elements $\{a,b\}$ of J, since an element of K is left over. So $(K \cup X) \not\sim J$, no matter which set X is. Hence $n(K \cup X) \neq n(J) = 2$; and the equation $3 + x = 2$ has no solution in W.

Finally, consider the equation $3 + x = 3$. A similar analysis shows that set X exists, but $X = \varnothing$ and so $x = 0$. Therefore the equation $3 + x = 3$ is satisfied by the *whole* number 0 but has no solution in *natural* numbers.

Perhaps you have spotted the difficulty in the last two cases, where no *natural* number x can be found. In the first case, by contrast, a natural number x *can* be found to add to 2 to yield a *larger* number, 5. Apparently $x \in N$ exists if the given addend is *less than* the given sum. We assess the results as follows:

1. Since 2 is less than 5, $2 + x = 5$ has a solution $x = 3 \in N$.
2. But since 3 is *not* less than 2, $3 + x = 2$ has no solution $x \in N$.
3. And since 3 is *not* less than 3, $3 + x = 3$ has no solution $x \in N$, though it does in W, since $x = 0$ satisfies the equation.

The above examples suggest the following:

Definition 4.7 Less Than If $a, b \in W$, then a is said to be *less than* b (in symbols, $a < b$) iff there exists a natural number x, such that $a + x = b$.

Parallel to Definition 4.7 is another:

Definition 4.8 Greater Than If $a, b \in W$, then a is said to be *greater than* b (in symbols, $a > b$) iff b is less than a.

As illustrations: $1 < 2$, since $1 + x = 2$ has a solution $x = 1 \in N$.

$2 \not< 2$, since $2 + x = 2$ has a solution $x = 0$, but $0 \notin N$.

$3 \not< 2$, since $3 + x = 2$ has no solution $x \in N$.

$2 > 1$, since $1 < 2$.

The order relations may be combined with equality. Thus $a \leq b$ (read, "a is *less than or equal to* b") means "$a < b$ or $a = b$." So we may write $5 \leq 7$, $9 \geq 8$, $6 \leq 6$, $6 \geq 6$, but not $6 < 6$.

At this point you should protest that we have already introduced an order relation by means of Definition 2.13. This is true: also, Definition 2.13 is more general than our new Definition 4.7, because the former applies to both finite and transfinite cardinals. We shall find, however, that Definition 4.7 is convenient and valuable, since it does not involve sets directly and it provides a useful link with subtraction, as given in Definition 4.9.

But before using these definitions simultaneously we must clear up a potentially serious logical difficulty; for each provides a different rule for determining, for any pair $a, b \in W$, whether or not $a < b$. Suppose the two rules give contradictory answers! We must prove that this cannot happen. Specifically, we must establish that the biconditional, "for any pair $a, b \in W$, $a < b$, satisfies Definition 4.7 iff $a < b$ satisfies Definition 2.13" is a logical equivalence according to Definition 3.11. Exercises 4.7, Problem 4 suggests how this may be done. Henceforth, we shall mostly use Definition 4.7, holding Definition 2.13 in reserve to deal with transfinite cardinals.

We shall now develop four basic order properties in W, beginning with *trichotomy*, a term which correctly suggests "cutting into three parts." Let us compare two given whole numbers, a and b, and list all the possibilities. We begin by choosing two sets, A and B, such that $n(A) = a$ and $n(B) = b$, and examining the possible relations between these sets. There are three:

1. $A \sim B$. In this case A and B can be matched one-to-one and have the same cardinal number; so, by Definition 4.3, $a = b$. 2. and 3. $A \nsim B$. Here the matching process fails in one of two ways and leaves at least one element unmatched in either A or B, but not in both.
2. Suppose the unmatched element(s) is (are) in B. Then a set $X \neq \varnothing$, disjoint from A, can be found such that $A \cup X = B$; hence $n(A) + n(X) = n(A \cup X) = n(B)$, or $a + x = b$, with $n(X) = x$ and $x \in N$. So by Definition 4.7, $a < b$.
3. Suppose, on the other hand, that the unmatched element(s) is (are) in A. Then a similar argument shows that $b + x = a$, with $x \in N$; hence $b < a$, or, by Definition 4.8, $a > b$.

It is also possible to reach these conclusions by use of Definition 2.13. At any rate, we observe that no two of the preceding three cases can occur simultaneously, and also that one of the cases *must* occur. We summarize the above argument—though we may not claim it as a rigorous proof—in the following:

THEOREM 4.23 Trichotomy If $a, b \in W$, then exactly one of the following holds: $a = b$, $a < b$, or $a > b$.

Earlier in this chapter we proved that equality is transitive. Is the "less than" relation transitive? You should check this by trying several examples; then you should state the theorem. It is

THEOREM 4.24 Transitivity for "Less Than" If $a, b, c \in W$ and if $a < b$ and $b < c$, then $a < c$.

How should we approach the proof of this theorem? Since the order concepts have just been introduced, the definitions and Theorem 4.23 are the only tools in W with which we can begin. Definition 4.7, however, leads immediately to an equation, and thus makes available all the properties in W which are associated with equalities. We shall give a sketch of the proof of Theorem 4.24 as a model of how inequalities may be handled. You can then supply the formal proof.

Sketch of Proof of Theorem 4.24: By Definition 4.7, $a < b$ means that there is a natural number x such that $a + x = b$. Similarly, $b < c$ means there is a $y \in N$ such that $b + y = c$. We need to show that there is a $z \in N$ such that $a + z = c$, so that $a < c$. To do this, we add y to both sides of the first equation, obtaining $(a + x) + y = b + y$ or $a + (x + y) = b + y$. Together with the second equation this yields $a + (x + y) = c$. But $(x + y)$ is a single natural number: let us put $x + y = z$. Then the equation becomes $a + z = c$, which is just what we needed to show $a < c$.

Two of the order theorems, involving addition and multiplication, respectively, are almost parallel to properties of equations. Numerical examples and proofs are left for you as starred exercises. They are

THEOREM 4.25 Addition for "Less Than" If $a, b, c \in W$ and if $a < b$, then $a + c < b + c$.

THEOREM 4.26 Multiplication for "Less Than" If $a, b, c \in W$ and $c \neq 0$ and if $a < b$, then $ac < bc$.

These two theorems can be modified by replacing $<$ with $>$ or $\leq$ or $\geq$. One of these modifications, for example, is

THEOREM 4.27 Addition for "Greater Than" If $a, b, c \in W$ and if $a > b$, then $a + c > b + c$.

We can now use Theorem 4.27 with Theorem 4.23 and Theorem 4.25 to provide a proof of Theorem 4.7, as promised at the end of Section 4.2. For reference we repeat

THEOREM 4.7 Cancellation for Addition If $a, b, c \in W$ and if $a + c = b + c$, then $a = b$.

Proof: Theorem 4.7 essentially states that $(a + c = b + c) \Rightarrow (a = b)$. We make an indirect proof by proving the contrapositive, $(a \neq b) \Rightarrow (a + c \neq b + c)$.[21] So our new hypothesis is $a, b, c \in W$ and $a \neq b$. Then, by Theorem 4.23, since $a \neq b$, either $a < b$ or $a > b$. Suppose $a < b$. Now, by Theorem 4.25, $a + c < b + c$, which, again by Theorem 4.23, implies that $a + c \neq b + c$. But this inequality is just what the conclusion of the contrapositive asserts. If we suppose instead that $a > b$, we get $a + c > b + c$ by Theorem 4.27, whence again $a + c \neq b + c$. Thus, in either case, $a \neq b$ implies $a + c \neq b + c$, or, equivalently, if $a + c = b + c$, then $a = b$.

---

[21]This indirect method, which is discussed at length in Section 4.10, is based on Definition 3.8 and Exercises 3.5, Problem 7, which proves that a conditional and its contrapositive are logically equivalent, so that proving that either is an implication automatically establishes the other as such.

This section also claims to deal with subtraction in W. How does the operation of subtraction relate to the order concepts? We recall that $2 < 5$ means there is a natural number x such that $2 + x = 5$. The solution, 3, is clearly the *difference* between 5 and 2: for $5 - 2 = 3$. Thus the equations $2 + 3 = 5$ and $5 - 2 = 3$ are two different ways of expressing the same basic relation among the three numbers. Some elementary schools introduce subtraction in just this fashion, as an addition problem in which the sum is given and one of the addends is missing. From this viewpoint, the following two problems are merely different ways of asking for the difference:

$$\begin{array}{r} 5 \\ -2 \\ \hline ? \end{array} \qquad \begin{array}{r} 2 \\ +? \\ \hline 5 \end{array}$$

$$\begin{array}{r} b \\ -a \\ \hline x \end{array} \qquad \begin{array}{r} a \\ +x \\ \hline b \end{array}$$

More generally, denoting the difference $b - a$ by the symbol x, the bottom two forms are equivalent.

Formally, the above approach to subtraction leads to the following:

Definition 4.9 Subtraction If $a, b \in W$, then the *difference* $b - a$ is a whole number x (if x exists) such that $a + x = b$.

A related definition with which the reader is undoubtedly familiar is

Definition 4.10 Minuend, Subtrahend In the difference $b - a$, b is called the *minuend* and a is called the *subtrahend*.

To distinguish between them, you may note that the *minuend* is the quantity "to be *diminished*" and the *subtrahend* the quantity "to be *subtracted*."

As an example of Definition 4.9, we consider $5 - 4$: this difference is a number x such that $4 + x = 5$; since $x = 1$, we have $5 - 4 = 1$. You should likewise explain the differences $9 - 6$, $5 - 0$, $8 - 8$, $0 - 0$, and $3 - 6$. We note that for $8 - 8$, $x = 0$, which is an acceptable answer, since Definition 4.9 permits a whole number. Does x in the equation $8 + x = 8$ fulfill the requirements of Definition 4.7, the "less than" definition? Why or why not? For the difference $3 - 6$ no whole number x exists: this is why the parenthetical expression is needed in Definition 4.9. Merely defining a term does not establish its existence! Is W closed under the operation of subtraction? Why or why not?

We should also investigate other properties. Is subtraction commutative? Is it associative, at least when the differences all exist, as in $(7 - 4) - 2$ and $7 - (4 - 2)$? Is there a distributive law for multiplication over subtraction? If the answer to any question is No, you should supply a counterexample; if it is Yes, you are invited to sketch a proof.

Two theorems on subtraction deserve special mention. The first states that subtraction "undoes" addition. The essential equation is $(a + b) - a = b$; it says that to add a to b and then subtract it off again leaves b. Since the proof illustrates a new technique, it is given in full. Formally, we have

THEOREM 4.28 If $a, b \in W$, then $(a + b) - a = b$.

Proof: Since $a, b \in W$, $(a + b) \in W$ by closure of addition. By the definition of subtraction, taking $(a + b)$ as the minuend, $(a + b) - a = b$ iff $a + b = (a + b)$.[22] But obviously the second equation, $a + b = (a + b)$, is true by definition of equality of whole numbers (Definition 4.3); hence the first equation, $(a + b) - a = b$, is true by Definition 3.11.

Though this proof is not difficult, the flow diagram in Figure 4.4 may provide further clarification and also eliminate a repetition.

G_1: $a \in W$

G_2: $b \in W$

T_3: cl. ad.

S_4: $(a + b) \in W$

D_5: def. equality in W

S_6: $a + b = (a + b)$

D_7: def. subtraction in W

C_8: $(a + b) - a = b$

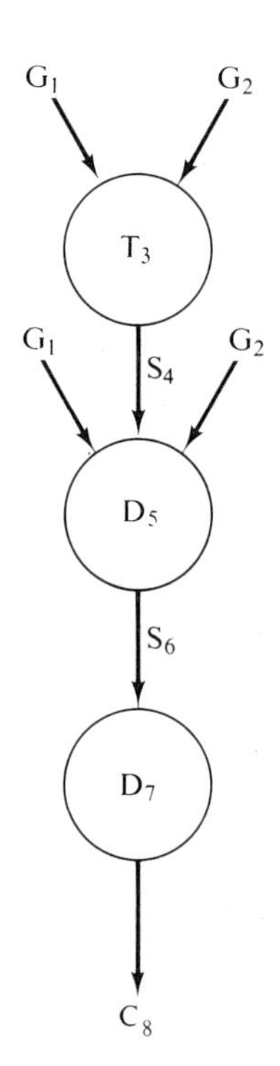

Figure 4.4

The second theorem states that multiplication distributes over subtraction. Its proof is sketched in the expectation that you can fill in the details. It is

[22] Here b plays the role of x in Definition 4.9; and the expression $(a + b)$ plays the role of the sum, b, in Definition 4.9. The expression "iff" means that the two equations it connects are logically equivalent, as is true with all definitions.

THEOREM 4.29 Distributivity of Multiplication over Subtraction If a, b, $c \in W$ and if $(b - c) \in W$, then $a(b - c) = ab - ac$.

Sketch of Proof: Since we know little about subtraction except its definition, we use the latter to change differences to sums. Since $(b - c) \in W$ by hypothesis, we let $b - c = x$, where $x \in W$. Then, using Theorem 4.13, we may multiply this equation by a to get $a(b - c) = ax$. Now, by the definition of subtraction, $b - c = x$ iff $c + x = b$. Multiplying the second equation also by a, we get $a(c + x) = ab$, or $ac + ax = ab$ (Why?), all terms being whole numbers. Using the definition of subtraction in the other direction, $ac + ax = ab$ iff $ax = ab - ac$. (Note that since ax exists as a whole number, so does its equal, the difference $ab - ac$.) It follows from $a(b - c) = ax$ and $ax = ab - ac$ that $a(b - c) = ab - ac$.

In summary, we note that the basic equation for the definitions of *both* $a < b$ and $b - a$ is $a + x = b$, where $x \in N$ in the first case, but $x \in W$ in the second. Understanding the relationships among these two concepts will help you to recall and use them without resort to rote memorization.

EXERCISES 4.7

1. (a) Explain, using appropriate definitions, why $3 < 7$ and $6 > 0$ are true, but $8 < 6$ and $4 < 4$ are false.

   (b) Explain the relationship of part (a) to the differences $7 - 3$, $6 - 0$, $6 - 8$, and $4 - 4$.

   (c) Using the results of part (b), decide whether or not W is closed under subtraction, and prove your answer.

2. Given the following pairs of whole numbers, insert between each pair the appropriate symbol from the set $\{<, >, =\}$: 7 9, 3 8, 8 3, 1 4, 5 5, 5 4, 0 3, 7 0, 0 0

3. (a) In the definition for $a \leq b$ following Definition 4.8, the phrase "$a < b$ or $a = b$" occurred. Is this "or" in fact inclusive or exclusive? Explain.

   (b) Write a definition for $a \geq b$.

4. (a) According to Definition 4.7, $4 < 6$ because $4 + x = 6$ is satisfied by $x = 2 \in N$. To satisfy Definition 2.13 we must find sets A and B, with $A \not\sim B$, such that $n(A) = 4$ and $n(B) = 6$, with A equivalent to a proper subset of B. Illustrate this sentence by finding such sets.

   (b) Let $C = \{m\}$ and $D = \{a,b,c\}$.

      (i) Use sets C and D to show why $n(C) < n(D)$ according to Definition 2.13.

      (ii) Explain why, according to Definition 4.7, $n(C) < n(D)$.

(c) Are the "greater than" definitions for whole numbers contained in Definitions 2.13 and 4.8 logically equivalent? Explain your answer.

*(d) Generalize the examples of parts (a) and (b) to obtain a proof that Definitions 2.13 and 4.7 are logically equivalent. (*Hint:* The proof is in two parts. One is, "If $a = n(A)$ and $b = n(B)$ are whole numbers and if $a < b$ according to Definition 4.7, then $n(A) < n(B)$ according to Definition 2.13.")

5. Which (if any) of the relations $>$, $\leq$, $\geq$ is (are) transitive? Illustrate your answers numerically.

6. The order properties may easily be interpreted geometrically on the number line, which has a marked point to represent each whole number. Draw a number line (as in Section 4.2) and answer the following questions:

   (a) If $a, b \in W$ and $a < b$, what is the geometric relationship between the points on the line which represent a and b?

   (b) Answer part (a) for the inequalities $a > b$ and $a \leq b$.

   (c) Interpret geometrically the trichotomy property (Theorem 4.23), beginning, "If a and b are any two points on the number line, then. . . ."

7. (a) Illustrate numerically each of Theorems 4.25, 4.26, 4.27, and 4.28.

   (b) In particular, verify that Theorem 4.25 holds when $c = 0$.

   (c) Is Theorem 4.26 true if "$c \neq 0$" is omitted? Explain.

8. (a) Is subtraction commutative? Explain and/or give examples.

   (b) Is subtraction associative? Consider the examples $a = 9$, $b = 7$, $c = 3$ and $a = 9$, $b = 5$, $c = 3$.

9. Prove the cancellation theorem for multiplication (Theorem 4.15). (*Hint:* See the proof of Theorem 4.7.)

10. Write out formal proofs of (a) Theorem 4.24 and (b) Theorem 4.29, along the lines suggested in the text.

11. Prove the following theorems: (a) If $a, b \in W$, then $(a + b) - b = a$. (b) If $a, b \in W$ and $(a - b) \in W$, then $(a - b) + b = a$.

*12. Write out formal proofs of (a) Theorem 4.25 and (b) Theorem 4.26.

*13. Rewrite Theorem 4.25 and Theorem 4.26, substituting $>$ for $<$, and compare each with Theorem 4.27. Then explain how you would adapt your proofs of Problem *12 to prove these new theorems.

*14. Shamus O'Leary has four daughters. One of them is a nurse. Can you tell which from these clues?

    1. If Kate is older than Mary, then Kate is the nurse.

2. If Kate is older than April, then April is the nurse.
3. If April is older than Erin, then Erin is the nurse.
4. If Erin is older than Mary, then Mary is the nurse.
5. Kate is older than Erin.
6. April is older than Mary.
7. The ages of the girls are 22, 24, 26 and 28.[23]

Section 4.8 Factors, Multiples, and Division

In Section 4.4 we drew attention to the many properties which addition and multiplication share in W. We now exploit these resemblances by developing this section in analogy to the preceding one. It turns out that many — but not all — of the definitions and theorems of Section 4.7 will remain true if the words *multiplication* and *division* are substituted for *addition* and *subtraction*, respectively, and if certain changes are made in the terminology regarding order.

We begin with the basic definition of order in W, Definition 4.7. Upon "translating" it, we get

Definition 4.11 Factor, Divisor If $a, b \in W$, then a is said to *be a factor of* or to *be a divisor of* or to *divide* b (in symbols, $a \mid b$)[24] iff there exists a *whole number* x such that $ax = b$.

The italicized passages of this definition constitute the changes made in Definition 4.7. Parallel to Definition 4.8 is

Definition 4.12 Multiple If $a, b \in W$, then a is said to be *a multiple of* b or to be *divisible by* b (symbols are not commonly used here) iff b is a *factor of* a.

Examples of these definitions follow: you should check that these statements are true: $2 \mid 10$, $5 \mid 245$, $1 \mid 5$, $3 \mid 3$, $2 \mid 0$, $4 \nmid 7$, $4 \nmid 2$, 10 is a multiple of 2, 8 is a multiple of 8, 20 is divisible by 5, 7 is not a multiple of 5, 3 is not a multiple of 12, 2 is not divisible by 3.

Although the preceding analogy between the pairs of definitions is striking, it cannot be pushed too far. We do not, for example, have any theorem regarding factors which corresponds to the trichotomy theorem for order. Possible parallels to Theorems 4.24, 4.25, and 4.26 are discussed in the exercises.

Very significant is the definition of division, which is perfectly analogous to Definition 4.9 for subtraction. You are urged to stop to construct this defini-

---

[23]*Table Talk*, 17:6 (April, 1971), 1. Washington: Marriott Corp. Reprinted by permission.

[24]The line separating a and b is *vertical*. This symbol should not be confused with "/", the fraction symbol, as in a/b, which has a quite different meaning.

tion before proceeding: you will be pleasantly surprised at your success! Here are two definitions:

Definition 4.13 Division If $a, b \in W$, then the *quotient* $b \div a$ is a whole number x (if x exists) such that $ax = b$.

Definition 4.14 Dividend, Divisor In the quotient $b \div a$, b is called the *dividend* and a is called the *divisor*.

As an example we consider $6 \div 3$. This quotient is a whole number x such that $3x = 6$. Since $x = 2$ satisfies $3x = 6$, the quotient is $6 \div 3 = 2$. You should likewise consider the quotients $10 \div 2$, $10 \div 5$, $3 \div 3$, $0 \div 3$, $3 \div 1$, $3 \div 4$, and $3 \div 2$. Is W closed under the operation of division? Why or why not? In the exercises we investigate commutativity, associativity, and (in a starred exercise) distributivity; we also explore whether division "undoes" multiplication.

We end this section with an examination of the peculiar and significant behavior of zero in division. We employ only two tools: the definition of division (Definition 4.13) and the zero product theorem (Theorem 4.16). Let us consider the examples $0 \div 1$, $1 \div 0$, and $0 \div 0$. It turns out that the first quotient is 0, but the remaining two quotients cannot be defined. Let us see why. By the definition of division, $0 \div 1 = x$ iff $1 \cdot x = 0$. Since $x = 0$ satisfies $1 \cdot x = 0$ and no other number does, we may claim that $0 \div 1 = 0$. More generally we have

THEOREM 4.30 If $a \in N$, then $0 \div a = 0$.

The proof is left as an exercise. You should note that $a \in N$ implies that $a \neq 0$.

Now we turn to $1 \div 0$. Here $1 \div 0 = x$ iff $0 \cdot x = 1$. Which whole number will satisfy this equation? None! For by the zero product theorem, $0 \cdot x = 0 \neq 1$ no matter which whole number replaces x. More generally, it is easy to show, by the same reasoning, that the quotient, $a \div 0$, with $a \neq 0$, does not exist. This demonstration is also left for the exercises.

With regard to the remaining quotient, $0 \div 0$, the situation seems quite different. Here $0 \div 0 = x$ iff $0 \cdot x = 0$. It is easy to satisfy this equation; for by the zero product theorem we may take $x = 0, 1, 2, 3, \ldots$ — indeed, *any* whole number will do! But this plethora of solutions is an embarrassment as serious as no answer at all: of what use is a quotient whose meaning is ambiguous?

In sum, the quotient of any whole number *by zero* is undefined: for if $a \neq 0$, there is no quotient (no answer), and if $a = 0$, the quotient is not unique (too many answers). Thus, regretfully, we have

THEOREM 4.31 If $a \in W$, then the quotient $a \div 0$ is undefined.

It is important to stress that division of a whole number by zero has been proved *impossible:* it is not merely a difficult operation which some day mathematicians may hope to perform. To put the matter another way, Theorem 4.31, like all the others we have stated, is a logical — and therefore inescapable — consequence of the definitions of multiplication and division and of Theorem 4.16, which in turn can be traced back to set theory. Thus the whole number system, and the extensions which we shall make later, based on the same foundation of set theory, inherently and unavoidably suffer from the "defect" that division by zero is not permitted. It might be possible to build a new system of some kind of numbers in which such quotients are permissible, but only at the cost of sacrificing some of the fundamental properties which we expect all numbers to enjoy. Such a weird new system, moreover, would probably have few, if any, useful applications in the real world. Fortunately, our conventional number systems are not seriously hampered by the prohibition of division by zero; on the contrary, they are rich, both in theoretical development and concrete application.

EXERCISES 4.8

1. (a) Explain, using appropriate definitions, why $4 \mid 8$, $6 \mid 6$, and "35 is a multiple of 7" are true; but $4 \mid 5$, $12 \mid 6$, and "8 is divisible by 15" are false.

   (b) Explain the relationship of part (a) to the quotients $8 \div 4$, $6 \div 6$, $35 \div 7$, $5 \div 4$, $6 \div 12$, and $8 \div 15$.

   (c) Using the results of part (b), decide whether or not W is closed under the operation of division, and prove your answer.

2. (a) Which (if either) of the relations "is a factor of" and "is a multiple of" are transitive? Illustrate your answers numerically.

   (b) *Prove* your answer concerning "is a factor of," either by producing a counterexample (if your answer to part (a) is No) or by stating and proving a theorem (if your answer is Yes).

3. Investigate, by using numerical examples, whether division is commutative and/or associative. To the extent possible, select numbers for which the relevant quotients all exist.

4. (a) Rewrite Theorems 4.25 and 4.26, substituting the symbol $\mid$ (is a factor of) for $<$, but making no other changes.

   (b) Investigate the truth or falsity of the results of part (a) by means of numerical examples.

   *(c) Prove or disprove (as the cases may be) the statements of part (a).

5. Prove Theorem 4.30.

6. Prove that, if $a \neq 0$, $a \div 0$ does not exist, thus completing the proof of Theorem 4.31.

7. (a) State and prove the division theorem analogous to Theorem 4.28, thus showing that division "undoes" multiplication.

   (b) Similarly state and prove the division analogs of Exercises 4.7, Problem 11.

8. Apply Definition 4.11 to the special cases $1 \mid 0$, $0 \mid 1$, and $0 \mid 0$. What relation, if any, do these cases have to Theorems 4.30 and 4.31?

*9. Investigate each of the following statements involving distributivity, and either disprove it by a counterexample or state and prove it as a theorem. Be careful when using numerical examples and stating theorems that the quotients actually exist.

   (a) Multiplication distributes over division.

   (d) Division distributes over addition. (*Hint:* Investigate both orders of division: $a \div (b + c)$ and $(b + c) \div a$.)

*10. There *is* a number system in which the commutative, associative, and distributive laws hold and in which subtraction and division are possible, without any exception. Find such a system and check that it has all these properties. (*Hint:* The system contains only one number. Which one is it?)

Section 4.9 A Taste of Number Theory

A very old, honorable, and still active, branch of mathematics is the theory of numbers. Euclid wrote about it in the third century, B.C.; and a number of mathematicians are currently engaged in research, some of it aided by the computer. Number theory has an appeal to the novice as well, in that some of its interesting conclusions and even some of its most intriguing unsolved problems can be stated in elementary terms. It is also true, on the other hand, that this branch of mathematics is quite abstract and has few applications outside of mathematics. Perhaps partly for this reason, as a research area it elicits the interest of relatively few mathematicians.

Briefly stated, the theory of numbers deals with the system of integers, which includes whole numbers and the negative integers. Sometimes, however, it is not necessary to use the negative integers. For instance, if it is a question of whether 6 divides 48, no important information is added by considering $^-6$ and $^-48$ also. This is so because $^-6$ divides $^-48$ iff 6 divides 48. Since we have not yet studied the negative integers, our "taste" will be confined to the whole numbers.

Our point of departure is to *partition* the whole numbers into four nonoverlapping subsets. The verb *partition* implies that every element of W is a member of exactly one of the four subsets. Now 0, as we have seen, is a very special integer: besides being the additive identity, it has other peculiar properties. Also, 1 is the multiplicative identity, and it serves as a "building block" for W, in the sense that by adding it successively to 0 we get 1, to 1 we get 2, etc.;

and thus we generate from it the entire set W. For these reasons, we take two of the subsets as $\{0\}$ and $\{1\}$. The other two are characterized by the divisibility property (Definition 4.11). We observe that every number $a \in W$, with $a > 1$, has itself and 1 as factors: thus $5 = 1 \cdot 5$ and $6 = 1 \cdot 6$. But whereas 5 has no additional factors, 6 has the additional factors 2 and 3, since $6 = 2 \cdot 3$. This observation makes appropriate the following definitions:

Definition 4.15 Prime A whole number > 1 whose only factors are itself and 1 is called a *prime* number.

Definition 4.16 Composite A whole number > 1 which is not prime is called a *composite* number.

Thus 5 is a prime, but 6 is a composite. Is 1 a prime? Do either of the above definitions apply to 0? In sum, we have partitioned W into four subsets:

1. $\{0\}$
2. $\{1\}$
3. The primes: $P = \{2,3,5,7,11, \ldots\}$
4. The composites: $C = \{4,6,8,9,10, \ldots\}$

You are advised to continue the list of primes, at least as far as 31, as practice in applying Definition 4.15.

One theoretically easy, but practically difficult problem, which has been greatly assisted by computers,[25] is determining whether a given large number is prime or composite. Of course, there are some guidelines. For example, it is obvious that the following two numbers are composite (Why?): 273,498,214; 668,572,300,415. But what about the number 134,023?[26]

Certain "divisibility tests" are an aid. You can assuredly formulate rules for telling when a number is divisible by 2 or 5. A rule for divisibility by 3 (or 9) is the following: "A whole number is divisible by 3 (or 9) iff the sum of its digits (when written in decimal notation) is divisible by 3 (or 9)." As an example: 258 is divisible by 3, but not by 9, because $2 + 5 + 8 = 15$, which is divisible by 3, but not by 9; on the other hand, 2,583 is divisible by both 3 and 9, and 2,581 is divisible by neither 3 nor 9. There are a number of more complicated tests for divisibility by 7, 11, etc.[27] You need only consider tests for divisibility by primes: for to ask for divisibility by 6 is equivalent to asking for divisibility by 2 *and* by 3, which are both primes. In other words, if a number is not divisible by any prime, then it certainly is not divisible by composites, which are made up of prime factors.

As an example, let us test whether 223 is prime or composite. Clearly, the

---

[25]See, for example, the Buchman reference.

[26]*Answer*: $134{,}023 = 223 \cdot 601$.

[27]Consult the chapter bibliography.

quick tests for 2, 3, and 5 show that 223 is not divisible by them. Trials show that it is not divisible by 7, 11, or 13. It is not necessary to test for divisibility by 17 or higher primes. Why not? So 223 is prime.

An exhaustive (and frequently exhausting!) method of obtaining all the first n primes is an ancient one called the *sieve of Eratosthenes*. For the first few primes it is quite effective; but for larger numbers it rapidly becomes impractical, as you can quickly show. We shall apply it to the first 50 numbers, beginning with 2. We start by listing every whole number from 2 to 50, inclusive: 2, 3, 4, 5, 6, . . . 50. Now, the first number in our list, 2, is a prime; but its multiples, 4, 6, 8, . . . , are not. We therefore *cross off* every second number after 2, beginning with 4, thus obtaining 2, 3, ~~4~~, 5, ~~6~~, . . . ~~50~~. In this way we have eliminated or "sieved out" all even numbers > 2. The next number after 2 which remains uncrossed is 3, also a prime. We leave it alone, but cross off every third number thereafter, starting with 6. (It happens that 6 has already been crossed out, but we cross it again — at least, mentally — in order to preserve our count.) Thus we "sieve out" all multiples of 3 which are greater than 3. Continuing in this manner, we next eliminate all multiples of 5 and 7, except the numbers themselves. The result is Table 4.2, in which all the composites have been eliminated, leaving only the set of primes less than 50, which are printed in boldface type.

TABLE 4.2

	2	**3**	~~4~~	**5**	~~6~~	**7**	~~8~~	~~9~~	~~10~~
11	~~12~~	**13**	~~14~~	~~15~~	~~16~~	**17**	~~18~~	**19**	~~20~~
~~21~~	~~22~~	**23**	~~24~~	~~25~~	~~26~~	~~27~~	~~28~~	**29**	~~30~~
31	~~32~~	~~33~~	~~34~~	~~35~~	~~36~~	**37**	~~38~~	~~39~~	~~40~~
41	~~42~~	**43**	~~44~~	~~45~~	~~46~~	**47**	~~48~~	~~49~~	~~50~~

Thus the primes less than 50 are 2, 3, 5, 7, 11, 13, 17, 19, 23, 29, 31, 37, 41, 43, and 47. You are invited to ponder why, in Table 4.2, it is not necessary to consider eliminating multiples of primes greater than 7, such as 11 and 13.

The property of "primeness" furnishes a good opportunity to apply inductive reasoning. Can you discover any rule or pattern of regularity among the primes? Do they occur as frequently among the larger numbers as among the smaller? In order to obtain answers or to further check any generalizations, you would probably wish to extend the table of primes. What is the most efficient way of doing so?

Inspection, aided by elementary divisibility rules such as those already mentioned, is quite feasible for the first 200 or so numbers, but becomes increasingly tedious as the numbers get larger. The sieve, while entirely routine and (theoretically) infallible, soon breaks down under the sheer weight of writing and crossing out numbers. Is there not a general formula, into which one might substitute various numbers, which produces all the primes, or at least only primes? At first glimpse, $n^2 - n + 41$ looks promising: for $n = 0$ or 1,

$n^2 - n + 41 = 41$, a prime; for $n = 2$, $n^2 - n + 41 = 43$, also a prime. You should try several more values for n. But alas, this formula eventually begins to produce composites. The truth is that, while there are formulas for primes, mathematicians have not yet discovered a simple and practical one. Frankly, the best way to obtain a list of primes is to consult a table which records the results of others' labors, formerly performed by mathematicians or human calculators, but now by electronic computers under the supervision of mathematicians. Large tables listing primes up to ten million have been published.

Let us use a modest table,[28] up to 2,000, to see whether the primes continue to be distributed irregularly and to become less frequent as they get larger. Some results are summarized in Table 4.3.

TABLE 4.3

Interval	*Number of Primes*	*Interval Density*[29]	*Cumulative Density*[29]
1–20	8	.400	.400
21–40	4	.200	.300
41–60	5	.250	.283
61–80	5	.250	.275
81–100	3	.150	.250
1–100	25	.250	.250
101–120	5	.250	.250
121–140	4	.200	.243
141–160	3	.150	.231
161–180	4	.200	.228
181–200	5	.250	.230
101–200	21	.210	.230
201–300	16	.160	.207
301–400	16	.160	.195
401–500	17	.170	.190
1–500	95	.190	.190
501–1000	73	.146	.169
1001–1500	71	.142	.160
1501–2000	64	.128	.152
1–2000	303	.152	.152

Table 4.3 suggests an interesting question. If the number of primes "thins out" as the numbers increase, will the primes cease altogether after a while? In other words, is there a *largest* prime, so that all whole numbers beyond it are

[28]*Standard Mathematical Tables*, 20th ed. (Cleveland, O.: The Chemical Rubber Co., 1972), pp. 37–45 or 681–88.

[29]The *interval density* is the ratio of the number of primes *in the given interval* to the number of whole numbers in the given interval; the *cumulative density* is the ratio of the number of primes *from 1* to the upper end of the interval, to the number of whole numbers from 1 to the upper end of the interval.

composites? Can we answer the above question by extending the table of primes? If so, how far must it be extended? If not, why not? Euclid answered the question as to whether the number of primes is infinite by proving a remarkable theorem, in a remarkable way. We shall present Euclid's theorem in the next section.

With regard to whole numbers greater than 1, an interesting question arises: In how many different ways can they be factored? For prime numbers, the answer is trivial: they can be "factored" into only one factor, except for 1's. Thus $5 = 5$ or $5 \cdot 1$. But for composites, the situation is different. For example, we note that $60 = 2 \cdot 30 = 3 \cdot 20 = 4 \cdot 15 = 5 \cdot 12 = 6 \cdot 10$; so there are 5 different ways of separating 60 into 2 factors, not counting mere differences in order, such as $10 \cdot 6$. But at least one of each pair of factors is itself composite, and could be further factored. If we carry out the factoring as far as possible — to primes — we obtain $60 = 2 \cdot 30 = 2 \cdot 2 \cdot 15 = 2 \cdot 2 \cdot 3 \cdot 5$, or $2^2 \cdot 3 \cdot 5$. We note also that $60 = 3 \cdot 20 = 3 \cdot 2 \cdot 10 = 3 \cdot 2 \cdot 2 \cdot 5$, which is the same as the preceding factorization into primes, except for the order of the factors. Similarly, we may try factoring $60 = 4 \cdot 15$ into primes, etc. This experiment makes plausible the following statement, which is called the *Fundamental Theorem of Arithmetic:*

THEOREM 4.32 Fundamental Theorem of Arithmetic Every whole number greater than 1 is either a prime or else it can be written as a product of primes, uniquely except for order.

Since the proof of this theorem rests on several subsidiary theorems which we have not discussed, we shall not give it here.[30] We shall, however, feel free to use it in subsequent proofs.

As indicated previously, easily stated unsolved problems abound in the theory of numbers. Several are discussed in the exercises. One which has intrigued historians of mathematics is Fermat's "Last Theorem." We consider an equation of the type $x^n + y^n = z^n$, with $x, y, z, n \in N$. Which solutions, if any, does it have? If $n = 1$, the equation becomes $x + y = z$, which obviously has an infinite number of solutions, one of which is $x = 3$, $y = 4$, $z = 7$. The case $n = 2$ is more interesting, because it is more difficult: here we require that $x^2 + y^2 = z^2$. You may recognize this equation as that of the Pythagorean Theorem for right triangles, but here the sides of the triangle must be natural numbers only: fractions or radicals are not allowed. The smallest set of so-called Pythagorean triples which satisfies this equation is $x = 3$, $y = 4$, $z = 5$ — the 3–4–5 right triangle. Any multiple of this triple is also a solution: e.g., 6–8–10. You may know of still other solutions, as 5–12–13. Indeed, a formula has been worked out which yields all sets of Pythagorean triples and shows that there is an infinity of them.

What, then, of the case $n = 3$: $x^3 + y^3 = z^3$? Strangely, it has been *proved* that there are no solutions at all in integers! But Fermat (1601?–1665), an out-

---

[30]See any of the references on number theory in the chapter bibliography.

standing French mathematician, claimed more than this: he asserted that for $n = 3,4,5, \ldots$, the equation $x^n + y^n = z^n$ has no solution in positive integers. He stated, moreover, that he had proved this, but that the margin of the book in which he wrote many of his notes was too small to contain the proof. Now, Fermat certainly knew what a proof is, and was not likely to be mistaken — though a mistake is always a possibility. On the other hand, in the three centuries since his death, mathematicians, despite many attempts, have never been able to discover a proof, nor disprove Fermat's theorem.[31] The theorem, to be sure, is not very important in itself; but mathematicians, in their so-far fruitless efforts to prove it, have made many important discoveries in number theory.

EXERCISES 4.9

1. (a) Name the smallest prime.
   (b) How many even primes are there?
   (c) Is it true that a number is prime iff it is odd? (*Hint:* This question has two parts; examine them both.)
2. Does Table 4.3 seem to show that the primes are irregularly distributed and/or that their frequency decreases as they become larger? Explain.
3. (a) Use the sieve of Eratosthenes to find all the prime numbers less than 100.
   (b) Name the largest prime whose multiples it is *necessary* to cross out in part (a). Explain why larger primes need not be considered in this process.
   (c) In order to use the sieve to find the primes less than 300, it would be necessary to cross out all multiples of 2, 3, 5, 7, . . . , k. Without actually constructing the sieve, determine the value of k.
4. (a) By inspection, find all the primes which lie between 100 and 150.
   (b) To perform part (a) it is *necessary* to determine whether a number is divisible by 2, 3, 5, 7, . . . k. Determine the value of k and explain why no larger primes need be considered as divisors.
   *(c) Generalize the result of part (b), by determining k when finding all primes less than n.
5. (a) Use the results of Problems 3(a) and 4(a) to construct the following density table, similar to Table 4.3:

---

[31] See the H. F. Duncan reference for some partial results.

Interval	Number of Primes	Interval Density	Cumulative Density
1–25			
26–50			
51–75			
76–100			
101–125			
126–150			

(b) Does the table of part (a) show irregularities in the distribution of primes? a generally decreasing frequency of primes between 1 and 150?

6. Test each of the following numbers to see if it is prime. If it is, so state; if not, factor it into its unique prime factors. For example, 73 is prime, but $72 = 2 \cdot 2 \cdot 2 \cdot 3 \cdot 3$ or $2^3 \cdot 3^2$. The numbers are: 255; 468; 387; 197; 812; 637; 953; 1007.

7. Pairs of primes which differ by 2 are called *twin primes*. List all twin primes less than 100. (It is an unsolved problem as to whether the total number of twin primes is finite or infinite.)

8. (a) Evaluate $n^2 - n + 41$ for $n = 0, 1, 2, \ldots, 7$, and determine whether the resulting numbers are prime.

   (b) Do the results of part (a) *prove* that $n^2 - n + 41$ produces *only* primes? *all* the primes?

   (c) Would your answer to part (b) change if it turns out (as it does!) that $n^2 - n + 41$ produces *only* primes for $n = 0, 1, 2, \ldots, 40$? Explain your answer.

   *(d) Can you find any value of n for which $n^2 - n + 41$ yields a composite number?

   *(e) What bearing has your answer to part (d) on the truth of the statement "The formula $n^2 - n + 41$ produces only primes"?

*9. (a) Goldbach's *conjecture* states that every even number greater than 2 is the sum of two (not necessarily different) primes. Thus $4 = 2 + 2$, $8 = 3 + 5$, etc. Why is $8 = 1 + 7$ *not* an example of Goldbach's conjecture?

   (b) Illustrate the conjecture for several two-digit numbers.

   (c) Is the representation of even numbers as sums of primes unique? Explain or illustrate your answer.

   (d) What does the fact that the preceding statement is called a conjecture rather than a theorem indicate about its truth or falsity? Explain.

*10. (a) Find three consecutive composite numbers. In particular, obtain, if possible, the smallest such numbers.

(b) Similarly find 4 or 5 consecutive composite numbers.

(c) Is it possible to have *exactly* 4 consecutive composite numbers — i.e., 4 consecutive composites with a prime number at each end? Justify your answer informally.

(d) Explore to what extent parts (a)–(c) can be generalized to permit the existence of 6, 7, 8, . . . consecutive composite numbers. Is there a limit on the number of consecutive composites? Also, can there be *exactly* 6, *exactly* 7, etc.? If possible, prove your answers.

Section 4.10 Indirect Proof

In the preceding section we claimed that Euclid gave an answer to the question, "Is the number of primes finite or infinite?" His proof is significant not only because it settles the question, but also because it is a very ingenious proof. It is frequently cast in the form of an indirect argument and thus can serve as a model for the indirect method of proof. Perhaps you have met a few examples of this method in high school geometry or algebra, though many texts tend to avoid it. Certainly the indirect method is more complicated logically than the direct method, which we have used so far in our proofs:[32] so the latter is preferable where it is feasible. But many important theorems — increasingly those in advanced mathematics — can only be proved by the indirect method.

This method, in its most general form, is based on the requirement that the axioms of a deductive system be *consistent*. Consistency of the axioms means that no two axioms contradict each other, nor are any two contradictory theorems deducible from the set of axioms. The second part of this demand is hard to establish if the system contains many theorems. The system of whole numbers, for example, is rich in theorems and undoubtedly has many which have not even been discovered. So far, no contradiction has been brought to light; and after several centuries of intensive study of this system, we can be fairly confident that none will appear (but of course we cannot be sure). A similar situation exists with respect to Euclidean geometry. The indirect method is based on the *assumption* that the system in which the theorem is stated is consistent. In the remainder of this section, this means that we are assuming that the system of whole numbers is consistent.

With this understanding, we proceed with the discussion of methods of proof. In the direct method, a chain of reasoning is forged, beginning with the hypotheses and leading to the conclusion. The flow-chart method of proof shows this most clearly. An example of the direct method is given in the proof of Theorem 4.33, followed by examples of two distinct indirect methods. These

[32]Except for Theorem 4.7, which was proved in Section 4.7.

examples are mostly theorems concerning odd and even (whole) numbers, and are helpful in their own right. We begin with a formal

Definition 4.17 Even Number A whole number n is said to be *even* iff it can be written in the form $n = 2k$, where k is a whole number.

Thus 18 is an even number, since $18 = 2 \cdot 9$ (i.e., $k = 9$); but 3 is not even, since there is no *whole* number k such that $3 = 2k$.

Definition 4.18 Odd Number A whole number n is said to be *odd* iff it is not even.

By definition, we know that an even number always has the form $2k$, with $k \in W$. It can be shown that an odd number always has the form $2k + 1$, with $k \in W$.

We now state Theorem 4.33 and prove it by the direct method.

THEOREM 4.33 If a whole number is even, then its square is even.

Proof: Let n be the given whole number; n is even by hypothesis. Now $n = 2k$ $(k \in W)$ by Definition 4.17. Hence $n^2 = (2k)(2k)$ by the multiplication law for equality (Theorem 4.14); and $(2k)(2k) = 2(2k^2)$ by associativity and commutativity and the definition of exponent. Also, $k \in W$ implies that $2k^2 \in W$, since W is closed under multiplication. Hence $n^2 = 2(2k^2)$ by transitivity. But since $2k^2 \in W$, the preceding equation shows that n^2 is even, because Definition 4.17 is satisfied.

THEOREM 4.34 If a whole number is odd, then its square is odd.

Proof: A starred exercise.

The first of the indirect methods uses the fact, which we have established by truth tables, that a conditional and its contrapositive are logically equivalent: i.e., $(p \to q) \leftrightarrow (\sim q) \to (\sim p)$ is a tautology. So, if it seems preferable, we may prove the contrapositive of a theorem instead of the original statement; then because we have proved a statement equivalent to the theorem, we have (albeit indirectly) proved the theorem itself. As a quick example of this method we prove

THEOREM 4.35 If the square of a whole number is even, then the number itself is even.

Proof: The contrapositive of the theorem is: "If a whole number is not even, then its square is not even." But, by the definition of an odd number, a number that is "not even" must be odd. So the contrapositive can be rephrased as: "If a whole number is odd, then its square is odd." But this rephrased statement is precisely Theorem 4.34, which has already been proved. Since the contrapositive has been established, the original statement also holds.

The second, and more general, indirect method uses the fact (which we can also prove by truth tables) that $\sim(p \rightarrow q)$ is logically equivalent to $p \wedge (\sim q)$. Again, let $p \rightarrow q$ be the theorem we wish to prove. If in some manner we can prove that $p \wedge (\sim q)$ is false, then $\sim(p \rightarrow q)$ (which is the *denial* of the theorem) is also false, so $p \rightarrow q$ is true, as desired. The manner of proving $p \wedge (\sim q)$ false is to show that accepting it as true would lead to a *contradiction*; and, since a mathematical system (such as W) is consistent, and contradiction cannot occur, $p \wedge (\sim q)$ cannot be true. As an example we prove a theorem which we have already established directly, but which is also easy to cast into indirect form. Since it is a portion of Theorem 4.31, we shall denote it by

THEOREM 4.31′ If $a \in W$ and $a \neq 0$, then $a \div 0$ does not exist.

Proof: Suppose $p \wedge (\sim q)$ is true. (We have no logical justification for this supposition, but it is a possibility.) This means that "a is a nonzero whole number" (p) and that "$a \div 0$ exists as a unique whole number x" ($\sim q$). Now, by the definition of division, $a \div 0 = x$ iff $0 \cdot x = a$. Since $a \neq 0$ by hypothesis, the product $0 \cdot x$, which equals a, is nonzero by definition of equality of whole numbers. But the equation $0 \cdot x \neq 0$ contradicts the zero product theorem (Theorem 4.16) in W, which says that $0 \cdot x = 0$ for all $x \in W$. Since W is a consistent system, this contradiction cannot be tolerated. But the argument above is deductive and therefore inescapable, except for the first step ($p \wedge (\sim q)$ is true), which has no logical justification. Hence this step, the supposition that $p \wedge (\sim q)$ or $\sim(p \rightarrow q)$ is true, must be false. Since $\sim(p \rightarrow q)$ is false, the theorem, $p \rightarrow q$, is true, or p implies $q (p \Rightarrow q)$.

Now the stage is set for the proof of Euclid's theorem on primes: it is an indirect proof of the second type.[33]

THEOREM 4.36 (Euclid) The number of primes is infinite.

Strategy of the Proof: For clarification, the theorem can be reworded: "If P is the set of primes, then P is an infinite set." From this it is apparent that the hypothesis (p) of the theorem ($p \rightarrow q$) is "P is the set of primes" and the conclusion (q) is "P is an infinite set." We now begin with the (unjustified) supposition, $p \wedge (\sim q)$, that P is the set of primes and P is finite — i.e., the cardinal number of P is m, a whole number. By some device (This is where Euclid's ingenuity showed itself!) we use deductive reasoning to arrive at a contradiction. Since the whole number system is consistent, the contradiction indicates a mistake in the reasoning. Because we have used valid reasoning throughout the argument *after* the first supposition, the "mistake" must be in this unsupported supposition, which is therefore false; hence the theorem is true.

---

[33]A number of proofs have been made, not all indirect. See Section 2.15 of the Archibald reference for five of them.

Proof:

1. Suppose P is the set of primes, and $n(P) = m$, a whole number. (This supposition is a possibility, but unsupported by any valid reason.)

2. Clearly, $P \neq \varnothing$, since $2 \in P$. So $n(P) \neq 0$.

3. Then the set $P = \{p_1, p_2, p_3, \ldots, p_m\}$ is a finite set with m elements, where we may assume $p_1 = 2 < p_2 = 3 < p_3 < \cdots < p_m$, so that p_m is the largest prime. These statements are justified by the definition of a finite set (in one-to-one correspondence with the first m natural numbers: notice the subscripts, running from 1 to m) and the fact that these numbers can be ordered.

4. The stroke of genius! Consider the number $R = (p_1 \cdot p_2 \cdot p_3 \cdot \ldots \cdot p_m) + 1$. Clearly, $R \in W$ by closure of multiplication and addition; so it is in exactly one of the four categories: $\{0\}$, $\{1\}$, primes, or composites, by definition of these terms.

5. Now $R \neq 0, 1$, because $R > p_1 + 1 = 2 + 1 = 3 > 1 > 0$ by the order properties for whole numbers.

6. So R is either a prime or a composite, because only these two categories are left.

7. But R is *not* a prime, because $R > p_m$, the *largest* prime, by the order properties.

8. We claim that R is not divisible by any prime whatsoever.
   (a) For $p_1 = 2$, $2 \nmid R$, since $2 \mid (2 \cdot p_2 \cdot p_3 \cdot \ldots \cdot p_m)$, so that $R/2$ leaves a remainder of 1.
   (b) Similarly, for $p_2 = 3$, $3 \nmid R$, since $3 \mid (2 \cdot 3 \cdot p_3 \cdot \ldots \cdot p_m)$, so that $R/3$ leaves a remainder of 1.
   (c) Similarly, $5(=p_3)$, $7(=p_4)$, $\ldots$, p_m do not divide R.
   (d) Hence, since $p_1, p_2, \ldots, p_m$ are *all* the primes, R is not divisible by any prime whatsoever.

9. In view of Step 8, R cannot be written as a product of primes, as required by the Fundamental Theorem of Arithmetic (Theorem 4.32); so R is *not* composite.

10. Together, Steps 7 and 9 contradict Step 6. CONTRADICTION!

11. In view of the contradiction, the consistency of the system of whole numbers, and the valid reasoning in all steps except Step 1, Step 1 must be false, and the theorem is true.

From the above theorem, it is clear that the set of primes is infinite: no matter how far one goes out in the set of whole numbers, primes continue to occur. Another theorem, called the prime number theorem and first proved independently in 1896 by two French mathematicians, shows that the frequency of the primes decreases, irregularly but more and more nearly predictably, as they become larger. Thus our conjecture suggested by Table 4.3 is correct. The significance of Euclid's theorem is that, though the primes become increasingly rare, they never cease altogether.

EXERCISES 4.10

1. (a) Using Definition 4.17, explain why each of the following is an even number: 12; 366; 0.

   (b) Let $n = 35$. Since $35 = 2(17\frac{1}{2})$, we have written n in the form $n = 2x$. Is 35, then, an even number according to Definition 4.17? Explain.

   (c) An odd number always has the form $2k + 1$, where $k \in W$. Illustrate this by determining k for each of the following odd numbers: 3, 27, 1, 1001.

2. Let $P^* = \{p_1, p_2, \ldots, p_5\} = \{2,3,5,7,11\}$.

   (a) Compute $R^* = (p_1 \cdot p_2 \cdot \ldots \cdot p_5) + 1$.

   (b) Verify that $R^* > p_5$ and that none of $p_1, \ldots, p_5$ divide R^*.

   (c) Note that, starting with the set P^*, we have been able to produce a prime *not* in P^*, because either

      (i) R^* is itself a prime; or else

      (ii) R^* is a composite containing a prime not in P^*.

   Explain in each of the two cases why R^* contains a prime not in P^*.

   *(d) Is R^* in fact prime or composite?

3. Using truth tables, establish that $p \wedge (\sim q)$ and $\sim(p \rightarrow q)$ are logically equivalent. (See Exercises 3.7, Problem 4(e)).

4. Prove the following theorem by using an indirect method:

THEOREM 4.37 If the square of a whole number is odd, then the number itself is odd.

5. Problem 3, in the presence of a contradiction, gives a logical explanation of the structure of an indirect proof. A more sophisticated logical expression, however, incorporates the contradiction in the form $r \wedge (\sim r)$. If accepting $p \wedge (\sim q)$ implies both the statement r and also its negation, $(\sim r)$, this is a contradiction; and so the theorem, $p \rightarrow q$, is established indirectly by rejecting $\sim(p \rightarrow q)$. Symbolically, this description is expressed as

$$(p \rightarrow q) \leftrightarrow \{[p \wedge (\sim q)] \rightarrow [r \wedge (\sim r)]\}$$

   Show that this expression is a tautology.

*6. (a) Give a direct proof of Theorem 4.34. (*Hint:* Let $n = 2k + 1$.)
(b) Prove that the sum of two odd numbers is even.

*7. Use a (partly) indirect argument to prove that in any gathering of six people there are either three who are mutual acquaintances or three who are strangers to one another.

*8. The Hallerberg reference explains how to make a flow diagram of an indirect proof. Apply this method to Theorem 4.36, as proved in the text.

Suggestions for Investigation

1. We have based our development of the whole numbers, including the natural numbers, on set theory. It is possible, however, to create the natural numbers independently, i.e., without reference to set theory, on the basis of five axioms. To see how this is done, study the first few pages of the Landau reference. Note how addition and multiplication of natural numbers are defined. In Chapters 2–5, Landau makes successive extensions of the natural numbers all the way to the complex numbers.

2. A second approach, mentioned in Section 4.1, permits building up the natural numbers more easily and rapidly than does Landau, but still independently of set theory. It is well explained in Chapter 3 of the Freund reference, which bases the natural number system on twelve axioms, most of which are theorems in our approach. Freund uses these axioms in Chapter 5 to prove theorems and further develop the system; in later chapters he makes extensions all the way to the complex numbers. Study of number theory in Chapter 4 would also be rewarding.

3. In Section 4.9, the set of whole numbers was partitioned into four classes, of which the two substantial ones were primes and composites. In a similar fashion, other sets of numbers, such as $\{1,2,4,6,8,10, \ldots\}$, can be partitioned and the concept of "prime" with respect to those sets can be defined. The results, however, may or may not correspond to those for the set of whole numbers. In particular, the Fundamental Theorem of Arithmetic need not hold. To find examples of different situations, consult the Brown and Stone references.

4. In the discussion in Section 4.9 on testing for primes, a few very easy divisibility tests were mentioned for 2, 3, 5, and 9. Many tests are available for higher numbers. References under the names of Bold, Jordan, Morton, Pruitt, and Tinnappel deal with this topic.

5. "Sundaram's sieve," discovered in 1934 by a young East Indian student, also filters prime numbers. For a description and proof, see the Honsberger reference, p. 75ff.

6. A thirteenth century mathematics text, written by Leonardo Fibonacci, contains the following problem: "How many *pairs* of rabbits can be pro-

duced from a single pair in a year if every month each pair begets a new pair, which from the second month on becomes productive?" If we begin on January 1 with a newborn pair, it will produce its first litter on March 1. The answer to this problem is the famous "Fibonacci sequence" (1,1,2,3,5, . . .), which is discussed in the references under Gardner, Sharpe, and Vorob'ev.

7. Other topics of interest are perfect numbers and amicable (friendly) numbers, which appear in the references under Gardner, Hanawalt, Rolf, and Schaaf.

Bibliography

Archibald, R. G. *An Introduction to the Theory of Numbers.* Columbus, Ohio: Charles E. Merrill Publishing Co., 1970*a*.

______. "How Many Primes?" *New York State Mathematics Teachers' Journal*, 20 (1970*b*), 23–25.

Barnett, I. A. *Some Ideas about Number Theory.* Washington: National Council of Teachers of Mathematics, 1961.

Birkhoff, Garrett and Saunders MacLane. *A Survey of Modern Algebra*, 3rd ed. New York: The Macmillan Co., 1965.

Bold, Benjamin. "A General Test for Divisibility by Any Prime (Except 2 and 5)." *Mathematics Teacher*, LVIII (1965), 311f.

Brown, S. I. "Of 'Prime' Concern: What Domain? " *Mathematics Teacher*, LVIII (1965), 402–407.

______. "Multiplication, Addition and Duality." *Mathematics Teacher*, LIX (1966), 543–50.

Buchman, A. L. "Patterns in Algorithms for Determining Whether Large Numbers Are Prime." *Mathematics Teacher*, LXIII (1970), 30–41.

Duncan, D. C. "Happy Integers." *Mathematics Teacher*, LXV (1972), 627–29.

Duncan, H. F. "Fermat's Last Theorem." *Mathematics Teacher*, LVIII (1965), 321f.

Freund, J. E. *A Modern Introduction to Mathematics.* Englewood Cliffs, N.J.: Prentice-Hall, Inc., 1956.

Gardner, Martin. "Perfect Numbers and Amicable Pairs." *Scientific American*, 218:3 (Mar., 1968), 121ff.

______. "The Multiple Fascinations of the Fibonacci Sequence." *Scientific American*, 220:3 (Mar., 1969), 116–120.

Hallerberg, A. E. "A Form of Proof." *Mathematics Teacher*, LXIV (1971), 203–14.

Hanawalt, Kenneth. "The End of a Perfect Number." *Mathematics Teacher*, LVIII (1965), 621f.

Herwitz, P. S. "The Theory of Numbers." *Scientific American*, 185:1 (Jul., 1951), 52–55.

Honsberger, Ross. *Ingenuity in Mathematics*. New York: Random House, Inc., 1970.

Jordan, J. Q. "Divisibility Tests of the Non-congruence Type." *Mathematics Teacher*, LVIII (1965), 709–12.

Landau, Edmund. *Foundations of Analysis*. Bronx, N.Y.: Chelsea Publishing Co., 1951.

Lieber, L. R. *Infinity*. New York: Rinehart & Co., Inc., 1953.

Morton, R. L. "Divisibility by 7, 11, 13 and Greater Primes." *Mathematics Teacher*, LXI (1968), 370–73.

Niven, Ivan and H. S. Zuckerman. *An Introduction to the Theory of Numbers*, 2nd ed. New York: John Wiley & Sons, Inc., 1966.

Pruitt, Robert. "A General Divisibility Test." *Mathematics Teacher*, LIX (1966), 31–33.

Rolf, H. L. "Friendly Numbers." *Mathematics Teacher*, LX (1967), 157–60.

Rosenbloom, P. C. "Recent Information on Primes." *Twenty-Eighth Yearbook*. Washington: National Council of Teachers of Mathematics, 1963. Chapter 3.

Schaaf, W. L., ed. *Prime Numbers and Perfect Numbers*, SMSG *Reprint Series*, 2. Pasadena, Calif.: A. C. Vroman, Inc., 1966.

Sharpe, Benjamin. "A Remarkable Research Area" (Fibonacci Numbers). *Mathematics Teacher*, LVIII (1965), 420f.

Sierpinski, Waclaw. *Pythagorean Triangles*. New York: Yeshiva University, 1962.

Sominskii, I. S. *The Method of Mathematical Induction*. New York: Blaisdell Publishing Co., 1961.

Spreckelmeyer, R. L. and K. Mustain. *The Natural Numbers*. Boston: D. C. Heath and Co., 1963.

Stein, M. L., S. Ulam, and M. Wells. "A Visual Display of Some Properties of the Distribution of Primes." *American Mathematical Monthly*, 71:5 (1964), 516–20.

Stone, E. J. "New Domains." *Mathematics Teacher*, LVIII (1965), 514–16.

Suppes, Patrick. *Axiomatic Set Theory*. New York: D. Van Nostrand Co., Inc., 1960. Chapter 4.

Tinnappel, Harold. "On Divisibility Rules." *Twenty-Seventh Yearbook*. Washington: National Council of Teachers of Mathematics, 1963*a*. Chapter 16.

———. "Using Sets to Study Odd, Even and Prime Numbers." *Twenty-Seventh Yearbook*. Washington: National Council of Teachers of Mathematics, 1963*b*, Chapter 19.

Topics in Mathematics for Elementary School Teachers, Twenty-Ninth Yearbook. Washington: National Council of Teachers of Mathematics, 1964. Chapter 2, "The Whole Numbers."

Uspensky, J. V. and M. A. Heaslet. *Elementary Number Theory*. New York: McGraw-Hill Book Co., 1939.

Vorob'ev, N. N. *The Fibonacci Numbers*. Boston: D. C. Heath and Co., 1963.

5

Systems of Numeration

Are numerals numbers? Which principles make the Roman and the Hindu-Arabic numeration systems work? What is base-five arithmetic? Why is the binary system important?

Section 5.1 Ancient Systems of Numeration

We turn from *number* systems to *numeral* or *numeration* systems, which are procedures for representing numbers by symbols. The distinction between numbers and numerals is stressed in many recent school texts. In Chapter 4 we discussed properties of whole *numbers*, referring to them, however, by the conventional Hindu-Arabic *numerals:* 0, 1, 2, 3, ... For example, we saw that 1 is the multiplicative identity or unit and that 2 is the smallest prime; in the Ionic Greek numeral system the preceding statement would become "α is the multiplicative identity or unit and β is the smallest prime." The meaning or content of the statement is the same for both sentences; only the symbols have changed. Consider, on the other hand, the statement "The numeral '9' is the numeral '6' turned upside down." This sentence describes numerals in our system; it has nothing to do with the properties of numbers. Indeed, in the Roman numeral system, the corresponding statement, "The numeral 'IX' is the numeral 'VI' turned upside down," is absurd.

Numerals, and other mathematical symbols, however, are important and worthy of study in their own right. The history of numeration, although imperfectly known, is rich and fascinating: civilizations in different parts of the world not only invented different symbols, but combined them according to different principles. Most of the ancient systems had the primary aim of merely recording numbers: there was little thought of performing calculations within the systems. Consequently, calculations which we can easily perform in our Hindu-Arabic system were done on an abacus or some other device, often very efficiently, and the results were "translated" into the numeral system.

One of the simplest and most ancient of the numeration systems is the Egyptian hieroglyphic system, which was used as far back as 3400 B.C. This system, like many others, was based on the number ten. The Egyptians used a separate symbol for each power of ten, as Table 5.1 shows.

TABLE 5.1

one | (a staff)	one thousand 𓆼 (a lotus flower)
ten ∩ (a heel bone)	ten thousand 𓂭 (a pointing finger)
one hundred 𓍢 (a scroll)	one hundred thousand 𓆐 (a burbot fish)
one million 𓁨 a man with his arms raised in astonishment)	

By combining these symbols according to two simple principles, the Egyptian could write any natural number from one up into the millions. The two principles are:

1. Repetition: ||| = 3; ∩∩ = 20. Note that *only one* symbol is used and that the value of this symbol is counted as many times as the symbol appears.
2. Addition: ∩| = 10 + 1 = 11; 𓍢∩∩||| = 123. When *different* symbols are juxtaposed, their values are added. The latter example illustrates both repetition and addition.

The Egyptian system is especially easy in that each symbol has exactly one value, regardless of its position. For example, ∩ has the value ten in 𓍢∩|, ∩𓍢|, |𓍢∩, etc. Is a comparable statement true of 5 in our Hindu-Arabic system?

It is assumed that you are familiar with the Roman numeral system. Which principles does it exhibit and how do they compare with those of the Egyptian system? A new one is the *principle of subtraction*, illustrated by IV and XLIX. Note that changing the order of writing the symbols changes the *operation* from addition to subtraction; but it does not change the *value* of the symbol: e.g., X stands for 10 in LX and in XL.[1]

You should now reexamine our familiar Hindu-Arabic system to see which principles it uses and which it does not. A new principle of the greatest importance is that of *position*, or *place value*. According to this principle, the value of a digit is determined, not only by its form, but also by its position. Thus in the numeral 321, the symbol 1, because of its position at the extreme right, has the value *one*. The next symbol, 2, is said to be in the tens' position and consequently has the value of two *tens*, or 20. Similarly, the 3 has the value of three *hundred* by virtue of its position to the left of the tens' digit.

The principle of position also occurs in a few other numeration systems, notably the Mayan and the Babylonian systems. More significantly, the *abacus*, which many peoples have used as a calculating device, is based on position. In a simple abacus the beads on one wire represent units, those on the adjacent wire to the left, tens, etc. Thus the abacus sketched in Figure 5.1 records the number five hundred two. Now, on the abacus it is clear that this number contains no tens. In recording the number, however, it would not do to write 5 2, for the 5 might be misconstrued as 5 *tens*. It is thus necessary to have a symbol to represent an empty wire. As we know, that symbol is 0. The invention of this symbol — a mark to represent "nothing" or emptiness! — was an intellectual breakthrough for mankind; for without it, no positional system of numeration can function efficiently.

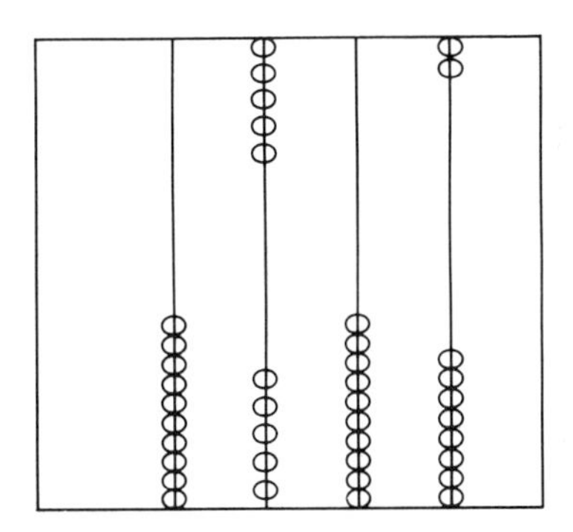

Figure 5.1

The principles of addition and position, together with the invention of 0, give to our Hindu-Arabic system its remarkable flexibility and utility. In contrast to the Egyptian and Roman systems, it is possible for us to express any number, no matter how large, by proper use and positioning of only the ten digits: it is not necessary to keep inventing new symbols for higher and higher powers of ten. And the positioning also facilitates the development of rules of computation *within* the system itself, so that there is no longer a need to resort to external calculating devices like the abacus.

---

[1]In the Middle Ages, a fourth principle, *multiplication*, was used: a bar over a numeral multiplied its value by one thousand, as in $\overline{\text{LX}}\text{CCXI}$ for 60,211.

With its obvious advantages for calculating, we might think that all civilizations would have welcomed eagerly the introduction of the Hindu-Arabic system. Indeed, the Moorish period in Spain amply demonstrated its benefits, especially in architecture. As testimony to the stubborn conservatism of many men, however, it has to be recorded that when this new system began to be introduced and gain adherents, the authorities tried to "stem the tide" by passing laws forbidding its use! But now Roman numerals are so little used that, it has been said, they are preferred for recording the date on a film so that, when it is rerun years later, audiences will not be able readily to determine its age!

EXERCISES 5.1

1. In the following statements, put quotation marks around the symbols which stand for *numerals;* leave unchanged those which represent *numbers:*

   (a) The numeral 13 has the digit 3 in the units' place; the number 13 is the sixth prime in order.

   (b) If a number written in the Hindu-Arabic notation has a 0 in the units' column, then it is divisible by 10.

   (c) This check is worth $2,000.00, because the line after the dollar sign contains 2,000.00.

   (d) 4 is half of 8, because $4 \cdot 2 = 8$. But 3 is half of 8, because a vertical line through 8 separates it into Ɛ and 3.

2. (a) Express in words the value of each of the following:

   36 4,125 7,011 80,015 1,000,240

   (b) Express in Hindu-Arabic notation each of the following:

   five hundred fifteen, twenty thousand sixty-one, twelve million four hundred ten, one less than two million fifty thousand

3. (a) Convert the following Egyptian numerals to Hindu-Arabic notation:

   (i) 𓍢𓎆𓎆𓎆𓎆𓏺𓏺 (ii) 𓆼𓍢𓏺

   (iii) 𓆼𓍢𓍢𓎆𓎆𓎆𓎆𓎆𓎆𓏺𓏺𓏺𓏺𓏺𓏺𓏺𓏺 (iv) 𓂭𓂭𓎆𓎆𓎆𓏺

   (b) The Egyptian numeration system is built on two principles. Name each and use one of the examples of part (a) to illustrate.

   (c) Convert the following Hindu-Arabic numerals to Egyptian notation:

   32 250 1,407 1,020,304

4. (a) Convert the following Roman numerals to Hindu-Arabic notation:

LXVI DCCXCIX MDLIV MCMXLVIII

(b) The Roman numeration system is built on three principles. Name each and use an example of part (a) to illustrate.

(c) Convert the following Hindu-Arabic numerals to Roman notation:

52 639 1,991 2,444

*5. The Attic Greek numeration system contains the following symbols:

one	|	one hundred	H
five	Γ	one thousand	X
ten	Δ	ten thousand	M

The symbol for 5 appears both alone and combined with others; thus is 50. As a larger example, 13,668 appears as

MXXX𐅅H𐅄ΔΓ|||

(a) Convert the following Attic Greek numerals to Hindu-Arabic notation:

HHΔΓ H𐅅ΔΔ|| M𐅆HHHH𐅄ΔΔ MM𐅆HHΔΔΔΓ|

(b) One of the principles of the Attic Greek system may be called *multiplication*. State this principle and illustrate it. Also identify and illustrate any other principles which you discover.

*6. The Babylonian system contains only three symbols, yet it represents numbers quite compactly. These symbols are:

one |

ten <

the subtractive symbol Γ

Thus 24 is <<|||| and 18 (= 20 − 2) is <<Γ||. This procedure continues until 60; thereafter, the principle of position also applies, as in | << |||, which is 83 (one sixty plus twenty-three). An example of a larger number is

<| <<||| <|||||

which is 40,995 (eleven sixty-times-sixties plus twenty-three sixties plus fifteen).

(a) Convert the following Babylonian numerals to Hindu-Arabic notation:

<<<<<|| <<<<┌┐ ||<|| <<<||<<<<|| <<┌┐<|||||<<<||

(b) Express the numbers of Problem 2(a) in Babylonian notation.

(c) Identify and illustrate the principles on which the Babylonian numeration system is based.

(d) The Babylonians did not have a symbol for zero. Provide illustrations to show how this lack leads to ambiguities in expressing large numbers in the Babylonian system.

*7. Investigate the Mayan numeration system, which has a symbol for zero. Use an encyclopedia or the Bidwell or Eves references. Work through the examples in your reference, then express the numbers of Problem 2(a) in Mayan notation.

Section 5.2 Exponents: A Shorthand

We discuss exponents at this point for several reasons. First, we need them for a full explanation of our numeration system and other systems related to it. Second, since exponents are a familiar and necessary part of elementary algebra, our discussion constitutes a good review, possibly from a different standpoint than that of an algebra text. Third, the "story of exponents" is an example of how *one simple* definition, introduced as a time-saving shorthand, can almost take on a life of its own, generating theorems and further definitions that carry it far beyond its original intent — and, indeed, ultimately lead to logarithms and to the exponential function which is so usefully applied in theoretical mathematics and in physical and economic situations involving growth and decay. While these "ultimate" ramifications are beyond the scope of this course, it will be useful to begin the development of exponents here.

You are familiar with the convenient shorthand which permits us to write the product $2 \cdot 2 \cdot 2 \cdot 2$ as 2^4, where the superscript, 4, means that 2 is to be taken as a factor 4 times. The expressions 3^7 and 10^{11} have corresponding meanings, the latter being a short expression for 100 billion. In expressing physical quantities like the diameter of the universe or the number of atoms in a cubic centimeter of iron, scientists use much larger exponents to the base 10. A definition is

Definition 5.1 Exponent The symbol x^m, where $x \in W$ and $m \in N$, means "x taken as a factor m times"; or

$$x^m = \underbrace{x \cdot x \cdot x \cdot \ldots \cdot x}_{m \text{ factors}}$$

In the expression x^m, x is called the *base* and m is called the *exponent*.[2]

Consider the product

$$8 \cdot (16) = 2^3 \cdot 2^4 = (2 \cdot 2 \cdot 2)(2 \cdot 2 \cdot 2 \cdot 2) = (2 \cdot 2 \cdot 2 \cdot 2 \cdot 2 \cdot 2 \cdot 2) = 2^7$$

By a generalization of the associative law for multiplication, the two groups of 3 and 4 factors each can be combined into one group, which can then be simply expressed by use of the exponent 7. But we note that 7, the exponent of the product, is the *sum* of the two given exponents:

$$2^3 \cdot 2^4 = 2^{3+4} = 2^7$$

The same kind of reasoning, generalized, leads to the following theorem:

THEOREM 5.1 If $x \in W$ and $m, n \in N$, then $x^m \cdot x^n = x^{m+n}$.

Is there an analogous theorem for division? Consider the example

$$4^5 \div 4^2 = \frac{4 \cdot 4 \cdot 4 \cdot 4 \cdot 4}{4 \cdot 4} = 4 \cdot 4 \cdot 4 = 4^3$$

This example suggests the theorem $x^m \div x^n = x^{m-n}$. Let us try it on three more troublesome examples: $3^5 \div 3^5 = 3^{5-5} = 3^0$; $3^4 \div 3^5 = 3^{4-5} = 3^{-1}$; and $3^3 \div 3^5 = 3^{3-5} = 3^{-2}$. The first example leads to 3^0, which is meaningless, since our definition of x^m requires that the exponent be a *natural* number. The second example is meaningless for the same reason, even if we assume (as we *shall* do during this section) that negative numbers exist.

There is, however, an alternative procedure, using the fraction form for division, and cancelling as appropriate. We compute, for example,

$$3^5 \div 3^5 = \frac{3 \cdot 3 \cdot 3 \cdot 3 \cdot 3}{3 \cdot 3 \cdot 3 \cdot 3 \cdot 3} = 1$$

Similarly,

$$3^4 \div 3^5 = \frac{3 \cdot 3 \cdot 3 \cdot 3}{3 \cdot 3 \cdot 3 \cdot 3 \cdot 3} = \frac{1}{3}$$

and

$$3^3 \div 3^5 = \frac{1}{3 \cdot 3} = \frac{1}{3^2}$$

Thus the division theorem must be stated in three parts in order to avoid meaningless exponents which are not natural numbers.

---

[2]A more formal, recursive definition is: $x^1 = x$; $x^m = x \cdot x^{m-1}$ for m any natural number greater than 1.

THEOREM 5.2 If $x, m, n \in N$, then

1. $x^m \div x^n = x^{m-n}$ if $m > n$.
2. $x^m \div x^n = 1$ if $m = n$.
3. $x^m \div x^n = 1/x^{n-m}$ if $m < n$ (so that $n - m \in N$).

Now since 3^0, 3^{-1}, 3^{-2}, etc., are undefined at this point, we are logically free to define them as we wish. It would be sensible, however, to choose a convenient definition, not a crazy one like $3^0 = 1{,}972$. Since we know that $3^5 \div 3^5$ *does* equal 1, and we also observe that Theorem 5.2(1), if it *could* be applied, would yield 3^0, let us assign 1 as the meaning of 3^0. Now we can use Theorem 5.2(1) for *any* division problem involving *equal* exponents to the base 3, since the result, 3^0, will, by our new definition, give the correct answer of 1. More generally, let us announce

Definition 5.2 $x^0 = 1$ $(x \neq 0)$.

Similarly, if we define $3^{-1} = 1/3$, $3^{-2} = 1/3^2 = 1/9$ etc., we shall be able to use Theorem 5.2(1) for cases when $m < n$. Accordingly, we announce in general

Definition 5.3 $x^{-m} = 1/x^m$, for $m \in N$, and $x \neq 0$.[3]

We have thus accomplished two things: we have given meanings to zero and negative exponents, and we have greatly simplified the statement of Theorem 5.2 by eliminating the need for equations (2) and (3). We state the simplified result as

THEOREM 5.3 If $x, m, n \in N$, then $x^m \div x^n = x^{m-n}$.

We now consider an easy application of exponents. In the Hindu-Arabic numeral 444.44, each 4 stands for a different value. Reading from left to right, the first 4 stands for 4 hundred, or $4(10)^2$; the last 4 stands for 4 hundredths, or $4(1/100)$, or $4(10)^{-2}$. In fact, we may write the numeral 444.44 in the following expanded form:

$$444.44 = 4(10)^2 + 4(10)^1 + 4(10)^0 + 4(10)^{-1} + 4(10)^{-2}$$

Note the way in which the exponents decrease regularly: this is rather neat! To save writing, we usually omit the exponent 1; and, since $(10)^0 = 1$ (Why?), we omit this factor altogether. Thus we write:

$$444.44 = 4(10)^2 + 4(10) + 4 + 4(10)^{-1} + 4(10)^{-2}$$

[3]Why is this restriction necessary? See Theorem 4.31.

Similarly, we may write, in expanded form,

$$2{,}036.75 = 2(10)^3 + 0(10)^2 + 3(10)^1 + 6(10)^0 + 7(10)^{-1} + 5(10)^{-2}$$

or

$$2(10)^3 + 3(10) + 6 + 7(10)^{-1} + 5(10)^{-2}$$

You should try writing in expanded form the numeral 3,141,592,654, which represents approximately the value "one billion pi."

Although the expanded form exhibits the meaning of our Hindu-Arabic notation, it is too cumbersome to use. There is, however, another use of exponents in expressing very large or very small numbers which is quite efficient. We refer to *standard*, or *scientific*, notation. The 1970 census, for instance, showed New York City with a population of 7,870,000, to the nearest ten thousand. In standard notation this number appears as 7.87×10^6. You may ascertain that, since 10^6 is one million, this form has the same value as the original one. Moreover, it can be converted back to the original, merely by moving the decimal point to the *right* the number of places indicated by the exponent (6), inserting zeros where needed. To take another example, the length of the poliomyelitis virus is about 0.0012 cm. Expressed in standard notation, this becomes 1.2×10^{-3} cm, the negative exponent indicating that, to recapture the original numeral, the decimal point should be moved 3 places to the *left*. The real convenience of this notation becomes evident when it is necessary to express distances in the universe or within the atom. Thus the nearest star is about 4.3 light-years, or 2.53×10^{13} miles, away; and the diameter of the copper atom is about 2.56×10^{-8} cm.

You will observe that, in all the above illustrations, the first part of the standard notation consists of a single nonzero digit, followed by a decimal point and further digits as necessary. This is the usual convention; it leads to

Definition 5.4 Standard Notation A number[4] r is represented in *standard notation* when it is expressed in the form $r = s \times 10^k$, where s is a single nonzero digit followed by a decimal point and (possibly) other digits, and k is an integer, positive, negative, or zero.

The use of standard notation is not confined to expressing quantities; it also extends to calculating with them. The task of multiplying 710,000 by 0.00063, for example, becomes easier and less subject to decimal errors if the numerals are first converted to standard notation. Thus

$$\begin{aligned} 710{,}000 \times 0.00063 &= (7.1 \times 10^5) \times (6.3 \times 10^{-4}) \\ &= (7.1 \times 6.3) \times (10^5 \times 10^{-4}) = 44.73 \times 10^1 = 447.3 \end{aligned}$$

[4]Here r is not necessarily a whole number: it may contain one or more digits to the right of the decimal point.

Can you justify some of the steps by reference to properties of whole numbers and of exponents? We are assuming here, of course, the rules for multiplying decimal expressions ($7.1 \times 6.3 = 44.73$) and for adding signed integers [$5 + (^-4) = 1$], neither of which we have yet discussed.

As a more complicated example we submit

$$\frac{32{,}000{,}000 \times 0.00005}{40{,}000} = \frac{(3.2 \times 10^7) \times (5 \times 10^{-5})}{(4 \times 10^4)}$$

$$= \frac{3.2 \times 5}{4} \times \frac{10^7 \times 10^{-5}}{10^4}$$

$$= \frac{16}{4} \times \frac{10^2}{10^4} = 4 \times 10^{-2} = 0.04$$

EXERCISES 5.2

1. Express the values of the following numbers in simplest form without exponents:

$2^5 \quad 3^4 \quad 4^3 \quad 5^0 \quad 0^5 \quad 10^9 \quad 1^{100} \quad 2^{-3} \quad 3^{-2} \quad 10^{-4} \quad 100^{-2}$

2. Simplify the following expressions, using exponents in your answers:

$3 \cdot 3 \cdot 3 \cdot 3 \cdot 3 \cdot 3 \quad 2 \cdot 3 \cdot 5 \cdot 5 \cdot 5 \cdot 9 \cdot 16 \quad 2^6 \cdot 2^5 \quad 5^7 \div 5^3 \quad 7^4 \div 7^6$

$10^8 \cdot 10^{-3} \quad 10^{16} \div 10^5 \quad 10^3 \div 10^{-1} \quad 5^4 \cdot 5^0 \quad 2^{17} \div 2^{17}$

$9 \cdot 27 \quad 64 \div 16$

3. In Theorems 5.2 and 5.3, why do we write $x \in N$ instead of $x \in W$? (*Hint:* See Theorem 4.31.)

4. (a) Write in expanded form as sums of powers of ten each of the following Hindu-Arabic numerals:

359, 60,215, 1,005,632, 28.34, 705.309, 0.01235

(b) Write in ordinary (compact) Hindu-Arabic notation each of the following:

$1 \cdot 10^3 + 9 \cdot 10^2 + 6 \cdot 10^1 + 9 \cdot 10^0 \quad 2 \cdot 10^5 + 3 \cdot 10^4 + 7 \cdot 10^2 + 8$

$4 \cdot 10^4 + 1 \cdot 10 + 6 + 3 \cdot 10^{-1} + 2 \cdot 10^{-2} \quad 8 \cdot 10^2 + 5 + 3 \cdot 10^{-3}$

$1 + 2 \cdot 10^{-1} + 5 \cdot 10^{-4}$

5. (a) Convert each of the following to standard notation:

2,400, 93,200,000, 0.00166, 0.000007

(b) Convert each of the following to ordinary Hindu-Arabic notation:

$$3.7 \times 10^4 \quad 6 \times 10^{10} \quad 1.28 \times 10^{-4} \quad 6.5 \times 10^{-12}$$

(c) Perform the following computations, both in ordinary notation and after converting to standard notation:

$$2{,}100{,}000 \times 15{,}000 \quad 21{,}000{,}000 \times 0.000015$$

$$60{,}000 \div 24{,}000{,}000 \quad 0.0018 \div 0.0000045$$

$$\frac{0.00745 \times 720{,}000}{0.015}$$

6. Some of the following physical constants are expressed in standard notation; others are expressed in "ordinary" notation.[5] Convert each form to the other form.

   (a) The constant of gravitation, 6.670×10^{-8} cm^3/g-sec^2, is the force in dynes between two 1-gram masses 1 cm apart.

   (b) The acceleration due to gravity (at sea level) is 32.172 ft/sec^2, or 980.665 cm/sec^2.

   (c) The velocity of light in a vacuum is 2.99776×10^{10} cm/sec, or 9.83514×10^8 ft/sec, or 186,272 mi/sec.

   (d) The mean wavelength of sodium light is 5893 Angstrom units (A), or 0.00005893 cm.

   (e) The half-life of carbon-14 is 5,700 years. Change this figure to standard notation, then find out how many days this is (multiplying by 365) expressed in standard notation.

   (f) The half-life of a certain isotope of polenium is 3.2×10^{-6}.

   (g) The charge on an electron is 4.80×10^{-10} e.s.u.

   (h) Avogadro's number, the number of molecules in a mole, is 6.02×10^{23}.

   (i) The mass of an electron is 9.1066×10^{-28} grams, while that of a proton is 1.67248×10^{-24} grams. *Without* converting these data, tell which mass is greater. Now calculate (still without converting) the ratio of the larger mass to the smaller.

*7. (a) Explore numerical examples such as $3^{-2} \cdot 3^3$, $4^3 \cdot 4^0$, and $2^3 \cdot 2^{-5}$ to see whether Theorem 5.1 can be extended to the cases where m and/or n are zero or negative integers.

   (b) Investigate a similar extension of Theorem 5.3.

---

[5] All these constants are approximate, since they are obtained by measurement.

*8. An exponential expression may be raised to a power, as

$$(5^2)^3 = (5^2)\cdot(5^2)\cdot(5^2) = 5^6$$

(a) Similarly expand $(3^2)^4$ and $(2^3)^5$.

(b) Generalize part (a) to create a formula for $(x^m)^n$, where $x \in W$ and $m, n \in N$. Consider further numerical examples, if needed.

(c) Prove the following:

THEOREM 5.4 $(x^m)^n = x^{mn}$, where $x, m, n \in N$.

(d) Explore numerical examples like $(3^4)^0$, $(2^5)^{-1}$, $(3^{-2})^2$, and $(5^{-2})^{-3}$ to see whether Theorem 5.4 can be extended to the cases where m and/or n are zero or negative integers.

*9. The exponent definitions for x^m ($m \in N$), x^0, and x^{-m} do not apply to expressions like $9^{1/2}$ or $8^{-2/3}$. We are therefore free to define these as we choose; but we should like to do so in such a way that these fractional exponents will obey the laws of operation for integral exponents.

(a) If $9^{1/2}$ is to obey Theorem 5.1 or Theorem 5.4, we must have $9^{1/2}\cdot 9^{1/2} = 9^{1/2+1/2} = 9^1 = 9$, or $(9^{1/2})^2 = 9^1 = 9$. Recall that $\sqrt{9}\cdot\sqrt{9} = 3\cdot 3 = 9$; then suggest a suitable definition for $9^{1/2}$. Does it also apply to $16^{1/2}$? to $5^{1/2}$? (*Hint:* $\sqrt{5}\cdot\sqrt{5} = 5$.)

(b) By considering $8^{1/3}$ and $5^{1/3}$, generalize your definition of part (a) to $x^{1/n}$, where $x, n \in N$.

(c) By considering suitable examples, make a further generalization, to $x^{m/n}$, where $x \in W$ and m and n are integers ($n \neq 0$), thus creating a meaning for *any* fractional exponents, including $27^{-1/3}$ and $4^{-3/2}$.

*10. (a) Write the largest number obtainable by rearranging the digits of 4,176, *viz.*, 7,641; also write the smallest number so obtainable; then find their difference.

(b) Use the procedure of part (a), starting with 6,345. After you have found the difference, repeat with it the same procedure. Repeat several times. What do you discover? (*Hint:* Compare your answer with that of part (a)).

(c) Try the repetitive procedure of part (b) also on 4,326, as well as on two other four-digit numbers of your own selection. (*Hints:* Do not make trivial choices like 4,444. Also, be patient: repeat the subtraction up to seven times.)

(d) State carefully a conjecture based on your observations in (a)–(c). (For a proof, see Essay 10 in the Honsberger reference.)

Section 5.3 Some Fundamental Arithmetic Algorithms

An *algorithm* is a mechanical computing rule or method. During our school careers we have all learned a number of arithmetic and algebraic algorithms. A simple illustration is finding the sum of 46 and 32. Though the definition of addition of whole numbers requires that we obtain the union of two disjoint sets, containing 46 and 32 elements, respectively, we do not generally find the sum that way. Instead, we split both numerals mentally into two pieces, add the units' digits and the tens' digits separately, and then combine the pieces to give the answer. This process is revealed by using the expanded notation:

$$\begin{array}{cc} 4\ 6 & 4\cdot 10 + 6 \\ 3\ 2 & 3\cdot 10 + 2 \\ \hline 7\ 8 & 7\cdot 10 + 8 \end{array}$$

But by what right do we perform this alchemy of fission and fusion? The theorem and proof which follow provide a justification in terms of the basic properties of our numeration system and the system of whole numbers.

Before giving the proof, we shall assume, to save time, that all the theorems represented by the *addition table* from $0 + 0$ to $9 + 9$ have been proved, either by applying the definition of addition (Definition 4.4) or by using other methods. Accordingly, we shall cite "addition table" as a reason in our proofs. The point of the algorithm is to help us add two numbers, as large as we please, without using sums greater than $9 + 9$ [or $(1 + 9) + 9$ if a "carry" occurs].

We now prove the following special theorem. The proof is not complete, because we omit the closure steps and, in the second line, condense several obvious steps into one. Thus the proof gains brevity and clarity. But you should be capable of supplying the omitted steps and reasons if asked.

THEOREM 5.5 $46 + 32 = 78$.

Proof:

1. $46 + 32 = (4\cdot 10 + 6) + (3\cdot 10 + 2)$ by the principles of Hindu-Arabic numeration.

2.[6] $= (4\cdot 10 + 3\cdot 10) + (6 + 2)$ by several steps, using the commutative and associative laws of addition.[7]

3. $= (4 + 3)\cdot 10 + (6 + 2)$ by the distributive law.

---

[6]The complete equation in Step 2 is $(4\cdot 10 + 6) + (3\cdot 10 + 2) = (4\cdot 10 + 3\cdot 10) + (6 + 2)$; but, to save space and copying, we will understand that the blank to the *left* of each equal sign is to be filled by the same expression which appears to the *right* in the *preceding* equation. Thus Steps 1–5 form a chain of equalities, which together imply Step 6.

[7]These steps are suggested by the chain $(4\cdot 10 + 6) + (3\cdot 10 + 2) = [(4\cdot 10 + 6) + 3\cdot 10] + 2 = [4\cdot 10 + (6 + 3\cdot 10)] + 2 = [4\cdot 10 + (3\cdot 10 + 6)] + 2 = [(4\cdot 10 + 3\cdot 10) + 6] + 2 = (4\cdot 10 + 3\cdot 10) + (6 + 2)$.

4. $= 7 \cdot 10 + 8$ by the addition table.
5. $= 78$ by the principles of Hindu-Arabic numeration.
6. So $46 + 32 = 78$ by the transitive law of equality.

As a second example of addition, this one involving carrying, we take $76 + 59$. Most of the reasons in the proof are omitted so that you may have the pleasure of supplying them. Again the proof is abbreviated in the interest of clarity. You are invited to pick out the step(s) in which the carrying actually occurs.

THEOREM 5.6 $76 + 59 = 135$.

Proof:

1. $76 + 59 = (7 \cdot 10 + 6) + (5 \cdot 10 + 9)$
2. $= (7 \cdot 10 + 5 \cdot 10) + (6 + 9)$
3. $= (7 + 5) \cdot 10 + (6 + 9)$
4. $= 12 \cdot 10 + 15$
5. $= (1 \cdot 10 + 2) \cdot 10 + (1 \cdot 10 + 5)$ by prin. H-A num.
6. $= [(1 \cdot 10) \cdot (10) + 2 \cdot 10] + (1 \cdot 10 + 5)$ by distr.
7. $= (1 \cdot 10^2 + 2 \cdot 10) + (1 \cdot 10 + 5)$ by assoc. ad. and def. exp.
8. $= 1 \cdot 10^2 + (2 \cdot 10 + 1 \cdot 10) + 5$ by gen. assoc. mult.
9. $= 1 \cdot 10^2 + (2 + 1) \cdot 10 + 5$
10. $= 1 \cdot 10^2 + 3 \cdot 10 + 5$
11. $= 135$

In multiplication we also have a useful algorithm. In computing the product of 32 and 27, for example, we arrange our work vertically, multiplying single digits and carrying where necessary. As part of the algorithm, we are taught to *indent* the second 4 (from $2 \cdot 2$), rather than placing it under the 4 in the first partial product.

```
   32
 × 27
 ----
  224
 ..4
 ----
 ....
```

Can you explain why? Again, a careful justification of the algorithm on the basis of the properties of whole numbers requires a lengthy proof. We assume, of course, both addition and multiplication tables up to $9 + 9$ and 9×9. Here is the proof, with most reasons omitted:

THEOREM 5.7 $32 \cdot 27 = 864$.

Proof:

1. $32 \cdot 27 = (3 \cdot 10 + 2) \cdot (2 \cdot 10 + 7)$
2. $= (3 \cdot 10)(2 \cdot 10) + (3 \cdot 10) \cdot 7 + 2 \cdot (2 \cdot 10) + 2 \cdot 7$ by binomial expansion (Theorem 4.22).
3. $= (3 \cdot 2)(10 \cdot 10) + (3 \cdot 7) \cdot 10 + (2 \cdot 2) \cdot 10 + 2 \cdot 7$ comm. and assoc. mult.
4. $= 6 \cdot 10^2 + 21 \cdot 10 + 4 \cdot 10 + 14$ def. exp. and mult. table
5. $= 6 \cdot 10^2 + (2 \cdot 10 + 1) \cdot 10 + 4 \cdot 10 + (1 \cdot 10 + 4)$
6. $= 6 \cdot 10^2 + (2 \cdot 10^2 + 1 \cdot 10) + 4 \cdot 10 + (1 \cdot 10 + 4)$
7. $= (6 \cdot 10^2 + 2 \cdot 10^2) + (1 \cdot 10 + 4 \cdot 10 + 1 \cdot 10) + 4$ gen. assoc. mult.
8. $= (6 + 2) \cdot 10^2 + (1 + 4 + 1) \cdot 10 + 4$ gen. distr.
9. $= 8 \cdot 10^2 + 6 \cdot 10 + 4$
10. $= 864$

It would be possible to extend the analysis to problems having three or more digits: but no new learning would be gained and the length would be forbidding and boring. To analyze the algorithms for subtraction and division would be more profitable. Since, however, it is not worthwhile to dwell on computation techniques in a text stressing the ideas of mathematics, we shall be content to introduce a starred exercise on subtraction and encourage you to create, for all the algorithms you meet, at least an informal explanation of their workings. Division, the most difficult of the elementary operations, would benefit most from such an analysis. The fact that we are omitting it here should not be taken as an indication of its lack of importance.

EXERCISES 5.3

1. Explain informally, using principles of the Hindu-Arabic numeration system, why the boldface digits are placed where they are in the following:

$$\begin{array}{r} 431 \\ \times\ 152 \\ \hline 862 \\ 215\mathbf{5} \\ 43\mathbf{1} \\ \hline 65512 \end{array} \qquad \begin{array}{r} 347 \\ \times\ 206 \\ \hline 2082 \\ \mathbf{694} \\ \hline 71482 \end{array}$$

2. Supply the omitted reasons in the proofs of Theorem 5.6 and Theorem 5.7, as given in the text.

3. Supply abbreviated proofs, as in the text, for

   (a) $35 + 24 = 59$ (b) $27 + 49 = 76$

   (c) $26 \cdot 5 = 130$ (d) $43 \cdot 25 = 1{,}075$

*4. Illustrate numerically, then prove the following:

THEOREM 5.8 If $a, b, c, d \in W$ and if $a > c$ and $b > d$, then $(a + b) - (c + d) = (a - c) + (b - d)$.

*5. Complete the proof of $63 - 21 = 42$, assuming the subtraction theorems from $0 - 0$ to $9 - 9$.

Proof: $63 - 21 = (6 \cdot 10 + 3) - (2 \cdot 10 + 1)$
$= (6 - 2) \cdot 10 + (3 - 1)$ by Theorem 5.8.
$=$ etc.

*6. Prove that $71 - 48 = 23$. [*Hint:* At some stage it will be necessary to rewrite $(7 \cdot 10 + 1)$ as $(6 \cdot 10 + 11)$. Why?]

*7. Solving *cryptarithms* is an enjoyable pastime which involves simple but careful reasoning. A cryptarithm is an arithmetic problem in which the digits are disguised as letters. An old example is the following sum (said to be a message from an impecunious college student to his parent):

```
  S E N D
  M O R E
---------
M O N E Y
```

The usual rule is that each letter represents one and only one digit. Thus if E should represent 5, then it would have that value wherever it occurred, and no other letter could have that value. Moreover, since M is an initial letter, we may assume $M \neq 0$; similarly, $S \neq 0$. Cryptarithms can be solved on two levels. As a minimum, you should find one solution: a set of numerical equivalents for the letters which yields a correct sum. As a more advanced exercise, you may find *all possible* solutions, or prove that the solution you first found is unique. Below are a few cryptarithms, all in base-ten. For further exercises, many more challenging, see the Brooke reference.

(a)
```
    S E N D
  + M O R E
  ---------
  M O N E Y
```

(b)
```
      T E N
    + T E N
  F O R T Y
  ---------
  S I X T Y
```

(c)
```
      A
    G O
    G O
  G A L
  -------
  L O O K
```
[8]

[8]J. A. H. Hunter, *Mathematics Magazine*, 43 (September, 1970), p. 225, Problem 768.

(d) Multiplication

L	Y	N	D	O	N
					B
J O	H	N	S	O	N[9]

(e) Multiplication

S	P	I	R	O
				T
A	G	N	E	W[10]

Section 5.4 Nondecimal Numeration Systems

Let us consider an easy example in addition, in expanded and condensed forms:

Example 1

$$\begin{array}{r} 2(10) + 6 \\ + \; 1(10) + 5 \\ \hline 3(10) + 1(10) + 1 \\ \text{or} \\ 4(10) + 1 \end{array} \qquad \begin{array}{r} 26 \\ +\;15 \\ \hline 41 \end{array}$$

We may compare with this addition example the one below, also in several forms:

Example 2

$$\begin{array}{c} \text{2 weeks 6 days} \\ + \text{ 1 week 5 days} \\ \hline \text{3 weeks + (1 wk + 4 da)} \\ \text{or} \\ \text{4 weeks 4 days} \end{array} \qquad \begin{array}{r} 2(7) + 6 \\ + \; 1(7) + 5 \\ \hline 3(7) + 1(7) + 4 \\ \text{or} \\ 4(7) + 4 \end{array} \qquad \begin{array}{r} 26 \\ +\;15 \\ \hline 44 \end{array}$$

Note that the same work is done in all three forms, though more explicitly written out in the first two. Note also that the same procedure is used in the second example, involving time units, as in the first, except that the "regrouping" or "carrying" is by sevens instead of by tens, since there are only seven days in a week; also, the left column represents sevens instead of tens. In the condensed form the two examples appear not to differ except in the right digits of their answers; but actually the left digits differ in *meaning*. The difference consists precisely in the fact that, in the first example, we are using decimal or base-*ten* numeration, whereas in the second example we are using septimal or base-*seven* numeration.

---

[9]Martin Gardner. *The Numerology of Dr. Matrix* (New York: Simon and Schuster, 1967), pp. 76, 108.

[10]Martin Gardner. "Mathematical Games," *Scientific American*, 221:4 (October, 1969), p. 127.

Here are two other familiar examples:

Example 3 (Multiplication)

2 ft + 5 in.	2(12) + 5	25
× 4	× 4	× 4
8 ft + (1 ft + 8 in.)	8(12) + 1(12) + 8	98
or	or	
9 ft + 8 in.	9(12) + 8	

Example 4 (Subtraction)[11]

3 gal + 1 qt + 2 c	$3(4^2) + 1(4) + 2$	312
− 1 gal + 2 qt + 1 c	$- 1(4^2) + 2(4) + 1$	− 121
[change 1 gal to 4 qt]	[change $1(4^2)$ to 4(4)]	
2 gal + 5 qt + 2 c	$2(4^2) + 5(4) + 2$	252
− 1 gal + 2 qt + 1 c	$- 1(4^2) + 2(4) + 1$	− 121
1 gal + 3 qt + 1 c	$1(4^2) + 3(4) + 1$	131

The preceding examples are not important in themselves; but they serve as an introduction to calculations in bases other than ten. Why should we wish to calculate in other bases? First, a modest amount of time spent in calculating in unfamiliar bases will serve to reinforce and call attention to the principles of our algorithms, which have become so automatic in base-ten that we no longer are aware of them. Second, two nondecimal systems are of special interest: the duodecimal (base-twelve) system is used in counting by dozens and has some advantages over the decimal system; and the binary (base-two) system and its derivatives are used in digital computer design, where they have great superiority over the decimal system.

Now that we have already added in base-seven, multiplied in base-twelve, and subtracted in base-four, the rest ought to be easy. There are three processes to be understood:

1. Conversion from any base to base-ten, and vice versa.
2. Counting in any base.
3. Performing the four arithmetic operations in any base.

Conversion of a numeral *from* a nondecimal base to base-ten is simple. For example, consider the numeral 174_{twelve} (174 in the base-twelve system).

Reading from right to left, the 4 represents units, the 7 represents "twelves," or "dozens," the 1 represents "twelve times twelve" or "$(\text{twelve})^2$" or "gross."

---

[11]Remark to noncooks: 4 cups = 1 quart; 4 quarts = 1 gallon.

Thus

$$174_{\text{twelve}} = 1 \text{ gross} + 7 \text{ dozen} + 4 \text{ units} = 1(12^2) + 7(12) + 4$$
$$= 1(144) + 84 + 4 = 232_{\text{ten}}$$ [12]

Generalizing from the above example, the method of conversion *to* base-ten consists merely in writing the numeral in expanded notation (using the appropriate base) and then performing the indicated operations of multiplication and addition in base-ten to get a single sum as a result. Other exercises to try are:

$$2431_{\text{five}} \quad \text{and} \quad 11011_{\text{two}}$$

Does 736_{six} make sense as a numeral? The answers are: 366; 27; No, because base-six notation uses only the digits 0, 1, 2, 3, 4, 5.

The last answer will become clearer if we attempt to count in the different bases. For example, in base-five the count goes 1, 2, 3, 4, *then* 10. Why does 10_{five} represent the number "five" in our base-ten system? The 10 means "one group of five, and no units." To continue the count, after 10 comes 11 ("a group of five, and one unit"), 12, 13, 14, *then* 20 ("two groups of five"). Continuing again (you should fill in for the dots), 21, 22, . . . , 40, 41, 42, 43, 44, the last numeral being equivalent to 24_{ten}. The next count, 44 + 1, is 100. You should check that 100_{five}, when converted to base-ten, yields the correct value. Continuing the count, eventually reaching 144, what is the next numeral? What is the next one after 444?

Let us now start counting in base-twelve: 1, 2, 3, 4, 5, 6, 7, 8, 9. Is the next numeral 10_{twelve}? No, for 10_{twelve} means "1 dozen and no units," which is not the next number after "nine." Note that the first two-digit numeral, in *any* base, is 10. Now *ten* and *eleven*, being smaller than a dozen (10_{twelve}) must be represented by *single* digits. So we invent the symbols T and E, respectively, for these digits. Now our count proceeds: 1, 2, 3, 4, 5, 6, 7, 8, 9, T, E, 10, 11, 12, . . . , 18, 19 (Are we ready yet for 20?), 1T, 1E, *now* 20 (Is this really the count of "two dozen"?), 21, ...

You should write in base-twelve the number which follows 29; 2E; 7E; 9E; EE; EEE. The answers are: 2T; 30; 80; T0; 100; 1000. The last two numerals, in fact, have names: 100 is a *gross*; 1000 is a *great gross*.

Let us now turn to a much smaller base, for which new symbols need not be invented. In counting in the base-two, or *binary*, system, we realize that only two digits may be used: 0 and 1. Thus the count goes 1, 10 ("a pair and no units"), 11, 100 ("a quadruple, no pairs, no units"), 101. . . . You should continue with the next ten numerals. The answers are: 110, 111, 1000, 1001, 1010, 1011, 1100, 1101, 1110, 1111. What is the next numeral? It is 10000.

We have discussed conversion in the direction *from* other bases *to* base-ten notation. Conversion in the other direction is almost as simple in practice,

---

[12]Hereafter, when base-ten is meant, we shall not generally mention the base.

but requires the exercise of reason to be understood. Can you *reason out*, before continuing, the base-five numeral which corresponds to 347_{ten}?

One method for converting 347 to base-five is to note that we need only fill in blanks in the following equation with appropriate digits from the set 0, 1, 2, 3, 4:

$$347 = \underset{5^3}{\_} \ \underset{5^2}{\_} \ \underset{5}{\_} \ \underset{u}{\_}{}_{\text{five}}$$

Note that, for convenience, the blanks are labeled, underneath, with u (for "units"), 5 (the "fives" place), etc. Then we check to see what is the largest power of 5 which we shall need in order to express 347. Since $5^3 = 125 < 347$, but $5^4 = 625 > 347$, it is clear that the blank spaces provided above are just enough for this problem. But how many 5^3's are wholly contained in 347? Since there are 2, we place a 2 in the leftmost blank. Of the value 347, we still have to account for $347 - 250 = 97$. We then ask how many 5^2's are contained in 97. There are 3, since $3 \cdot 5^2 = 75 < 97$, but $4 \cdot 5^2 = 100 > 97$. This leaves $97 - 75 = 22$ unaccounted for; but 22 is 4 fives + 2. So the answer is 2342_{five}.

A second method, quicker and more mechanical, also needs discussion if it is to be well understood. To make it "come alive," suppose we have 347 pennies on the table. Since we are asked to convert 347 to base-five, let us exchange, so far as possible, a nickel for every five pennies. How many piles of 5 pennies each can we get from 347 pennies? Obviously, to find out, we divide:

$$5\overline{)347}$$
$$69 + \text{remainder of 2 pennies}$$

Thus we obtain 69 piles of pennies, or 69 nickels, with 2 pennies left over. This tells us already that the number of *units* in the base-five numeral we seek is 2: for these 2 pennies cannot be grouped by fives.

But we do not stop here. We must now stack the nickels in sets of five and exchange each stack for a quarter. From the 69 nickels, how many stacks of 5 can we make? Again we divide:

$$5\overline{)69}$$
$$13 + \text{remainder of 4 nickels}$$

What does the 13 represent? It represents 13 stacks of 5 nickels each; these are equivalent to 13 quarters. Note that we have 4 nickels left over — not enough to make a full stack: so the "fives"'-place digit is 4.

In the next step, of course, we group the 13 quarters by fives: let us say, we put five such groups (containing altogether the equivalent of 5^3, or 125 pennies) into a small paper bag. Then how many bags will we have? Again we divide:

$$5\overline{)13}$$
$$2 + \text{remainder of 3 quarters}$$

We have now reached the end of the road, since we cannot form from the last quotient, 2, any sets of five. Indeed, further division yields a quotient of 0 and a remainder of 2; but we could predict as much without actually performing the division. Now, looking at the remainders, we see that 347 contains two units, four 5's, three 5^2's, and two 5^3's; so the numeral is 2342_{five}. For efficiency we can speed up the process, performing the divisions and noting the remainders in sequence.

$$\begin{array}{l} 5\overline{)347} \\ 5\overline{)\ 69}, \text{ remainder } 2 \\ 5\overline{)\ 13}, \text{ remainder } 4 \\ 5\overline{)\ \ 2}, \text{ remainder } 3 \\ \quad\ \ 0, \text{ remainder } 2 \end{array} \quad \Bigg\uparrow \text{ (read up)}$$

As a further exercise, you should convert 57 to base-four, using the above-described methods. The answer is 321_{four}.

We have already seen some examples of arithmetic operations in other bases. If it were necessary to do much work in a given base, it would be possible to set up and memorize the basic addition and multiplication tables of that base. Since computational facility is not our goal, however, we shall merely figure out addition and multiplication combinations as we need them. For example, if in a base-twelve addition problem we need 7 + E, we immediately recognize that the numerical value is 18 (in base-ten) and so ask how many *dozen* this value is. The answer is, "one dozen, with 6 units left over": hence $7 + E = 16_{\text{twelve}}$. Similarly, the product $7 \cdot 9$, which we call 63 (in base-ten), is "5 dozen, 3 units" or 53_{twelve}.

Since the commutative, closure, associative, and distributive properties apply to *numbers* — no matter which symbols or numeration schemes are used to represent them — we would expect them to be applicable to any computational algorithms. Moreover, since the numeration systems in other bases also share with the decimal system the principles of addition and place value, the analysis of these algorithms parallels very closely that shown in the preceding section. In fact, the only difference is that the grouping of units is done in terms of the other base, not by tens. For example, the analysis of the base-seven addition problem at the beginning of this section, 26 + 15 = 44, would begin with the expansion of 26 to $2 \cdot 10 + 6$ where *here*, however, the 10 would refer to one *seven* and no units, and the base-seven addition table would yield sums like 6 + 5 = 14.

Because *duodecimal* (base-twelve) arithmetic has special interest and employs two additional symbols, examples will now be given in that base. Thus $1T5_{\text{twelve}} + 476_{\text{twelve}}$, written vertically for convenience, yields the sum:

$$\begin{array}{r} 1\ T\ 5 \\ +\ 4\ 7\ 6 \\ \hline 6\ 5\ E \end{array}$$

The units column does not involve carrying or regrouping, since the sum of 5 and 6 is less than a dozen and can therefore be expressed by the single digit E. In the dozens' column, however, the sum of 10 dozen and 7 dozen is 17 dozen, which is 1-dozen-and-5 dozen: so a 5 is placed in the dozens' column and the 1 dozen-dozen (or one gross) is recorded or remembered as a carry in the third column. Thus this carry, added to the 1 gross and the 4 gross in the third column, produces a sum of 6 gross. We may check the result by converting both addends and the sum to base-ten notation. Thus

$$1T5_{\text{twelve}} = 1 \cdot 12^2 + 10 \cdot 12 + 5 = 144 + 120 + 5 = 269$$

Similarly,

$$476_{\text{twelve}} = 666 \quad \text{and} \quad 65E_{\text{twelve}} = 935$$

We note that $269 + 666 = 935$ is a correct sum in base-ten.

As a subtraction example we take $5T3 - 128$.

$$\begin{array}{r} 5\,T\,3 \\ -\;1\,2\,8 \\ \hline 4\,7\,7 \end{array}$$

Since we cannot subtract 8 units from 3 units to get a nonnegative result, we convert the 10 dozen in the second column of the minuend to 9 dozen and 12 units. Now in the units column we have 12 plus 3, or 15 units, from which we can subtract 8 units to get 7 units as a remainder. The second column now requires subtracting 2 (dozen) from 9 (dozen), which gives 7 (dozen). The third column presents no problem. Thus the result is 477. This difference may be checked either by converting the entire problem to base-ten or by applying the definition of subtraction, adding (in base-twelve!) the difference to the subtrahend to get the minuend. You should make both checks for practice.

The following is an example of multiplying in base-twelve:

$$\begin{array}{r} 3\,E\,6 \\ \times \quad 4\,9 \\ \hline 2\,E\,7\,6 \\ 1\,3\,T\,0 \\ \hline 1\,6\,9\,7\,6 \end{array}$$

In multiplying 9 units by 6 units, we ask how many dozen are contained in 54. Since the answer is 4 dozen and 6 units, we record the 6 units in the first partial product and remember the 4 dozen so as to add it to the product of 9 units and 11 dozen. Since the latter product is 99 dozen, or 8 gross and 3 dozen, adding the remembered 4 dozen yields 8 gross and 7 dozen so far in the partial product. So we record the 7 dozen in the second column and remember the 8 gross to combine it with the next product, 9 units by 3 gross, to get a total of 35 gross, or 2 great-gross, and 11 gross, and thus complete the first partial product. The second partial product is completed in the same way. Finally, the two partial products are added in base-twelve to get the answer. The work may be checked by conversion to base-ten, yielding $570 \times 57 = 32{,}490$.

Since we have as yet provided no example of division, we shall start with a simple one in base-five, namely, $3{,}023 \div 4$. Because it will be necessary to ask how many times 4 is contained in various portions of the dividend, a base-five table of products of 4 will be helpful:

$$\begin{array}{l} 4\cdot 0 = 0 \\ 4\cdot 1 = 4 \\ 4\cdot 2 = 13 \\ 4\cdot 3 = 22 \\ 4\cdot 4 = 31 \end{array} \qquad \begin{array}{rcccc} & & 3 & & \\ 4) & \overline{3} & \overline{0} & \overline{2} & \overline{3} \\ & 2 & 2 & & \\ \cline{2-3} & & 3 & 2 & \end{array} \qquad \begin{array}{rcccc} & & 3 & 4 & 2 \\ 4) & \overline{3} & \overline{0} & \overline{2} & \overline{3} \\ & 2 & 2 & & \\ \cline{2-3} & & 3 & 2 & \\ & & 3 & 1 & \\ \cline{3-4} & & & 1 & 3 \\ & & & 1 & 3 \\ \cline{4-5} & & & & 0 \end{array}$$

Using this table, we note that 4 is contained in 3 zero times, but in 3 *fours* and no units (30) three times, since $4\cdot 3 = 22$. Writing down 22 and subtracting it (in base-five) from 30, we obtain 3. (Of course, we have really been stating that 4 is contained $3\cdot 5^2$ times in $3\cdot 5^3 + 2\cdot 5 + 3$, plus remainder; but correct placement of the digits automatically records this without further attention on our part.) Continuing, we find that 4 is contained in 32 four times, etc., yielding the completed solution shown. It may be checked by conversion to base-ten or by multiplying 342_{five} by 4_{five} and adding the remainder (0 here).

We present without comment another example in base-five,

$$\begin{array}{rccccc} & & & 2 & 1 & 3 \\ 41) & \overline{1} & \overline{4} & \overline{3} & \overline{3} & \overline{3} \\ & 1 & 3 & 2 & & \\ \cline{2-4} & & 1 & 1 & 3 & \\ & & & 4 & 1 & \\ \cline{3-5} & & & 2 & 2 & 3 \\ & & & 2 & 2 & 3 \\ \cline{4-6} & & & & & 0 \end{array}$$

this one with two digits in the divisor. We omit the (possibly helpful) construction of a table of multiples of 41.

This section closes with discussions of some special advantages of the binary (base-two) and duodecimal systems. The former, and its powers, the octal (base-eight) and hexadecimal (base-sixteen) systems, are used in computers. In fact, their advantages are so great that it is far better to build a computer operating in binary and include a routine for converting between decimal and binary than to attempt to build one that operates directly in decimal notation, as does a desk calculator, and avoids the double conversion. Let us see why. The most important reason is that the binary system has only two values, which can be represented electrically by tubes or transistors which are "on" (1) or "off" (0). Thus the current either flows in a transistor, or it does not — two states which are readily distinguished. (One can imagine what might happen if the ten digits of the decimal system were represented electrically by ten different strengths of electric current: a change in the voltage of the electrical supply might change the values of the digits!) Another advantage of the binary system appears in multiplication. In base-ten multiplication, one may obtain carries from 0 to 8, inclusive: thus $2 \cdot 4 = 8$ (no carry), but $9 \cdot 9 = 1$ (with a carry of 8). Carries enormously complicate the mechanics of computers. But in the binary system, multiplication never produces a carry, since the largest product is $1 \cdot 1 = 1$ (no carry). This makes multiplication in binary very easy, as is shown in:

$$\begin{array}{r} 1\,1\,1 \\ \times\,1\,0\,1 \\ \hline 1\,1\,1 \\ 1\,1\,1 \\ \hline 1\,0\,0\,0\,1\,1 \end{array}$$

To get a partial product with the multiplier 1, we merely copy down the multiplicand in the appropriate position; we omit partial products with the multiplier 0.

One unexpected advantage of the binary system requires lengthier discussion. Despite the number of digits required to express large numbers in base-two, this system is actually more efficient in terms of "machine positions" than is our decimal system. A desk adding machine, for example, can express with only two digits any number from 0 to 99, inclusive. Mechanically, this requires two wheels, each containing the ten digits from 0 to 9. Altogether, therefore, there are 2 times 10, or 20, different machine positions needed to record numbers under a hundred. Now, $99_{\text{ten}} = 1100011_{\text{two}}$; so 99_{ten} uses seven places in binary notation. But since each binary place is in one of only two states (0 and 1), the total number of machine states is $7 \cdot 2 = 14$, which is considerably less than 20. Moreover, the largest binary seven-digit number is $1111111_{\text{two}} = 127_{\text{ten}}$. Thus 14 binary states suffice for numbers from 0 to 127, inclusive, whereas 20 decimal states are needed to represent numbers up to only 99. Actually, it can be proved that base-three is slightly more efficient, in regard to machine states, than is base-two;[13] but the other advantages of the binary system far outweigh this slight advantage for base-three.

[13]Theoretically, a base of $e = 2.71828\ldots$ is the most efficient; and 3 is the nearest *integer* to e.

What of the duodecimal system? It has enough advantages, particularly in regard to handling fractional parts, that it is actually used, in effect, in some of our systems of measurement. Thus 12 inches make a foot; $60 = 12 \cdot 5$ minutes make an hour (and seconds, a minute); 60 minutes make a degree (and seconds, a minute); 12 *Troy* ounces make a pound; 12 months make a year; 12 shillings of English money made a pound until Britain's recent conversion to decimal money. The metric system, of course, uses units related by powers of ten; it was invented precisely *because* our numeral system is a decimal one. But if we were to convert to a duodecimal numeral system, it would be possible to set up a system of physical units of measurement based throughout on twelve, and so just as efficient within the new system as the metric system is within ours.

One interesting advantage of the duodecimal system emerges when we compare the decimal and the duodecimal equivalents of common fractions. These expressions are obtained, of course, by dividing (in the appropriate base) the numerator of the fraction by its denominator. The results are given in Table 5.2.

TABLE 5.2

Common Fraction	*Decimal Fraction*	*Duodecimal Fraction*
1/2	.5	.6 (6 twelfths = 1/2)
1/3	.333 . . .	.4
1/4	.25	.3
1/5	.2	.24972497 . . .
1/6	.1666 . . .	.2
1/7	Horrible in either system!	
1/8	.125	.16
1/9	.111 . . .	.14

We note that the duodecimal column has a larger proportion of *terminating* decimals, and those that terminate usually do so in fewer digits. The fact that fifths (and tenths) are awkward in the duodecimal system is of little consequence; for these fractions would scarcely occur in practice, except that our decimal system forces us to use them. In fact, the advantages of a duodecimal system have seemed so compelling to some that an organization, the Duodecimal Society of America,[14] has been formed with the specific aim of promoting such a change in our system of numeration. Such, however, is human inertia, and so costly would be a change of this magnitude, that little is likely to come of these efforts. We should have been born with six fingers on each hand!

[14]Duodecimal Society of America, 20 Carleton Place, Staten Island, N.Y.

EXERCISES 5.4

1. Beginning with 1, write the numerals for sixteen consecutive whole numbers in each of the following bases:

   ten eight twelve eleven five four two

2. In base-ten, the largest two-digit numeral is 99; in base-eight it is 77_{eight}, and the next consecutive base-eight numeral is 100_{eight}, which is equivalent to 64_{ten}. For each of the following bases, write the largest two-digit numeral, the next numeral in succession, and the latter's base-ten equivalent:

   five, twelve, three, eleven, two, b (where b is any whole number greater than 1, and the digits used are, in order, 0,1,2, . . .)

3. IBM is using a *hexadecimal* (base-sixteen) numeral system.
   (a) How many different symbols are needed to express numbers in this system?
   (b) The symbols used by IBM are 0,1,2, . . . , 9,A,B, . . . , F. Thus A is the single-digit numeral for "ten," "B" for "eleven," etc. Beginning with 1, write the numerals for twenty consecutive whole numbers in hexadecimal notation.
   (c) Work Problem 2 in the hexadecimal system.
   (d) How many hexadecimal digits are needed to express the number one thousand (1000_{ten})? Explain your answer.

4. Write each of the following numerals in expanded notation: $2134_{ten} = 2 \cdot 10^3 + \ldots$ etc.;

   631_{eight} 4213_{five} $15TE_{twelve}$ 12121_{three} 1101011_{two}

5. Convert each of the numerals, except the first, of Problem 4 to base-ten notation.

6. Convert each of the following numerals expressed in base-ten notation to each of the systems mentioned in Problem 1:

   35 209 1972

7. (a) Construct an addition table for the base-five system, showing all sums from $0 + 0$ to $4 + 4$.
   (b) Use this table to add 123_{five} and 341_{five}.
   (c) Check your answer to part (b) by converting all numbers to base-ten notation.

8. Adapt Problem 7 to the binary system, then answer it, using in part (b) the sum of 11110_{two} and 1011_{two}.

9. Perform, in the bases given, the following operations:

(a) *base-five*

$$\begin{array}{r} 1220 \\ +\ 3124 \\ \hline \end{array}$$

(b) *base-five*

$$\begin{array}{r} 1302 \\ +\ 3414 \\ \hline \end{array}$$

(c) *base-seven*

$$\begin{array}{r} 214 \\ +\ 625 \\ \hline \end{array}$$

(d) *base-twelve*

$$\begin{array}{r} 20E3 \\ +\ 952T \\ \hline \end{array}$$

(e) *base-two*

$$\begin{array}{r} 1010111 \\ +\ 1100111 \\ \hline \end{array}$$

(f) *base-five*

$$\begin{array}{r} 3412 \\ -\ 2401 \\ \hline \end{array}$$

(g) *base-five*

$$\begin{array}{r} 2023 \\ -\ 1414 \\ \hline \end{array}$$

(h) *base-eight*

$$\begin{array}{r} 10000 \\ -\ \ 6201 \\ \hline \end{array}$$

(i) *base-twelve*

$$\begin{array}{r} 310T \\ -\ 1E45 \\ \hline \end{array}$$

(j) *base-two*

$$\begin{array}{r} 1000101 \\ -\ \ 101011 \\ \hline \end{array}$$

10. (a) Construct a multiplication table for the binary system.

(b) Use it to compute the following products:

(i)
$$\begin{array}{r} 1011 \\ \times\ 101 \\ \hline \end{array}$$

(ii)
$$\begin{array}{r} 1101011 \\ \times\ \ 11010 \\ \hline \end{array}$$

(c) Check your answers in part (b) by converting to base-ten.

11. Compute the following products, working in the bases given:

(a) *base-five*

$$\begin{array}{r} 2143 \\ \times\ \ \ 2 \\ \hline \end{array}$$

(b) *base-five*

$$\begin{array}{r} 3014 \\ \times\ 104 \\ \hline \end{array}$$

(c) *base-eight*

$$\begin{array}{r} 3725 \\ \times\ 364 \\ \hline \end{array}$$

(d) *base-twelve*

$$\begin{array}{r} 2E59 \\ \times\ \ T4 \\ \hline \end{array}$$

12. Compute the following quotients, working in the bases given:

(a) *base-five*

$432 \div 3$

(b) *base-eight*

$2570 \div 4$

(c) *base-twelve*

$3896 \div T$

(d) *base-five*

$34103 \div 14$

13. Analyze the algorithms for addition in the manner of Theorem 5.5, thus proving the following theorems:

    (a) base-twelve: 36 + 54 = 8T (b) base-five: 42 + 31 = 123

*14. In which base is 297 a factor of 792?[15]

*15. (a) Convert 1/5 to a duodecimal numeral by dividing 1 by 5 in base-twelve.

    (b) Similarly convert 1/9, 1/10, and 1/11 to duodecimal form.

    (c) Compare the *duodecimal* result of 1/11 with the *decimal* result for 1/9. Try also 2/11 in base-twelve and 2/9 in base-ten.

    (d) Similarly try 1/5 and 2/5 in base-six.

    (e) Can you generalize the above results to *any* base ($b \in W$ and $b > 1$)?

*16. In base-ten a numeral represents an even number iff the last digit is even (0,2,4,6,8).

    (a) Discover whether this statement also holds in base-twelve, in base-five, and in several other bases. Explain.

    (b) Write a general statement describing the connection, if any, between the property of a number's being even and the last digit of its representation in base-*b*.

*17. In which base was each of the following sums performed? Could there be more than one answer? Explain.

    (a) 64 + 13 = 110 (b) 2T1 + 356 = 647

    (c) 851 + 326 = 1277 (d) 251 + 326 = 577

*18. (a) Convert 327_{eight} to base-ten, then the result to base-two.

    (b) Examine part (a) and its answers to see if you can find a method of converting from base-eight directly to base-two, without going through base-ten.

    (c) Use the method of part (b) to convert 635_{eight} directly to base-two.

*19. Cryptarithms occur in bases other than ten. Try the following two examples:

    (a) Base-six

```
  A G O
+   G O
-------
  R U M
```
[16]

---

[15] *Mathematics Magazine*, 38 (March, 1965), p. 124, Problem Q358.

[16] Charles W. Trigg, *Mathematics Magazine*, 39 (May, 1966), p. 187, Problem 622.

(b)

```
     S E N D
   + M O R E
   ---------
   M O N E Y
```
[17]

(i) Find three different solutions of this cryptarithm in base-eleven;

(ii) Prove that there is no solution in base-nine.

Suggestions for Investigation

1. A simple number-guessing game, based on a topic in this chapter, utilizes four cards containing the numbers given below:

Card #1		Card #2		Card #3		Card #4	
1	3	2	3	4	5	8	9
5	7	6	7	6	7	10	11
9	11	10	11	12	13	12	13
13	15	14	15	14	15	14	15

To play the game, ask another person to think of a whole number between 1 and 15, inclusive. Without telling you what the number is, the victim is to examine the four cards and hand to you just the cards on which the number appears. After making appropriate magical incantations over these cards, you announce the number he or she was thinking of. This "magic" is accomplished simply by adding the numbers which appear in the upper left corners of the cards handed to you.

(a) Explain why the above "magic" works, using a topic in this chapter.

(b) By the introduction of two more cards and correct placement of the additional numbers on all six cards, this game can be extended to numbers from 1 to 63, inclusive. Which numbers will appear on each of the six cards? The game can obviously be still further extended.

2. The game of Nim is simple to play; moreover, its outcome is completely determined if both players use a precise strategy based on the binary number system. Learn how to play Nim and try it with other persons. One reference is B. W. Jones (pp. 74–77) in the bibliography.

3. It is possible to use numeration systems based on negative integers. How would this be done, and what are some of the consequences? See the Jordan, Nelson, and Twaddle references. Could fractional bases be used?

---

[17]Adapted from E. P. McCravy, *Mathematics Magazine*, 44 (March, 1971), p. 105, Problem 789.

4. This chapter has some interesting historical background. You may wish to explore the following topics, among others:
   (a) The history of numeration.
   (b) Simon Stevin and the invention of decimals.
   (c) The abacus: its history, forms, and operation.
   (d) The history of computers.
5. Exercises 2.3, Problem *8, leads to a formula for the number of subsets of a set U having n elements. Look up the neat proof of this formula, using the binary system, in the Lindstrom reference.

Bibliography

Andrews, F. E. *New Numbers.* New York: Harcourt, Brace and Co., 1935.

Bidwell, J. K. "Mayan Arithmetic." *Mathematics Teacher*, LX (1967), 762–68.

Boyer, C. B. *A History of Mathematics.* New York: John Wiley & Sons, Inc., 1968.

Brooke, Maxey. *One Hundred Fifty Puzzles in Crypt-arithmetic*, 2nd rev. ed. New York: Dover Publications, Inc., 1969.

Cajori, Florian. *A History of Mathematical Notations*, 2 vols. LaSalle, Ill.: Open Court Publishing Co., 1928–29.

Davis, H. T., et al. "The History of Computation." *Thirty-First Yearbook.* Washington: National Council of Teachers of Mathematics, 1969. Chapter 3.

Duodecimal Society of America, 20 Carlton Place, Staten Island, N.Y.

Eves, Howard. *An Introduction to the History of Mathematics*, 3rd ed. New York: Holt, Rinehart and Winston, 1969.

Gardner, Martin. "How To Turn a Chessboard Into a Computer and To Calculate with Negabinary Numbers." *Scientific American*, 228:4 (Apr., 1973), 106–11.

Glaser, Anton. *History of Binary and Other Non-decimal Bases.* Southampton, Pa.: Anton Glaser, 1237 Whitney Road, 1971.

Gross, H. I. "A Problem, a Solution and Some Commentary." *Mathematics Teacher*, LXIV (1971), 221–24.

Gundlach, B. H., et al. "The History of Numbers and Numerals." *Thirty-First Yearbook.* Washington: National Council of Teachers of Mathematics, 1969. Chapter 2.

Honsberger, Ross. *Ingenuity in Mathematics.* New York: Random House, Inc., 1970.

Jones, B. W. *Elementary Concepts of Mathematics*, 3rd ed. New York: The Macmillan Co., 1970.

Jordan, J. H. "Small-digit Representation of Real Numbers." *Mathematics Teacher*, LXI (1968), 36–38.

Kovach, L. D. "Ancient Algorithms Adapted to Modern Computers." *Mathematics Magazine*, 37 (1964), 159–65.

Lindstrom, P. A. "The Number of Subsets of a Finite Set." *New York State Mathematics Teachers' Journal*, 21:2 (1971), 75.

Lloyd, D. B. "Recent Evidences of Primeval Mathematics." *Mathematics Teacher*, LVIII (1965), 720–23.

National Council of Teachers of Mathematics. "Numeration Systems for the Whole Numbers." *Twenty-Ninth Yearbook*. Washington: National Council of Teachers of Mathematics, 1964. Chapter 3.

Nelson, A. H. "Investigation to Discovery with a Negative Base." *Mathematics Teacher*, LX (1967), 723–26.

Payne, J. N. "Numeration Systems." *Twenty-Seventh Yearbook*. Washington: National Council of Teachers of Mathematics, 1963. Chapter 17.

Rudd, L. E. "Non-decimal Numeration Systems." *Twenty-Seventh Yearbook*. Washington: National Council of Teachers of Mathematics, 1963. Chapter 2.

Schurrer, A. L., ed. *Numeration*, SMSG *Supplementary and Enrichment Series*, 15. Pasadena, Calif.: A. C. Vroman, Inc., 1965.

Smith, D. E. *A Source Book in Mathematics*. New York: McGraw-Hill Book Co., 1929.

Twaddle, R. D. "A Look at Base Negative Ten." *Mathematics Teacher*, LVI (1963), 88–90.

6

Mathematical Systems

How sure is $2 + 2 = 4$? Are there operations which are not commutative? What are groups and relations, and how do they apply to number systems?

Section 6.1 A Bizarre but Familiar Arithmetic

Consider the following examples of a kind of arithmetic which diverges in the second and third lines below from the result we have been trained to expect:

$$2 + 3 = 5 \qquad 8 - 2 = 6 \qquad 2 \cdot 5 = 10$$

$$9 + 5 = 2 \qquad 2 - 8 = 6 \qquad 4 \cdot 5 = 8$$

$$12 + 12 = 12 \qquad 12 - 12 = 12 \qquad 12 \cdot 3 = 12$$

Perhaps you may recognize that this arithmetic is familiar from elementary school and is used by any traveler who plans a trip including more than half a day. In fact, it is the arithmetic of a twelve-hour clock. Thus, to explain

$9 + 5 = 2$, if we start with the hour hand at the "12" position and move it 9 hours in a clockwise direction, then 5 hours more, we arrive at the "2" position. If we interpret multiplication as repeated addition,[1] $4 \cdot 5$ becomes $5 + 5 + 5 + 5$; starting again at the "12" position, after 4 clockwise motions of 5 hours each, we reach the "8" position. You should verify the remaining equations.

The "clock arithmetic" just described is one example of a *finite* arithmetic system, so called because it uses only a finite set of numbers, here $\{1,2,3,4,5,6,7,8,9,10,11,12\}$. In order to study it conveniently, we may construct addition and multiplication tables, portions of which are given as Table 6.1.

TABLE 6.1

Arithmetic Tables (mod 12)[2]

+	1	2	3	4	...	12
1	2	3	4	5	...	1
2	3	4	5	6	...	2
3	4	5	6	7	...	3
4	5	6	7	8	...	4
...	...	...	...	...	...	...
12	1	2	3	4	...	12

·	1	2	3	4	...	12
1	1	2	3	4	...	12
2	2	4	6	8	...	12
3	3	6	9	12	...	12
4	4	8	12	4	...	12
...	...	...	...	...	...	...
12	12	12	12	12	...	12

The above tables are significant, not because they provide "instant answers" to specific problems, but because a study of them will reveal patterns and properties of the "clock arithmetic," hereafter called the *arithmetic* (*mod 12*).

Several important properties of the arithmetic (mod 12), such as closure, associativity, commutativity, and distributivity, are revealed by Table 6.1 or suggested by exercises. We call attention here to the multiplicative and additive identities. Specifically, for multiplication is there an element u in the system such that $1 \cdot u = 1$, $2 \cdot u = 2$, . . . , $12 \cdot u = 12$; or, in general, $xu = x$ for every number x in the set $\{1,2,3, \ldots 12\}$? The first row and first column of the multiplication table indicate that the element 1 behaves like u in $xu = x$ — i.e., the multiplicative identity of the system is 1.

Turning now to the addition table, we seek an element z such that $1 + z = 1$, $2 + z = 2$, . . . , $12 + z = 12$. Can you determine the value of z, the additive identity of the system? You should get rid of any preconceived notions of what the symbol "ought" to be and focus instead on finding an element z which behaves as $x + z = x$ for every number x.

Logically, there is no reason why a clockface should be divided into twelve, rather than some other number of equal parts. Astronomers and the United

[1]See Theorem 4.20.

[2]*mod* is an abbreviation for *modulo* 'according to the measure of.'

States Navy, for example, number the hours up to twenty-four. Mathematicians, on the other hand, are not satisfied unless they can generalize by dividing the clockface into n parts, where n is any whole number greater than 1. For example, with $n = 5$ we obtain the arithmetic (mod 5), based on Figure 6.1. Usually we label the number at the top 0 rather than n (here, 5), though the exact symbol used is not important. Indeed, we could even choose five nonnumerical symbols to represent the five positions on the dial: e.g., #, $, %, &, and *. Now the reader is invited to construct addition and multiplication tables (mod 5) in the same manner as those (mod 12), and compare the results with Table 6.2. It is easy to pick out the additive and multiplicative identities: the results are hardly surprising.

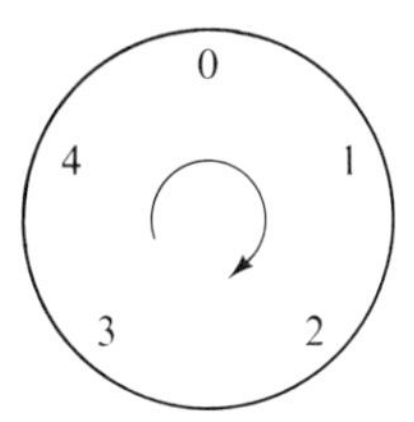

Figure 6.1

The first thing to notice about Table 6.2 is that it contains one and only one entry in each position: in other words, all sums and products are unique elements of the set $\{0,1,2,3,4\}$. An efficient way to summarize these facts is to state that the arithmetic (mod 5) is closed under addition and multiplication.

TABLE 6.2

Arithmetic Tables (mod 5)

+	0	1	2	3	4
0	0	1	2	3	4
1	1	2	3	4	0
2	2	3	4	0	1
3	3	4	0	1	2
4	4	0	1	2	3

·	0	1	2	3	4
0	0	0	0	0	0
1	0	1	2	3	4
2	0	2	4	1	3
3	0	3	1	4	2
4	0	4	3	2	1

For a finite system it is easy to check whether a given operation is commutative by glancing at its table. We note, for example, that $2 + 3 = 0$ (mod 5) and that $3 + 2 = 0$ (mod 5). Instead of tediously checking the sums of every pair of integers, we may note and compare the locations of the sums $a + b$ and $b + a$. The sum $2 + 3$, for example, is recorded in the third row, fourth column, of the addition table; while the sum $3 + 2$ is recorded in the fourth row, third column. If we draw a diagonal line from the upper-left to the lower-right corner in the body of the table, the two sums are symmetrically placed with respect to this diagonal. The same statement can be made for *any* corresponding sums $a + b$ and $b + a$. Now, the two sums will be equal iff the operation is commutative. A glance at the operation table is enough to determine whether each pair of squares that are symmetrically placed with respect to the diagonal has the same contents. To put the matter another way, if, when we fold an operation table along its diagonal, all pairs of squares *and their contents* coincide, then the operation is commutative. Evidently, both addition and multiplication (mod 5) are commutative, because their tables are symmetric about the diagonal.

The task of checking associativity is more complicated. To check each triple of numbers, with both groupings, is of course possible in any finite system; but it is very laborious. In the arithmetic (mod 5) there are 5^3, or 125, triples, each grouped two ways — a discouraging number even if certain short cuts are employed.[3] Fortunately, because all the modular systems are closely related to the system of whole numbers — which we have already proved to be associative under addition and multiplication[4] — it is possible to prove that any arithmetic (mod n) is associative under addition and multiplication. A proof is requested in a later optional exercise.[5] In our subsequent work with modular arithmetics let us accept associativity of addition and multiplication as theorems which we do not take time to prove.

The modular arithmetics illustrate another pair of properties which are very important in many mathematical systems, including arithmetic and high school algebra. These are the *additive inverse* property and the corresponding *multiplicative inverse* property. We illustrate the former, using the addition table (mod 5). We select any element and search for another element in the addition table which will produce with it a sum of 0, the additive identity. Thus if we choose 2 as the first element, we seek an element x' such that $2 + x' = 0$. From the table we see that $2 + (3) = 0$, and that 3 is the only replacement for x' in the equation $2 + x' = 0$ which makes that equation true. So we write $x' = 3$, and call 3 the additive inverse of 2 (mod 5). Similarly, the additive inverse of 4 is 1, since $x' = 1$ satisfies $4 + x' = 0$. More generally, we have

Definition 6.1 Additive Inverse Let S be a system which has a closed, binary operation of addition $(+)$ and an additive identity (0). Then, if x is any element of S, an element $x' \in S$ (if x' exists) such that $x + x' = 0$ or[6] $x' + x = 0$ is called an *additive inverse* of x.

You should note that Definition 6.1 is stated in terms of an abstract system, S, although we intend to apply it, for the present, to arithmetics (mod n).

---

[3]One can, for example, use the identity properties to advantage. See also the Hammel reference.

[4]See Theorem 4.3 and Theorem 4.11.

[5]See Exercises 6.5, Problems *11(c) and (d).

[6]If S is commutative, only one of these two equations is necessary; and this is also true of certain noncommutative systems.

The concept may also be applied to the whole number system, taking $S = W$. If we attempt to do so, for $x = 0$, we obtain $x' = 0$, since $x + x' = 0 + 0 = 0$ is true. For $x = 1$, however, the defining equation for x' becomes $1 + x' = 0$, an equation which has no solution in W.[7] This example makes clear the reason for including the parenthetical phrase, "(if x' exists)," in Definition 6.1. Merely defining a concept does not guarantee its existence. Indeed, the lack of additive inverses in W (except for $0'$) is a deficiency that we will seek to remedy in the next chapter.

It is encouraging to discover, on the other hand, that in the arithmetic (mod 5), *every* element has an additive inverse. We have already seen that $2' = 3$ and $4' = 1$. That $0' = 0$ is obvious: in fact, it is a theorem that, in any system, the additive identity is its own inverse.[8] Continuing, you can easily find $1'$ and $3'$ from the addition table (mod 5), and will discover, in fact, that each element has a *unique* inverse. In fact, it can be proved that in any arithmetic (mod n) — and, indeed, in more general systems called *groups* — an additive inverse exists and is unique.[9] For this reason, we shall henceforth speak of *the* additive inverse of an element.

It is natural to turn to the parallel concept of a multiplicative inverse. The obvious changes from Definition 6.1 are to require that the *product* of a selected element with another element be the *multiplicative* identity, 1. Thus if we select 2 (mod 5), we seek an element x'' such that $2 \cdot x'' = 1$. A glance at the multiplication table (mod 5) shows that $x'' = 3$. Similarly, $4 \cdot x'' = 1$ is satisfied by 4, so that $4'' = 4$. The formal definition is

Definition 6.2 Multiplicative Inverse Let S be a system which has a closed, binary operation of multiplication ($\cdot$) and a multiplicative identity (1). Then, if x is any element of S, an element $x'' \in S$ (if x'' exists) such that $x \cdot x'' = 1$ or[10] $x'' \cdot x = 1$ is called a *multiplicative inverse* of x.

Having found that $2'' = 3$ and $4'' = 4$ (mod 5), we try to complete the task by obtaining $1''$, $3''$, and $0''$. Now $1'' = 1$, as expected (Why?); and $3'' = 2$. But we look in vain for $0''$, since $0 \cdot x'' = 0 \neq 1$, no matter how we choose x''. This result again shows the necessity of inserting the parenthetical phrase "(if x''exists)." Indeed, it indicates that any system for which the zero product theorem[11] holds cannot contain a *multiplicative* inverse for the *additive* identity, 0. Finally, it is evident from the preceding example, and can be proved,[12] that a multiplicative inverse, when it exists, is unique: so we may speak henceforth of *the multiplicative* inverse of an element.

---

[7]Though $^{-}1$ satisfies the equation, it is not a *whole number*.

[8]See Problem *9.

[9]See Exercises 6.2, Problem *9, which applies here if the operation "∘" is interpreted as addition.

[10]See note 6.

[11]For the whole numbers, this is Theorem 4.16.

[12]See again Exercises 6.2, Problem *9.

EXERCISES 6.1

1. Copy and complete Table 6.1.
2. (a) Is the arithmetic (mod 12) closed under addition? Under multiplication? Justify your answers.
   (b) Are addition and multiplication (mod 12) commutative?
   (c) Give a numerical example or counterexample of part (b) for each operation.
   (d) Explain your answers to part (b) in terms of the appearance of the respective operation tables.
3. Give examples (mod 12) illustrating the associative properties of addition and multiplication.
4. For each element of the arithmetic (mod 12) obtain its additive inverse, if it exists; if it does not, so state. (Recall that 12 is the additive identity in Table 6.1.)
5. Answer Problem 4 for multiplicative inverses.
6. In Problems 4 and 5, what is the relation between the additive identity and its additive inverse? The multiplicative identity and its multiplicative inverse?
7. (a) Draw a "clock" for the arithmetic (mod 4); then construct addition and multiplication tables for this system.
   (b) Is the arithmetic (mod 4) closed under addition? Under multiplication? Explain.
   (c) Are these operations (mod 4) commutative? Explain.
   (d) These operations *are* associative. Provide an example for each operation.
   (e) For each element find, if possible, its additive inverse; if not possible, so state.
   (f) Answer part (e) for the operation of multiplication.

*8. Perhaps you have discovered a shorter way of constructing tables for arithmetics (mod n) than moving around clockfaces. If so, describe your method and illustrate it by finding $40 + 45$ (mod 60) and $15 \cdot 10$ (mod 60).

*9. Prove the following:

THEOREM: Let S be a system which has a closed binary operation and contains an identity element for that operation. Then the identity element is its own inverse.

(*Hint:* Consider the operation to be addition and the identity to be 0. Then prove that $0' = 0$. How may your proof be adapted to multiplication? Finish by generalizing to any operation $(\circ)$ and its identity (e)).

Section 6.2 An Introduction to Groups

In the preceding section, we developed operation tables for several of the arithmetics (mod n) and then examined some important properties of these systems. Most of these properties were similar to those we met in set theory and logic: closure, associativity, commutativity, and identity. Three of these properties, together with that of inverses, comprise a new and useful concept, that of a *group*. Though the formal definition is a bit lengthy, the parts are all familiar. It is

Definition 6.3 Group Let G be a nonempty set and $\circ$ be a binary operation defined on G. Then the system $(G,\circ)$ is called a *group* iff each of the following properties holds:

1. The operation $\circ$ is closed in G.
2. The operation $\circ$ is associative in G.
3. There exists an element $e \in G$ such that, for any element $a \in G$, $a \circ e = e \circ a = a$. The element e is called an *identity* for G.
4. For each element $a \in G$ there exists an element $a^* \in G$ such that $a \circ a^* = a^* \circ a = e$. The element a^* is called an *inverse* for a in G.

We recall that Property 1 means that, for any pair of elements $a, b \in G$, $a \circ b$ is a unique element of G. The requirement of Property 2 can be expressed by the equation $(a \circ b) \circ c = a \circ (b \circ c)$, where a, b, c are any three elements of G. The requirements of Properties 3 and 4 are expressed in Definition 6.3 by equations analogous to those of the preceding section.

As an example of Definition 6.3 we choose for the nonempty set, $G = \{0,1,2,3,4\}$. Since a group consists of a set, together with an operation, we must now specify the operation. Suppose we let $\circ$ represent the operation of addition (mod 5). Then the system $(G,\circ)$ has its behavior completely specified by the addition table (mod 5) displayed in the preceding section. In order to claim that this system is a group, we need only verify that Properties 1–4 of Definition 6.3 hold. This we did in the text of the preceding section, where we checked that the system is closed under addition, addition is associative, 0 is the additive identity, and every element has an additive inverse. As a bonus, we note that addition is commutative, though Definition 6.3 does not require this property. The efficiency of the definition of a group is indicated by the fact that all of the discussion of this paragraph can be wrapped up in a single sentence: "The arithmetic (mod 5) forms a commutative group under addition."

Since we have just used the very natural phrase, "commutative group," we may as well write down

Definition 6.4 Commutative Group A group whose operation is commutative is called a *commutative* group, or an *abelian*[13] group.

---

[13]Niels Abel (1802–1829), a Norwegian mathematician, was one of the founders of group theory.

You may be wondering what happens if we take the same set, G, but interpret $\circ$ as multiplication instead of addition. The result is entirely different and quite instructive. This time we must use the multiplication table (mod 5) in checking the four properties of Definition 6.3. Clearly, the first three are satisfied, with 1 as the identity. We earlier found, however, that the element 0 does not have a multiplicative inverse. But Property 4 of Definition 6.3 is adamant: *every* element must have an inverse. So the arithmetic (mod 5) fails to be a group under multiplication, solely because $0''$ does not exist.

Can we perhaps eliminate this frustrating source of failure? Let us try "surgery": let us simply remove the element 0 and see whether the reduced set, $\{1,2,3,4\}$, forms a group under multiplication. The operation table for this new set is obtained from the original multiplication table (mod 5) by merely deleting all references to 0. What remains is Table 6.3. We now test anew each of the four group properties. Table 6.3 exhibits closure, since each entry is one of the numbers in the reduced set; no zeros occur. Obviously, deletion of 0 does not affect the behavior, with respect to associativity (or commutativity), of the remaining elements. The identity, 1, is still present. And now every element, without exception, has a multiplicative inverse, since the only failure has been eliminated. Thus the arithmetic (mod 5), *with zero deleted*, forms a commutative group under multiplication.

TABLE 6.3

$\cdot$	1	2	3	4
1	1	2	3	4
2	2	4	1	3
3	3	1	4	2
4	4	3	2	1

In the examples of groups just drawn from the arithmetic (mod 5), we note that there is *exactly one* identity of each type, additive and multiplicative. Each element of the set, moreover, has only one inverse of each type, if it has any at all. In fact, in Section 6.1, we found that each of the arithmetics (mod n) had unique identities (0 and 1, respectively) and unique inverses (x' and x'', respectively), when they existed. Definition 6.3, on the other hand, speaks vaguely of "*an* identity" for the group and "*an* inverse" for each element. Why the difference in phraseology?

The explanation is that, in stating definitions or axioms in mathematics, we usually try to be as brief and simple as possible and to write down only what is necessary. So we do not claim uniqueness in Definition 6.3, because it is possible later to *prove* uniqueness. Indeed, proofs of the uniqueness of the identity and of inverses are requested in optional exercises. So it is legitimate hereafter to speak always of *the* identity of a group and of *the* inverse of a given element. From now on we shall do so.

Let us return to examples. The arithmetic (mod 4) is interesting to analyze.

In the exercises, you are asked to test this system for group properties under addition. To investigate its behavior under multiplication, we examine Table 6.4. An attempt to find $0''$, the multiplicative inverse of 0, fails, as with (mod 5). So (mod 4) does not form a group under multiplication. Can we "rescue" the system by deleting 0? You should check the closure property for the reduced system, $\{1,2,3\}$, and also seek inverses for 1, 2, and 3.

TABLE 6.4

$\cdot$	0	1	2	3
0	0	0	0	0
1	0	1	2	3
2	0	2	0	2
3	0	3	2	1

We now turn to a familiar system, W, to check whether $(W,+)$ or $(W,\cdot)$ is a group. First, taking addition as the operation, we check Property 4 of Definition 6.3. As expected, 0 has an additive inverse: in fact, $0' = 0$. But what of $1'$, $2'$, . . . ? Clearly, equations such as $1 + x' = 0$, $2 + x' = 0$, etc., have no solutions in W. Moreover, the failure of Property 4 is so sweeping that any attempt at "surgery" would kill the system rather than rescue it. Second, taking multiplication as the operation, you are invited to check for the group properties. Is deletion of any element(s) necessary here? Is it helpful?

Partly because of the difficulties just uncovered with respect to the whole number system, it is instructive to try to modify or enlarge W so that the new system will enjoy all the group properties with respect to both fundamental operations. This effort, which is of great significance in the theory and practical use of number systems, will be carried forward in the next chapter. In the meantime, we must recognize and accept the deficiencies of the whole-number system.

We now produce, to illustrate the generality of the group concept, two examples which appear quite different from all the previous ones. These examples have elements which are not numbers at all, but geometric motions; and the operation, likewise nonnumerical, is "composition" of motions—i.e., merely *following* one motion *by* another.

Example 1 Symmetries of the rectangle.

We consider the rectangle in Figure 6.2, with its four corners numbered as shown. The numbers are only labels used to identify the corners; we do not perform arithmetic operations with them. Since the rectangle may have to be flipped over, we shall also number the corners in the same way on the back side of the rectangle. The rectangle, when cut out of paper or cardboard and placed on a flat surface, such as a blackboard or another piece of paper on a table, will cover a rectangular region of the flat surface. You are advised actually to cut out such a rectangle. The boundaries of this region are shown above as *dashed* line segments surrounding the rectangle: actually, of course, the boundaries of the rectangle and those of the region coincide.

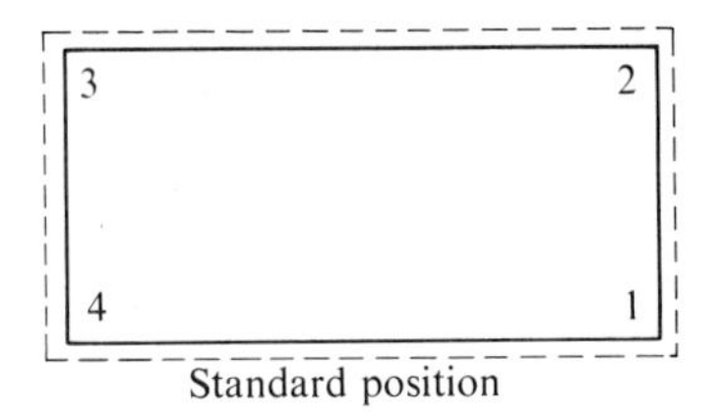

Figure 6.2

The position of the rectangle shown in Figure 6.2 is called the *standard* position. Figure 6.3 shows again the standard position (omitting the dashed segments); in it the numbers are also placed, for reference, outside the rectangle, close to their respective corners.

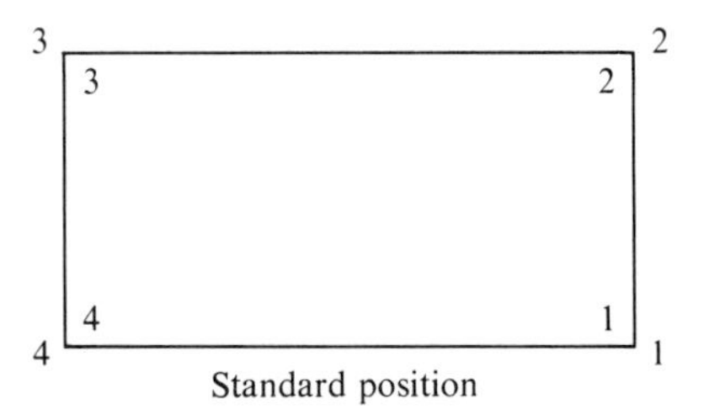

Figure 6.3

We shall now apply certain simple *motions* to the rectangle in such a way that, when a given motion is completed, the rectangle will still be covering the same region of the flat surface; but it will not necessarily have its corners in standard position. The change in position of the corners of the rectangle can be noted by comparing the numbers *on* the rectangle (which moved) with those *outside* the rectangle (which did not move). For example, Figure 6.4 shows

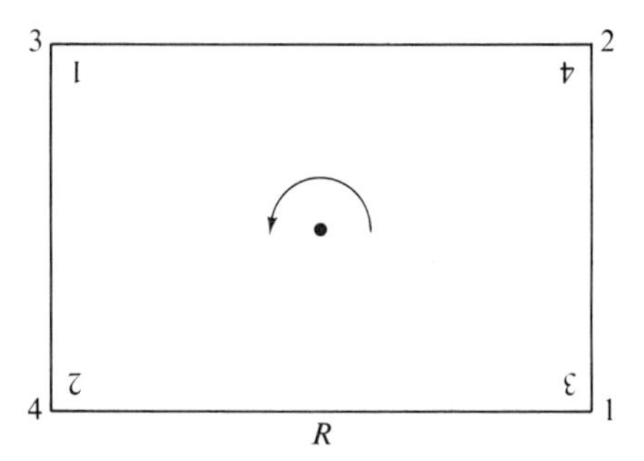

Figure 6.4

the result of a simple rotation of the rectangle 180° about its center. Note that the rectangle still covers the same region; but it is no longer in standard position: for example, Corner 1 has moved to a position opposite its original one.

The 180° rotation is one *element* of the set G, which we hope to show is a group. Let us designate this element by R (for "rotation"); thus

R stands for a 180° *rotation*.

Which other simple motions leave the rectangle covering the same region? Does a 90° rotation (counterclockwise, say) do this? A 360° rotation? Figure 6.5 depicts three more simple motions:

H stands for a flip about the *horizontal* axis of the rectangle.

V stands for a flip about the *vertical* axis of the rectangle.

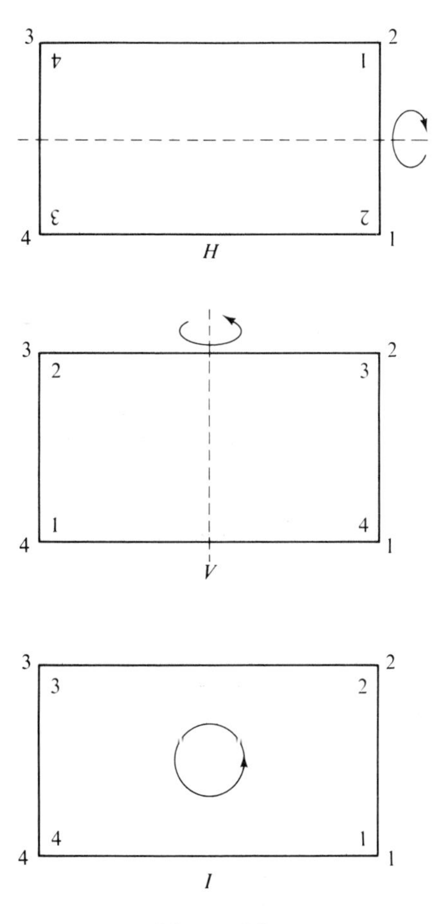

Figure 6.5

Finally, the 360° rotation is an acceptable simple motion which, however, returns the rectangle to standard position. Let us designate this motion by I; thus

I stands for a 360° rotation.

The preceding motions are the only ones which leave the rectangle covering its initial region — they are the *symmetries* of the rectangle.

In sum, the set G has the above four elements: $G = \{R,H,V,I\}$. What, then, is the binary operation? Let us take it as "followed by." Thus $R \circ R$ is an instruction to rotate the rectangle through 180°, and then, continuing from the new position, to perform another rotation of 180°. If we perform these two motions in succession, either mentally or with a cardboard rectangle, we find that the rectangle has been returned to standard position (see Figure 6.6). In other words, the two motions in $R \circ R$ had the same effect on the rectangle as the *single* motion I would have had. Briefly, we have shown that $R \circ R = I$.

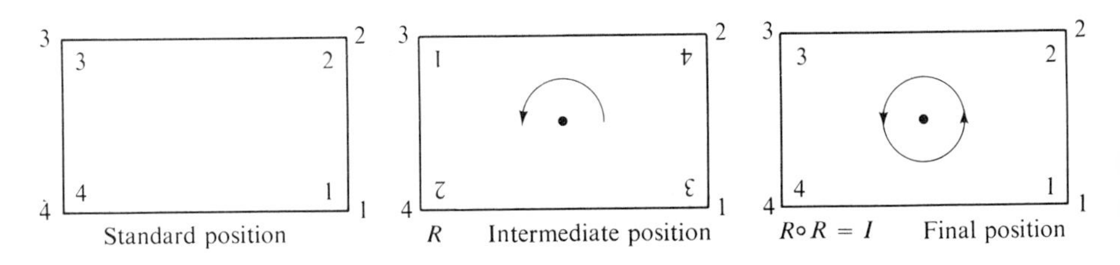

Figure 6.6

Let us now try $R \circ H$. The two stages are shown in Figure 6.7. Note that they leave the rectangle in the same position it would have assumed if the *single* motion V had been applied. Hence we write $R \circ H = V$. Similarly, we may find $R \circ V$, $R \circ I$, and $V \circ R$.

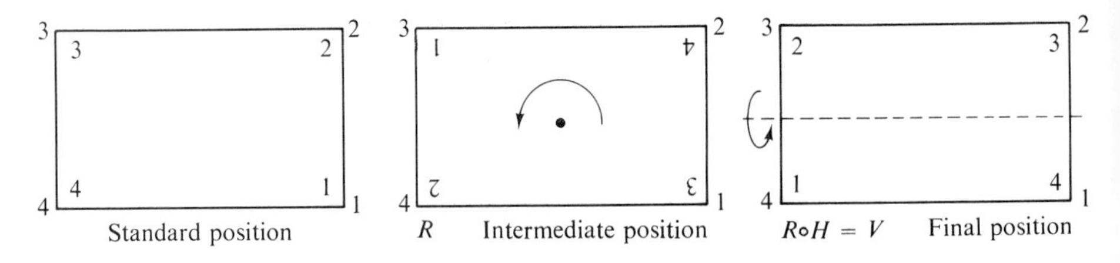

Figure 6.7

We can summarize the results thus far and provide space for further experimentation by setting up an "operation table" for $\circ$, Table 6.5. In entering the table, we shall agree to look up the first element in the leftmost vertical column and the second element in the topmost row. Thus, to record $R \circ H = V$, we find R in the leftmost column (here, R is marked by a single *) and H in the topmost row (here, H is marked by **); we then record the answer, V, at the intersection of the R* row and the H** column: the answer is in boldface type. The completion of this table is left as an exercise. It will turn out that this system satisfies the four properties, or axioms, for a group. So the symmetries of the rectangle, subject to the binary operation "followed by," form a group.

TABLE 6.5

$\circ$	I	R	H**	V
I				
R*	R	I	V	H
H				
V		H		

Example 2 Symmetries of the equilateral triangle.

We label the equilateral triangle in standard position as at the top of Figure 6.8. The elements of G will be the following six motions (see Figure 6.8):

R_1: 120° counterclockwise rotation about the triangle's center

R_2: 240° counterclockwise rotation

I: 360° (or 0°) rotation

F_1: flip about vertical axis (Corner 1 *location* when in standard position)

F_2: flip about axis slanted up to left (Corner 2 location when in standard position)

F_3: flip about axis slanted up to right (Corner 3 location when in standard position)

We set up the operation table as in Table 6.6, so that it may be checked easily. Our next task will be to fill it in. To do so, most people find it necessary to cut a triangle out of cardboard, number the vertices on both sides, and physically perform the motions, comparing the results of the motions with a fixed triangle in standard position. As a challenge, however, if you have a strong visual imagination, you may attempt to perform the motions mentally! Since I is the identity element, it is easy to fill in the first row and first column. The upper left-hand corner is hardly more difficult, since it concerns only the rotations of the triangle in the plane. We get, for example, $R_1 \circ R_2 = I$, be-

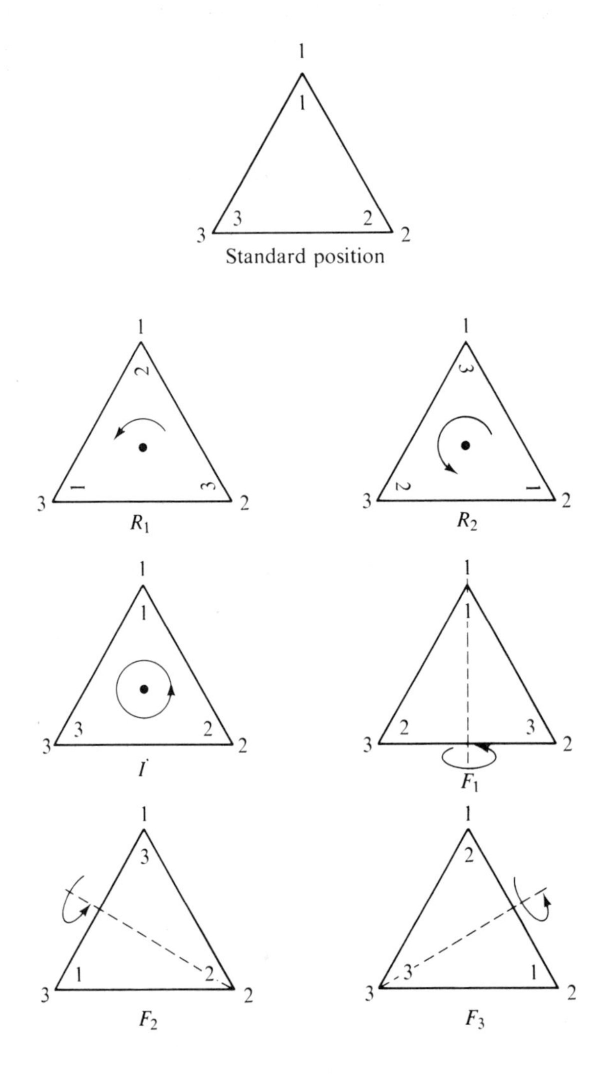

Figure 6.8

cause a 120° counterclockwise rotation, followed by one of 240°, causes the triangle to rotate through 360°, bringing it back to standard position. As a second easy observation, any flip of the triangle out of the plane, performed twice, brings the triangle back to standard position; this implies that $F_1 \circ F_1 = I$, and similarly for the other two flips.

TABLE 6.6

$\circ$	I	R_1	R_2	F_1	F_2	F_3
I	I	R_1	R_2			
R_1	R_1					
R_2	R_2					
F_1				I		
F_2						
F_3						

Let us now try a combination of a rotation and a flip, e.g., $R_1 \circ F_1$. The stages are shown in Figure 6.9. You should note that F_1 is a flip about the

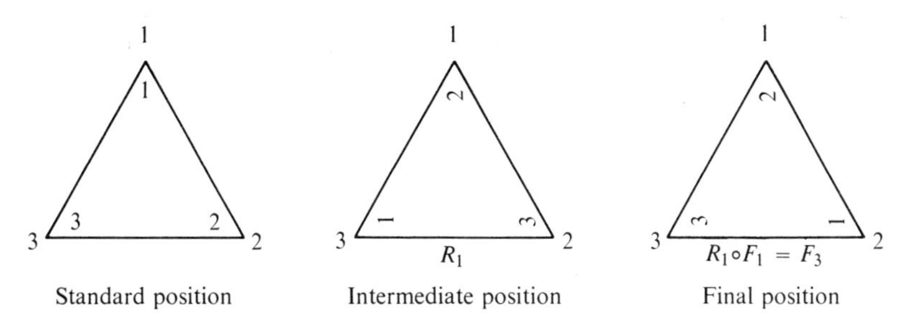

Figure 6.9

vertical axis (the axis passing through Vertex 1 *in standard position*); it has nothing to do with the location of Vertex 1 at the time the flip is made. To determine $R_1 \circ F_1$, we now ask: "Which *single* motion (of the six composing the system) transforms the triangle directly from standard position to the final position?" We observe that, in the final position, Vertex 3 is in standard position, while Vertices 1 and 2 have been interchanged from standard position. The single motion which accomplishes this is F_3: so $R_1 \circ F_1 = F_3$.

To complete the investigation, we try $F_1 \circ R_1$. The stages are given in

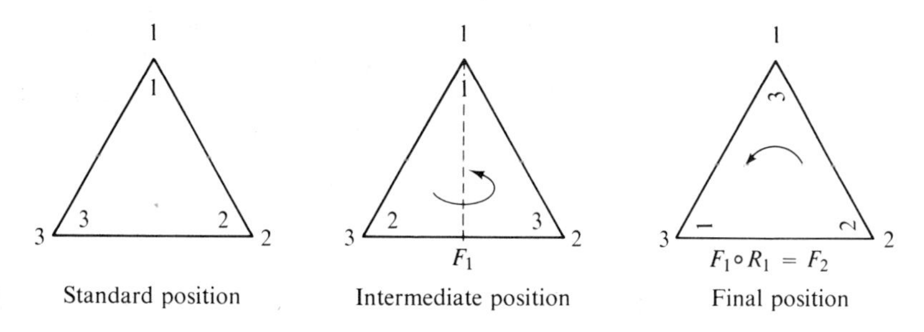

Figure 6.10

Figure 6.10. Comparing the final with the original position, we note that this time Vertex 2 is unmoved, while Vertices 1 and 3 have been interchanged. This implies that $F_1 \circ R_1 = F_2$. You should note that $R_1 \circ F_1 \neq F_1 \circ R_1$. Thus we have our first example of a noncommutative or nonabelian system. The completion of Table 6.6 is an exercise, as is also the determination that the system forms a group. In this connection, we accept without proof the theorem that composition is an associative operation.

By way of contrast we cite a wholly abstract, finite group $(G,\circ)$ having three undefined elements, a, b, c, and an undefined operation, $\circ$. The operation table, Table 6.7, completely categorizes the system. Let us check that this system satisfies Definition 6.3 and so forms a group. Certainly $G \neq \varnothing$, and the construction of the table shows that $\circ$ is a binary operation. Since every position in the body of the table contains one and only one element of G, the operation is closed. It is also associative, though we shall not tediously check the 27 different equations which would confirm this. The identity is a, as an examination of the a row and the a column indicates. As for inverses, $a^* = a$ (as expected), $b^* = c$, and $c^* = b$. Thus $(G,\circ)$ is a group. Is it commutative?

TABLE 6.7

$\circ$	a	b	c
a	a	b	c
b	b	c	a
c	c	a	b

The preceding example shows that a group is a system containing *abstract* elements, a, b, c, . . . , and *one* abstract operation, $\circ$. If we apply the group concept to a particular system, the elements may be numbers, vectors, or even geometric motions; and the operation may be called *addition*, *multiplication*, *composition*, or some other term. In fact, the beauty and utility of an abstract concept is that it can be applied in many concrete situations which, at first glance, seem to have little in common with each other. With respect to a number of mathematical systems we have all experienced in school, the group concept furnishes a common bond. The advantage of such a bond becomes evident when we realize that any theorem proved in the abstract system may be applied immediately in all concrete situations which exhibit the properties of a group. In the next section we shall encounter a striking instance of this power of the abstract to make sense out of the concrete.

EXERCISES 6.2

1. (a) Construct addition and multiplication tables for the arithmetic (mod 3).

   (b) Determine whether this system forms a commutative group

(i) under addition.

(ii) under multiplication.

(iii) under multiplication, with 0 deleted.

2. (a) Show that the arithmetic (mod 4) is a group under addition. Is it commutative?

   (b) Explain why the arithmetic (mod 4) is not a group under multiplication.

   (c) Show that the "reduced" system (mod 4) — with 0 deleted — is not a group under multiplication. (*Hint:* Consider the product $2 \cdot 2$; also $2''$.)

3. Answer Problem 1 for the arithmetic (mod 2).

4. Complete the operation table for the symmetries of a rectangle (Table 6.5) and show that it forms a group. Is it commutative?

5. Complete the operation table for the symmetries of an equilateral triangle (Table 6.6) and show that it forms a group. What does the pattern of the table indicate about commutativity?

6. If an arithmetic (mod n) has two *nonzero* elements whose product is zero, we call either element a *divisor of zero*. As an example, since $2 \cdot 2 = 0$ (mod 4), 2 is a divisor of zero (mod 4). More generally, we have the following:

Definition 6.5 Divisors of Zero Let S be a system which has two closed, binary operations, *addition* $(+)$ and *multiplication* $(\cdot)$, and contains an additive identity (0). If there are elements $x, y \in S$, with $x \neq 0$ and $y \neq 0$, such that $x \cdot y = 0$, then x and y are called *divisors of zero*.

   (a) Does the arithmetic (mod 5) have divisors of zero? If so, list all such pairs.

   (b) Answer part (a) for the arithmetic (mod 12). (Recall that the additive identity here is "12" on the clockface.)

   (c) Answer part (a) for the arithmetic (mod 6). (*Hint:* A little thought will make it unnecessary to construct the entire multiplication table.)

   (d) Does the system of whole numbers have divisors of zero? Justify your answer. (*Hint:* See Theorem 4.17.)

*7. Prove that the identity, e, of a group $(G, \circ)$ is unique. (*Hint:* Suppose $(G, \circ)$ contained two different identities, e_1 and e_2. Now use the equations $a \circ e_1 = a$ and $e_2 \circ b = b$, letting $a = e_2$ in the first and $b = e_1$ in the second.)

*8. State and prove that when $(G, \circ)$ is a group, then the conclusion of Exercises 6.1, Problem *9 holds.

*9. Prove that if x is any element of a group $(G,\circ)$, then its inverse, x^*, is unique. (*Hint:* Suppose $x \in G$ has two different inverses, $x_1{}^*$ and $x_2{}^*$. "Multiply" the equation $x \circ x_1{}^* = e$ *on the left* by $x_2{}^*$, obtaining the equation $x_2{}^* \circ (x \circ x_1{}^*) = x_2{}^* \circ e$; then simplify this equation to get $x_1{}^* = x_2{}^*$.)

*10. (a) By further experiments along the line of Problem 6, discover and state a rule giving the values of n for which an arithmetic (mod n) possesses divisors of zero. In particular, explain how the arithmetics (mod 7), (mod 9), and (mod 10) illustrate your rule.

(b) Discover, state, and illustrate a connection between the presence or absence of divisors of zero in an arithmetic (mod n) and the group properties of this arithmetic under multiplication, with zero deleted.

(c) Using the results of parts (a) and/or (b), explain for which values of n the arithmetic (mod n), with zero deleted, forms a group under multiplication.

*11. Consider the set of rotations *in the plane* of a square about its center. Consult Figure 6.11. Let R_1 be the 90° rotation counterclockwise; let R_2 be the 180° rotation counterclockwise; etc.

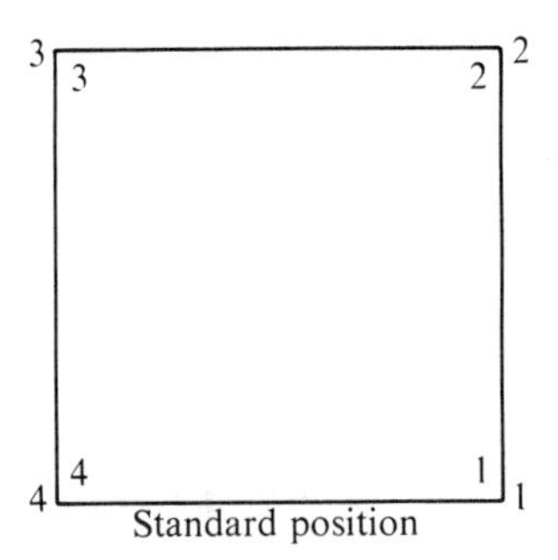

Figure 6.11

(a) How many distinct rotations are there? (Do not consider flips *out of* the plane.)

(b) Which rotation (if any) returns the square to its original position?

(c) Set up an operation table for this system, taking composition (one motion followed by another) as the operation.

(d) Does this system form a group? A commutative group? Justify your answer.

*12. Extend Problem *11 by considering *all* the symmetries of the square, including flips out of the plane about a horizontal axis, a diagonal axis, etc. You should obtain eight motions in all, including rotations and flips. Describe each of these motions and invent appropriate symbols for them (as R_1, R_2). Then answer questions similar to parts (b), (c), and (d) of Problem *11.

Section 6.3 Inverse Operations

In Section 6.1 we constructed tables of addition and multiplication for several arithmetics (mod n). These operations have many of the same properties as do the corresponding ones for the whole number system. But in W we also defined two other operations: subtraction (the inverse of addition) and division (the inverse of multiplication). Can we do the same for the arithmetics (mod n)? Certainly: we need only carry over Definition 4.9 and Definition 4.13, replacing W by "the arithmetic (mod n)." For these inverse operations (mod n) we shall use the same terms and symbols as in W.

As an example of subtraction we have $2 - 4 = 3$ (mod 5), because, by Table 6.2, $4 + 3 = 2$. Similarly, $1 - 2 = x$, such that $2 + x = 1$ (mod 5); from the table we see that $x = 4$. In the arithmetic (mod 12), the search for the value of x, as in $4 - 9 = x$, may take a little longer; but since the addition table has only a finite number of entries, it is possible to determine a value of x — if there is one — such that $9 + x = 4$ (mod 12). You, in fact, may check that $x = 7$ satisfies the equation. You should also note that, though the above differences exist in these arithmetics (mod n), the corresponding differences do not exist in W!

It is likewise possible, by reference to Table 6.2, to investigate division problems. Thus $1 \div 2 = 3$ (mod 5), because $2 \cdot 3 = 1$. Similarly, $2 \div 4 = x$ such that $4 \cdot x = 2$; from the table, $x = 3$. We look in vain, however, for the quotient $3 \div 0 = x$ (mod 5): there is no such element x. The quotient $8 \div 2$ (mod 12), on the other hand, is apparently not unique; for $2 \cdot x = 8$ is satisfied by $x = 4$ and $x = 10$. It is natural to ask: "Under which circumstances are quotients in the arithmetic (mod n) ill defined, either because they do not exist or because they are not unique?" Moreover, do differences also behave in this strange fashion?

A more important question is the following: "Are there more efficient methods of computing differences and quotients than resorting to the definitions, which require a search for a number x to satisfy the defining equations?" To cite a parallel, in the study of fractions in elementary school, finding the quotient $1/3 \div 4/5$ was assuredly not done by seeking a fraction x/y such that $(4/5)\cdot(x/y) = 1/3$. Indeed, it is not at all obvious from the equation that $x/y - 5/12$! To be honest, the writer obtained 5/12 by the usual rule of "invert and multiply."

We can deal with the questions of the two preceding paragraphs and also learn much more about the nature of inverse operations if we exploit the group concept. We propose to develop a sequence of abstract definitions and theorems on groups which we can then apply to subtraction and division in a variety of important contexts, including ordinary arithmetic and the arithmetics (mod n). We begin by abstracting from Definitions 4.9 and 4.13 a new, seemingly "off-beat" definition.

To make clear how this *abstracting* process occurs, we repeat these two definitions in a form which exhibits their parallelism, then employ the identical pattern with "neutral" symbols which refer to abstract group operations rather than to the four specific arithmetic ones. The three definitions are

$$\text{If}\left\{\begin{matrix} a, b \in W \\ a, b \in W \\ a, b \in G \end{matrix}\right\}, \text{then the}\left\{\begin{matrix} \textit{difference } b - a \\ \textit{quotient } b \div a \\ \ldots\ldots\ldots b \,\square\, a \end{matrix}\right\} \text{is}$$

$$\left\{\begin{matrix} \text{a whole number} \\ \text{a whole number} \\ \text{an element of } G \end{matrix}\right\} x \text{ (if } x \text{ exists) such that} \left\{\begin{matrix} a + x = b \\ ax = b \\ a \circ x = b \end{matrix}\right\}$$

Summarizing, the first two lines express a difference in terms of a sum, and a quotient in terms of a product, respectively. The third line expresses some abstract, so far nameless operation $\square$ in terms of the basic group operation, $\circ$. What shall we call this new operation? Let us call it "quifference."[14] Then the third line can be formally stated as

Definition 6.6 "Quifference" If $(G,\circ)$ is a group with the operation $\circ$, and if a, b are any elements of G, then the *quifference*, $b \,\square\, a$, is an element $x \in G$ (if x exists) such that $a \circ x = b$.

Another way to describe the relations among the operations discussed in the preceding paragraph is to say that subtraction is the operation *inverse to* addition, division is *inverse to* multiplication, and quifference is *inverse to* the basic group operation. Thus, associated with the *direct* operations, $+$, $\cdot$, and $\circ$, are their *inverse* operations, $-$, $\div$, and $\square$, respectively. Please be warned that this use of the word "inverse" to describe operations is different from its use with respect to elements or conditionals.

We have already illustrated the first two inverse operations for the whole numbers in Sections 4.7 and 4.8 and in the first part of this section for the arithmetics (mod n). For abstract groups we can illustrate the inverse operation $\square$ by reference to the group $(\{a,b,c\},\circ)$ defined by Table 6.7. By Definition 6.6, the quifference, $a \,\square\, c$, is an element x such that $c \circ x = a$. Inspection of the table for G reveals that, for $x = b$, $c \circ x = c \circ b = a$; hence $a \,\square\, c = b$.

We now have the machinery with which to answer the question of the fourth paragraph of this section: how to devise efficient methods for computing differences and quotients. We answer it in a way so general as to apply, not only to arithmetic and algebra and to "clock" arithmetics, but also to geo-

---

[14]"Quifference" is a homemade term, compounded from "quotient" and "difference."

metric systems such as the symmetries of the rectangle — indeed, to any concrete application of the group concept. Our approach is to use Definition 6.6 and the properties of a group contained in Definition 6.3 to prove a theorem which provides an explicit formula for finding $b \square a$ for $a, b \in (G,\circ)$, thus implying that $\square$ is a closed operation in $(G,\circ)$. Moreover, use of this formula avoids the necessity of hunting through a table for an element x to satisfy the equation $a \circ x = b$. The theorem is

THEOREM 6.1 Inverse Operation Theorem If a, b are any elements of a group $(G,\circ)$, then the quifference, $b \square a = a^* \circ b$.

In thinking through the proof, we recall that a^* is the element inverse to a, so that $a \circ a^* = e$, the group identity. It is also helpful to realize that one possibly effective tool in proving a theorem is the definition of any concept which it involves. Actually, in the case of this theorem, Definition 6.6 is all we really know about $\square$. This definition says, in effect, that $b \square a = x$ iff $a \circ x = b$. Now let us see what plays the role of x in the conclusion of the theorem: $b \square a = a^* \circ b$. Is it not the expression $(a^* \circ b)$? So $b \square a = a^* \circ b$ will be true iff $a \circ (a^* \circ b) = b$. Let us see if we can manipulate the expression $a \circ (a^* \circ b)$ so that it reduces to b, using only group laws in doing so. We get the following chain:

$$a \circ (a^* \circ b) = (a \circ a^*) \circ b = e \circ b = b$$

The formal proof goes as follows:

Proof:

1. $b \square a = a^* \circ b$ iff $a \circ (a^* \circ b) = b$, by definition of $\square$.
2. Now $a \circ (a^* \circ b) = (a \circ a^*) \circ b$, by associativity in $(G,\circ)$.
3. And $(a \circ a^*) \circ b = e \circ b$.(Why?)
4. And $e \circ b = b$.(Why?)
5. Hence $a \circ (a^* \circ b) = b$, by transitivity of equality.
6. So $b \square a = a^* \circ b$, by definition of logical equivalence.

The significance of Theorem 6.1 becomes apparent when we realize that (1) $(a^* \circ b)$ is a unique element of G (Why is this so?) and (2) this theorem and the next two apply to several familiar number systems. In fact, Theorem 6.2 is a restatement of (1):

THEOREM 6.2 Closure of Quifference In a group $(G,\circ)$ the inverse operation $\square$ is closed.

Proof: Closure of $\square$ requires that, for any elements $a, b \in (G,\circ)$, $b \square a$ be a unique element of G. But by Theorem 6.1, $b \square a = a^* \circ b$, which *is* a unique element of G. The latter statement is so, because $a \in G$ implies that a^* is a unique element of G, as is proved in Exercises 6.2, Problem *9; then the closure of $\circ$ in G implies that $a^* \circ b$ is a unique element of G.

THEOREM 6.3 If $(G,\circ)$ is a *commutative* group, then $b \square a = b \circ a^*$.

Proof: This follows directly from Theorem 6.1, since, if G is commutative, $b \square a = a^* \circ b = b \circ a^*$.

With the aid of Theorems 6.1–6.3, we return to the examples with which we began this section. Since the arithmetic (mod 5) forms a commutative group under addition, we interpret $\circ$ in Theorem 6.3 as $+$, $\square$ as $-$, and a^* as a', the additive inverse of a. Then Theorem 6.3 translates into $b - a = b + a'$. In particular, if $b = 2$ and $a = 4$, then $a' = 1$; and we get $2 - 4 = 2 + 1 = 3$ (mod 5), which agrees with our answer in the second paragraph of this section. Similarly, $1 - 2 = 1 + 2' = 1 + 3 = 4$. As another example, in the arithmetic (mod 12), $4 - 9 = 4 + 9' = 4 + 3 = 7$. You should note that all these results were obtained directly, without any "guessing." Why is it permissible to use Theorem 6.3 in performing subtraction in (mod 12)?

In order to use Theorem 6.1 or Theorem 6.3 for division, we merely interpret $\circ$ as $\cdot$, $\square$ as $\div$, and a^* as a'', the multiplicative inverse of a. But is the arithmetic (mod 5) a group under multiplication? No; for 0 has no multiplicative inverse. We have seen, however, that after deleting 0, the reduced system (mod 5) forms a commutative group under multiplication. So Theorem 6.3 may be used to obtain quotients containing any *nonzero* elements (mod 5). Thus $1 \div 2 = 1 \cdot 2'' = 1 \cdot 3 = 3$; and $2 \div 4 = 2 \cdot 4'' = 2 \cdot 4 = 3$. But Theorem 6.3 does not apply to $0 \div 3$ or $3 \div 0$ (mod 5), nor yet to $8 \div 2$ (mod 12). Why not? Only Definition 6.6 is available for these proposed quotients — and it may not produce a unique result or any result at all.

In summary, Theorem 6.2 guarantees that, whenever a system forms a group under a given operation, the inverse operation is closed. In addition, Theorem 6.1 or Theorem 6.3 (if applicable) provides an efficient method for performing the inverse operation when it is closed. Regrettably, the proposed theory cannot be used to compute differences or quotients in the whole number system. Why not? Chapter 7, however, will provide remedies for this lack.

EXERCISES 6.3

1. Find the answer (if it exists) to each of the following problems (mod 5) in two ways: (1) *First*, use the appropriate definition to write an equation in x, then determine x by inspecting the appropriate operation table (mod 5) (2) *Second*, use Theorem 6.1 or Theorem 6.3, if either applies. If the answer does not exist, or is not unique, so state.

   (a) $4 - 1$ (b) $3 - 2$ (c) $1 - 4$ (d) $3 - 3$

   (e) $2 - 0$ (f) $0 - 2$ (g) $3 \div 1$ (h) $4 \div 2$

(i) $3 \div 2$ (j) $1 \div 3$ (k) $0 \div 2$ (l) $2 \div 0$

(m) $0 \div 0$

2. As in Problem 1, find the answer to each of the following problems (mod 12):

   (a) $8 - 6$ (b) $3 - 7$ (c) $5 - 12$ (d) $12 - 5$

   (e) $5 \div 1$ (f) $6 \div 2$ (g) $2 \div 6$ (h) $3 \div 7$

   (i) $3 \div 8$ (j) $12 \div 5$ (k) $12 \div 6$ (l) $6 \div 12$

   (m) $12 \div 12$

3. Use the tables you made in Exercises 6.1, Problem 7.

   (a) Evaluate (mod 4) by any convenient method:

      (i) $3 - 1$ (ii) $3 - 0$ (iii) $0 - 3$ (iv) $1 - 1$

      (v) $1 - 3$

   (b) Which (if any) of the following are well-defined quotients (i.e., exist and are unique)?

      (i) $3 \div 1$ (ii) $2 \div 2$ (iii) $0 \div 1$ (iv) $1 \div 3$

      (v) $1 \div 2$ (vi) $0 \div 0$

   (c) Find all the quotients (mod 4) which are not well defined.

4. Supply the missing reasons in the proof of Theorem 6.1.

5. In ordinary algebra a quadratic equation never has more than two roots. The equation $x^2 - 3x + 2 = 0$, for example, can be factored into $(x - 1)(x - 2) = 0$ and then solved by applying Theorem 4.17, which implies that $x - 1 = 0$ or $x - 2 = 0$, so that the roots are 1 and 2. What happens when we attempt to solve a quadratic equation (mod n)?

   (a) If n is such that the arithmetic (mod n) has no divisors of zero, then Theorem 4.17 holds, and we get two (or fewer) roots. Consider $x^2 - 3x + 2 = 0$ (mod 5).

      (i) By factoring, obtain $x = 1$ and $x = 2$ as roots.

      (ii) Check that 1 and 2 *are* roots by actually substituting them in the equation. For instance, for $x = 2$ we get $2^2 - 3 \cdot 2 + 2 = 4 - 1 + 2 = (4 + 1') + 2 = (4 + 4) + 2 = 3 + 2 = 0$, which checks.

      (iii) By actual trial, show that $x = 0, 3$, and 4 do not satisfy the equation.

   (b) But if the arithmetic (mod n) has divisors of zero, then we cannot apply Theorem 4.17. Consider $x^2 - 3x + 2 = 0$ (mod 6). Then $(x - 1)(x - 2)$ may have the value zero even when neither of its factors is zero.

(i) Perform parts (i) and (ii) of (a) to obtain two roots of the equation.

(ii) By actual substitution, show that 4 is also a root.

(iii) Search for still other roots, checking $x = 0, 3$, and 5.

(c) It is possible, on the other hand, for a quadratic equation (mod n) to have no roots. By actual trial, show that the following have no solution:

(i) $x^2 - 3x + 3 = 0 \pmod 4$ (ii) $x^2 - 3x + 3 = 0 \pmod 5$

6. (a) Determine whether or not each of the following arithmetics is closed under subtraction. Justify your answers by means of Theorem 6.2. The arithmetics are

(i) (mod 5) (ii) (mod 12) (iii) (mod 4)

(b) Answer part (a) with respect to division, instead of subtraction.

*(c) Determine the values of n for which the arithmetic (mod n) is closed under subtraction; under division; under division, with quotients not involving zero. (*Hint:* See Exercises 6.2, Problem *10.)

*7. (a) Restate Theorem 4.28 in abstract terms (using G, $\circ$, $\square$, $*$).

(b) Prove the theorem you stated in part (a). (*Hint:* To what extent can you use the pattern of the proof of Theorem 4.28, with abstract notation?)

(c) Explain the relationship of parts (a) and (b) to Exercises 4.8, Problem 7(a).

Section 6.4 Relations

We have already met a number of instances of relations. In Chapter 2, the symbols $\subseteq$, $\subset$, $=$, and $\sim$ denote certain relations between two sets. Thus the statement "$A \subseteq B$," where A and B are sets, signifies that A and B are related as specified by Definition 2.1. In Chapter 3 we introduced the relations $\Rightarrow$ and $\Leftrightarrow$ between two propositions; and in Chapter 4 we met the relations $=$, $<$, $>$, and others between two natural numbers. In everyday conversations we also use relations, such as "Jack *is taller than* Jill" or "My uncle is *a brother of* my mother." In geometry we may assert, of two lines m and n in a plane, that *m is parallel to n* (in symbols, $m \parallel n$); thus parallelism is a relation between two lines. In all the preceding cases, we have a collection of elements and a relation — which we may, in general, designate by R — such that for certain pairs of elements we can assert that "element a is related to element b by the relation R": in symbols, $a\ R\ b$.

But there is another way of looking at relations which yields a precise, though abstract definition of a relation in terms of the cartesian product of two sets.[15] The definition itself is brief; it is

Definition 6.7 Relation Let A and B be sets; then a *relation* R *from* A *to* B is any subset of $A \times B$ (in symbols, $R \subseteq A \times B$).

It is not obvious that Definition 6.7 has any connection with the instances of relations given above. The following example, however, illuminates this connection. Let us consider the set U of all people attending a certain musical concert. Then $U = M \cup W$, where M is the set of all men (and boys) and W is the set of all women (and girls) in attendance. These three sets are shown in Figure 6.12. Note that, since there are only two sexes, M and W together occupy the entire rectangle, U. In M and W, certain individuals are indicated by name. Now, since M and W are sets, a cartesian product — say $M \times W$ — can be formed from them. This cartesian product includes elements such as (JT, MD), where we use initials to abbreviate (John Talbot, Mary Dixon). In fact, $M \times W = \{$(JT,MD),(JT,SF),(JD,MD),(SW,SF),(SW,JW), . . . $\}$. Now, let us define a relation between the sets M and W; specifically, let the relation R be "is the husband of." Then it is clear that some pairs of names do *not* have that relation: for example, according to social convention JT is not the husband of MD because their last names differ. Also, other pairs presumably *do* have that relation, for example, (JD,MD). Thus the pairs (JT,MD), (JT,SF), etc., do not satisfy the relation, "is the husband of"; while the pairs (JD,MD) and (SW,JW) do satisfy it. To put the matter another way, the set of pairs that *do* satisfy the relation is a *subset of all* the elements in $M \times W$. Thus $R = \{$(JD,MD),(SW,JW), . . . $\} \subseteq M \times W$.

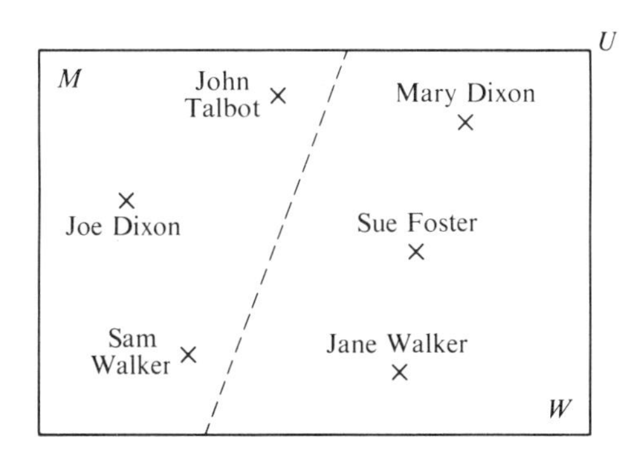

Figure 6.12

An important special case of Definition 6.7 occurs when A and B are the same set. This was so for most of the examples in the first paragraph of this section. In this case, a natural modification leads to

[15] See Section 2.8.

Definition 6.8 Relation (Same Set) Let A be a set; then a *relation R on A* is any subset of $A \times A$ (in symbols, $R \subseteq A \times A$).

As an example of Definition 6.8, let A be the set of whole numbers and R^* be the relation "is less than," as given by Definition 4.7. $A \times A$ contains all possible ordered pairs of whole numbers:

$$A \times A = \{(0,0),(0,1),(1,0),(0,2),(1,1),(2,0),(0,3),(1,2),(2,1),(3,0), \ldots\}$$

Some of these ordered pairs belong to R^* and others do not. Thus $(0,1) \in R^*$, since $0 < 1$ and, similarly, $(1,2) \in R^*$; but $(2,0)$ and $(1,1) \notin R^*$. So $R^* = \{(0,1),(0,2),(0,3),(1,2), \ldots\} \subseteq A \times A$.

We have now introduced two notations for expressing relations. For the statement "Joe Dixon is the husband of Mary Dixon," we may write either $(\text{JD,MD}) \in R$ or JD R MD. Similarly, we may write either $(1,2) \in R^*$ or $1\ R^*\ 2$. Hereafter, we shall use the latter notation, since it is often more convenient and closer to your previous mathematical experience.

Relations, like operations, may possess — or fail to possess — certain properties. We shall define and use three of them: the reflexive, symmetric, and transitive properties. To be fully applicable, these properties should be defined for a relation on A (i.e., satisfying Definition 6.8 rather than merely Definition 6.7). In the following discussion, therefore, we shall assume that $R \subseteq A \times A$. We shall introduce the three properties, one at a time, and follow each with an example and a counterexample.

First we have

Definition 6.9 Reflexivity A relation R on A is *reflexive* iff $a\ R\ a$ for every element $a \in A$.[16]

As an example, if $A = W$ (the set of whole numbers) and R is the relation of equality, then R is reflexive, since $a = a$ is true for every whole number. As a counterexample, if $A = W$ and R is the relation "less than," then R is not reflexive, since $2 < 2$ is false.[17]

Second, we have

Definition 6.10 Symmetry A relation R on A is *symmetric* iff $a\ R\ b$ implies $b\ R\ a$.

As an example, if $A = W$ and R is the equality relation, then R is symmetric, because whenever $a = b$, then $b = a$. Note that Definition 6.10 does not require that every two elements of A be related by R, but only that *if $a\ R\ b$*,

---

[16]In the cartesian product notation, we may say that R is reflexive iff $(a,a) \in R$ for every $a \in A$.

[17]As a matter of fact, $a < a$ is *never* true; but a single failure of $a\ R\ a$ is enough to show that R is not reflexive.

then R must also hold in the reverse order, $b\ R\ a$. As a counterexample, if $A = W$ and R is the relation "is a divisor of," or "divides,"[18] then R is not symmetric: for although $3 \mid 12$ holds, $12 \mid 3$ does not.

Third, we have

Definition 6.11 Transitivity A relation R on A is *transitive* iff $a\ R\ b$ and $b\ R\ c$ together imply $a\ R\ c$.

As an example, if $A = W$ and R is the "less than" relation, Theorem 4.24 asserts that $a < b$ and $b < c$ guarantee that $a < c$. As a counterexample, if A is the set of teams in a basketball league and R is the relation "wins a game against," then R is not transitive. In fact, the possibility that Team c may win a game against Team a after Team a has beaten Team b and Team b has beaten Team c is one factor which makes athletic contests interesting!

Many relations, of course, enjoy some, but not all, of the three properties we have been discussing. Some relations, however, satisfy all three. An example is the equality relation on W. Can you think of others? This type of relation is so important that it warrants

Definition 6.12 Equivalence Relation A relation R on a set A which is reflexive, symmetric, and transitive is called an *equivalence relation.*

Thus equality in W is an equivalence relation; but the "less than" relation in W is not (Why not?). What about equality of *sets*, as given by Definition 2.2?

It is worth pointing out that the behavior of a relation depends not only on what the relation is, but also upon the set on which it is defined. To emphasize this point, consider the following two examples: in each case, let U be the set of children at a certain neighborhood playground at a specified time. Also, let B be the set of boys, G the set of girls.

Example 1

Let R be the relation "is a full brother of," defined on the set U. Is R reflexive? symmetric? transitive? The first answer is clearly No. So also is the second answer. To see why, consider Jim Brown, who is a full brother of Susie Brown. Is there any difficulty about answering the third question?

Example 2

Let R be the same relation as in Example 1, *except that* R is defined on the set B. One of the answers is different from that of Example 1. Which is it?

Why are we especially interested in equivalence relations? For one thing, they occur significantly in many branches of mathematics. In geometry, for example, "congruence" is an equivalence relation on the set of triangles

---

[18]See Definition 4.11.

in the plane. For another, the following important theorem applies to equivalence relations. We shall not prove it but we will provide an example.

THEOREM 6.4 If R is an equivalence relation defined on a set A, then R *partitions* the set A into subsets in such a way that each element of A is a member of one and only one equivalence class.

Definition 6.13 Equivalence Classes The subsets into which R, in Theorem 6.4, partitions A are called *equivalence classes*.

More informally, what the theorem is saying is that an equivalence relation functions as a kind of "sorter" of the elements of a set, with some elements dropping into one "bin" (equivalence class), others into a second "bin," etc., with no elements left unsorted and no element dropping into two different bins.

As an example of Theorem 6.4, let us return to the set U of persons attending the musical concert,[19] this time taking as the relation R the phrase "is the same age in years as." You are invited to check that R is an equivalence relation defined on U. Suppose U contains at least one 21-year-old person: let us denote him or her by p_{21}. Now let us consider the set, S_{21}, of all persons in U who have the relation R to p_{21}. Formally, $S_{21} = \{x \mid x \in U \text{ and } x\,R\,p_{21}\}$; so S_{21} is the set of all 21-year-olds attending the concert. This set is a subset of U, as shown in Figure 6.13. Similarly, we can define S_{22}, S_{23}, S_{24}, . . . and S_{20}, S_{19}, . . . S_0, continuing this process until we have defined a set to which every person in U can belong. There will, of course, be some sets that are empty — like S_1 (perhaps) and S_{120}: these may be ignored and need not be shown in the diagram. But surely every person in U is a member of *exactly one* set, $S_i \subseteq U$; and the sets S_i are the equivalence classes into which R partitions U. In other words, R has "sorted" the elements of U — by age, in this instance — into disjoint subsets whose union is U: and this is just what is meant by a *partition* of U.

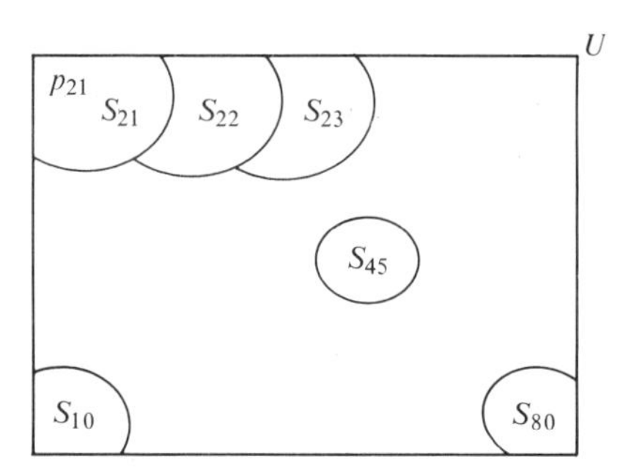

Figure 6.13

[19]See the paragraph following Definition 6.7.

EXERCISES 6.4

1. Explain why the immediately preceding relation, "is the same age in years as," defined on the set U of concert-goers, is an equivalence relation.

2. Turn back to the cartesian-product example at the beginning of Section 2.8. Let U be the set of all cities and towns represented on the map of New York State, and let R be the relation "is in the same region as," in terms of the 180 keyed regions on the road map. Assume that no city or town lies exactly on the boundary of two or more regions. Does this example illustrate Theorem 6.4? Explain your answer.

3. Each of the following relations is defined on a specified set, A. For each, answer the questions: (1) Which (if any) of the three properties (reflexivity, symmetry, transitivity) does it satisfy? (2) Is it an equivalence relation?

   (a) R: is taller than A: the set of concert-goers in the text

   (b) R: resides in the same county as A: as in part (a)

   (c) R: greater than A: W

   (d) R: $\leq$ A: W

   (e) R: $\neq$ A: W

   (f) R: is perpendicular to A: the set of lines in a given plane

   (g) R: is parallel to or coincident with A: as in part (f)

   (h) R: is the mother of A: the set of all people who ever lived

   (i) R: is an ancestor of A: as in part (h)

   (j) R: is directly east of A: all points within continental United States

   (k) R: as in part (j) A: the set of all points on the earth

   (l) R: lives within 5 miles of A: the set of all people living in the state of Arkansas

   (m) R: lives on the opposite side of the street from
   A: the set of all people living on 42nd Street in New York City

   (n) R: $\subset$ A: the set of all subsets of W

   (o) R: $\subseteq$ A: as in part (n)

   (p) R: $=$ A: as in part (n)

   (q) R: $\Rightarrow$ A: the set of all propositions

   (r) R: $\Leftrightarrow$ A: as in part (q)

4. (a) Explain why the relation "is equivalent to," $\sim$, defined on the set of all sets by means of Definition 2.8, is an equivalence relation in the sense of Definition 6.12.

(b) After rereading the first part of Section 2.8, explain how the concepts of equivalence ($\sim$) of sets and cardinal number (Definition 2.9) are related to Theorem 6.4 and Definition 6.13.

*5. Let A be a given set of people and R_1 and R_2 be two relations defined on A as follows: R_1: "is the same age in years as"; R_2: "is the same height in inches as."

(a) Are R_1 and R_2 equivalence relations? Explain.

(b) Since these relations are subsets of $A \times A$, they may be combined by means of set operations. Thus $R_1 \cap R_2$ is the relation "is the same age in years and the same height in inches as," defined on A. What is the meaning of $R_1 \cup R_2$?

(c) One of $R_1 \cap R_2$ and $R_1 \cup R_2$ is an equivalence relation, but the other is not. Explain which is which, and why.

(d) For the relation in part (c) which is not an equivalence relation, show by a counterexample that Theorem 6.4 does not hold.

(e) The example of part (d) illustrates the inverse of Theorem 6.4, which is also a theorem. State the inverse of Theorem 6.4, and also its converse. Is the converse of Theorem 6.4 a theorem? Why or why not?

Section 6.5 Arithmetics (mod *n*) As Equivalence Classes

Surely a major theme of mathematics is the degree of interrelationship of its parts, including some which at first glance seem to have little connection. As a minor, but quite worthwhile, instance of the coherence of mathematics, we use the concept of equivalence classes to develop the arithmetics (mod n) from a purely algebraic point of view, dispensing with clockfaces. We do this first for the familiar arithmetic (mod 5) and then generalize the procedure.

Let us begin with the set of whole numbers. We note that some elements of W are divisible by 5; by Definition 4.12, these comprise the subset $\{0,5,10,15, \ldots\}$. We shall denote this subset by [0], because any member of it, when divided by 5, produces a remainder of 0. Thus

$$[0] = \{0,5,10,15, \ldots\}$$

Some other elements of W, however, when divided by 5, yield a remainder of 1. Examples are 6 and 21. Such elements also form a subset of W, which we shall denote by [1]. Thus

$$[1] = \{1,6,11,16,21, \ldots\}$$

Similar considerations induce us to define (mod 5)

$$[2] = \{2,7,12,17,22, \ldots\}$$

$$[3] = \{3,8,13,18,23, \ldots\}$$

$$[4] = \{4,9,14,19,24, \ldots\}$$

Now all the elements of any one of the subsets [0], [1], [2], [3], and [4] share a common divisibility property. We can express this property as a relation R, defined on W. In words, R is "has the same remainder on division by 5 as." Thus $10 \, R \, 5$ is true, because 10 and 5 both yield a remainder of 0 on division by 5; similarly, $23 \, R \, 8$ and $6 \, R \, 21$ are true; but $14 \, R \, 7$ is false. It is easy to reason out informally that R is an equivalence relation. But if this is true, then Theorem 6.4 insists that R induces a partition of W into equivalence classes. Which are they? A brief examination shows that they are precisely the five sets, [0], [1], [2], [3], and [4]. We observe that, though W is an infinite set, the set of equivalence classes (mod 5) is finite, having exactly 5 members. We must be careful to distinguish between the whole number, 4, and the equivalence class, [4], to which it belongs.

In the theory of numbers, the relation R is more briefly expressed by the phrase, "is congruent to (mod 5)." Thus we may state that 23 is congruent to 8 (mod 5): in symbols, $23 \equiv 8 \pmod 5$. Similarly, $10 \equiv 5 \pmod 5$ and $6 \equiv 21 \pmod 5$; but $14 \not\equiv 7 \pmod 5$. Alternatively, we may describe the relation R by asserting that two whole numbers are congruent (mod 5) iff their difference, taken in whichever order it exists, is divisible by 5. In particular, $23 - 8 = 15$, $10 - 5 = 5$, and $21 - 6 = 15$[20] are all divisible by 5, whereas $14 - 7 = 7$ is not. Thus $a \equiv b \pmod 5$ iff $5 \mid (a - b)$ or $5 \mid (b - a)$, whichever difference exists. More generally, we have

Definition 6.14 Congruence (mod n) If $a, b \in W$ and $n \in N$, then $a \equiv b \pmod n$ iff $n \mid (a - b)$ or $n \mid (b - a)$, whichever difference exists.

With the aid of Definition 6.14, we can formally prove that congruence is an equivalence relation on W. We are going to prove reflexivity; symmetry and transitivity are left as exercises. Summarizing these efforts, we shall have

THEOREM 6.5. Congruence (mod n), defined on W, is an equivalence relation.

Partial Proof (Reflexivity only): We must show that, for any element $a \in W$, $a \equiv a \pmod n$. But by Definition 6.14, $a \equiv a \pmod n$ iff $n \mid (a - a)$ or $n \mid 0$, where $n \in N$. Now any natural number divides zero, since $n \cdot x = 0$ for $x = 0$ by the zero product theorem in W.[21] Thus by Definition 4.11, $n \mid 0 = (a - a)$, and $a \equiv a \pmod n$ because Definition 6.14 is satisfied.

But we need not stop here; for it is possible to define binary operations on the equivalence classes (mod 5). In particular, we can add two *classes* by

---

[20]Note the reversal of the numbers, since $6 - 21$ is not a whole number, but $21 - 6$ is.

[21]Theorem 4.16.

selecting an element from each class, adding these whole numbers in the usual way (in W), and defining the *class* which contains the sum of the two elements as the *sum* of the given classes. For instance, the sum of [3] and [4] may be found by taking $8 \in [3]$ and $14 \in [4]$, obtaining $8 + 14 = 22$ by addition in W, and concluding that, since $22 \in [2]$, $[3] + [4] = [2]$.

But what would happen if we made another choice of elements from [3] and [4] — e.g., $23 \in [3]$ and $9 \in [4]$? We would get a different sum from the elements, since $23 + 9 = 32$. But since $32 \in [2]$, the *class* to which the sum belongs is the same one as before: so that, in terms of classes, the sum, $[3] + [4] = [2]$, is unchanged. You should investigate still other choices, including the simplest pair, $3 \in [3]$ and $4 \in [4]$, which yield $3 + 4 = 7 \in [2]$, as before. Fortunately, it turns out that, whichever elements are selected from [3] and [4], the sum of these elements is always contained in the class [2]. Indeed, it can be proved in general that the sum of two classes (mod n) is a *unique class* (mod n): briefly, addition of classes (mod n) is closed.

For the record, we now formally state

Definition 6.15 Addition (mod n) If $[a]$ and $[b]$ are any two equivalence classes (mod n), then $[a] + [b] = [a + b]$, where $[a + b]$ signifies the equivalence class containing $a + b$.

Though Definition 6.15 may have a different appearance from our examples, it really expresses the same meaning, quite compactly. According to it, $[3] + [4] = [7]$. But [7] is the equivalence class containing the element 7, namely, [2]: hence $[7] = [2]$, where the symbol $=$ refers to the equality of classes or sets in the sense of Definition 2.2.

To round out the discussion, there are now two tasks to pursue. The first is to use Definition 6.15 to construct an addition table (mod 5) for the classes $[0], \ldots, [4]$, compare it with the arithmetic table (mod 5) of Section 6.1, and examine its properties. The second task is to invent a definition for multiplication of classes (mod n) similar to Definition 6.15 and use it to construct a multiplication table (mod 5) and make corresponding comparisons with Section 6.1. As an example of multiplication of classes, we cite $[2]\,[4] = [8] = [3]$. For further examples you should study the following exercises.

EXERCISES 6.5

1. Definition 6.14 tacitly assumes that, for any two numbers a and $b \in W$, at least one of the differences, $a - b$ or $b - a$, exists as a whole number.
   (a) Illustrate this assumption, using first $a = 9$, $b = 5$; then $a = 11$, $b = 17$.
   (b) Show that, for some choices of a and b, both differences exist.
   (c) Explain, by reference to material in Section 4.7, why the assumption about the existence of at least one of these differences is justified.

2. (a) Apply Definition 6.14 to show that (mod 5) $8 \equiv 3, 24 \equiv 9, 9 \equiv 24, 7 \equiv 7, 16 \equiv 31, 35 \equiv 0$.

   (b) For each congruence of part (a), name the equivalence class (mod 5) containing the pair of whole numbers.

   (c) Answer part (a) for $9 \equiv 6$ (mod 3), $4 \equiv 10$ (mod 3), $23 \equiv 37$ (mod 7), $20 \equiv 8$ (mod 12).

3. Using Definition 6.15, find the following sums (mod 5), selecting (1) only whole numbers greater than 10 from the equivalence classes, (2) only the smallest possible whole numbers from these classes. The sums requested are

   (a) $[1] + [3]$ (b) $[4] + [4]$

   (c) $[2] + [0]$ (d) $[1] + [4]$

4. (a) Set up a complete addition table for the equivalence classes (mod 5) and fill in all the entries, using the results and methods of Problem 3.

   (b) Compare your table with that in Section 6.1 for the arithmetic (mod 5).

5. Write a definition of multiplication (mod n) which parallels Definition 6.15 for addition (mod n). Call it

Definition 6.16 Multiplication (mod n)

6. Using Definition 6.16, compute the following products (mod 5) selecting (1) only whole numbers greater than 10 from the equivalence classes, (2) only the smallest possible whole numbers from these classes.

   (a) $[2]\cdot[4]$ (b) $[3]\cdot[3]$ (c) $[4]\cdot[1]$

   (d) $[3]\cdot[0]$ (e) $[3]\cdot[2]$

7. Answer Problem 4 for multiplication (mod 5).

8. (a) Define each of the equivalence classes (mod 4) and list several elements in each class, as was done for (mod 5) at the beginning of this section.

   (b) Construct addition and multiplication tables for the arithmetic of equivalence classes (mod 4).

   (c) Compare the tables of part (b) with those you constructed in Exercises 6.1, Problem 7.

9. Using Definition 6.14, finish Theorem 6.5 by proving that congruence (mod n) is

   (a) a symmetric relation

   *(b) a transitive relation,

   hence completing the proof that it is an equivalence relation.

10. (a) Using the definitions of subtraction and division in Chapter 4 as models, write out definitions for

    (i) $[a] - [b] \pmod{n}$ (ii) $[a] \div [b] \pmod{n}$

    (b) How could Exercises 6.3, Problems 1, 2, be changed so as to illustrate part (a) above?

    (c) Is subtraction of equivalence classes (mod n) a closed operation? Justify your answer. (*Hint:* Invoke group properties.)

    (d) Answer part (c) for division.

*11. Now that the arithmetics (mod n) have acquired a precise algebraic foundation in terms of equivalence classes, it is possible to prove that addition and multiplication are commutative and associative. In the following theorems, we are given that $[a]$, $[b]$, $[c]$ are any three equivalence classes (mod n). Prove the theorems.

(a)

THEOREM Commutativity of Addition $[a] + [b] = [b] + [a]$.

(*Hint:* By Definition 6.15 $[a] + [b] = [a + b]$. But $[a + b] = [b + a]$. Why?)

(b)

THEOREM Commutativity of Multiplication $[a] \cdot [b] = [b] \cdot [a]$.

(c)

THEOREM Associativity of Addition $([a] + [b]) + [c] = [a] + ([b] + [c])$.

(d)

THEOREM Associativity of Multiplication Similar to the theorem for associativity of addition.

*12. Congruences share a number of properties with ordinary algebraic equations; but the analogies, although suggestive, are also dangerous.

    (a) The analog of Theorem 4.5 is "If $a \equiv b \pmod{n}$, then $a + c \equiv b + c \pmod{n}$.

        (i) Illustrate this analog, using (mod 5).

        (ii) This statement is a theorem. Give a proof, or at least an informal argument in support of it.

    (b) (i) Write the analog of Theorem 4.13.

        (ii) Either prove it (at least, informally) or else find a counterexample to show that it is not a theorem.

    (c) (i) Write the converse of part (a). Which theorem is its analog in Chapter 4?

        (ii) Use the cancellation theorem to *solve* the congruence (i.e., find values of x for which it is true), $x + 3 \equiv 7 \pmod{5}$, which can be rewritten as $x + 3 \equiv 4 + 3 \pmod{5}$.

(iii) Similarly solve $x + 12 \equiv 5 \pmod{9}$.

(d) The converse of part (b)(i) would be the cancellation law for multiplication (mod n). In unrestricted form it is not true. To see this, try solving the following congruences, being sure to check each answer:

(i) $2x \equiv 5 \pmod{7}$ (*Hint:* $[5] = [2]\cdot[6] \pmod{7}$.)

(ii) $2x \equiv 5 \pmod{8}$

(iii) $2x \equiv 6 \pmod{7}$

(iv) $2x \equiv 6 \pmod{8}$ (*Hint:* This congruence has two solutions, only one of which is given by the unrestricted cancellation law.)

*Section 6.6 Isomorphic Groups

In mathematics we sometimes discover that apparently different symbols, or even different concepts, behave in the same way. In the arithmetic of fractions, for instance, we frequently find it advantageous to substitute 4/1 for 4, and vice versa, as in the equations

$$\frac{8}{9} \div 4 = \frac{8}{9} \div \frac{4}{1} = \frac{8}{9}\cdot\frac{1}{4} = \frac{2}{9} \quad \text{and} \quad \frac{7}{3} + \frac{5}{3} = \frac{12}{3} = \frac{4}{1} = 4$$

Although 4 is a whole number and 4/1 is, at least in form, a fraction, we do not hesitate to replace either expression by the other in performing the arithmetic operations. We can do this because 4 and 4/1 behave in the same way when subjected to arithmetic operations. In this section we investigate, on a more abstract level, the conditions under which two *groups* which appear to differ may actually have the same structure, which is merely disguised by different notations. We begin with a familiar, intuitive example.

It is easy to prove the following properties for odd and even integers: The sum of two even numbers is even; the sum of two odd numbers is even; the sum of an odd and an even number is odd. If we denote the set of even numbers by E and the set of odd numbers by D, we can put the preceding facts in the form of an addition table (see Table 6.8).

TABLE 6.8

$+$	E	D
E	E	D
D	D	E

$+$	0	1
0	0	1
1	1	0

For comparison purposes, let us now develop the addition table for the integers (mod 2). It is also given in Table 6.8. We compare the two addition tables. Do they seem to have the same structure? Certainly each has one element occurring on the diagonal slanting down to the right: these are the elements E and 0, respectively. In both tables, moreover, the second element occupies the other diagonal. In fact, if we merely replace E and D, wherever they occur in the first table, by 0 and 1, respectively, we obtain the second table. It appears, then, that the two tables are indistinguishable, except for notation.

We can make the preceding ideas more precise by setting up a certain one-to-one correspondence between the sets $\{D,E\}$ and $\{0,1\}$, namely,

$$E \leftrightarrow 0$$

$$D \leftrightarrow 1$$

and noting that all entries in the two tables match under this correspondence. In this way we clearly demonstrate that the two groups, $(\{D,E\}, +)$ and $(\{0,1\}, +)$, have the same structure. If we say "of the same structure" in classical Greek, the word comes out as "isomorphic." Thus we now assert that the systems $(\{D,E\}, +)$ and $(\{0,1\}, +)$, with addition as the operation, are *isomorphic:* or, since they are both groups under addition, we say that they are *isomorphic groups*. More formally, we have the definition

Definition 6.17 Isomorphic Groups Two groups $(G,\circ)$ and $(H,\not\circ)$ are said to be *isomorphic* iff there is a one-to-one correspondence between the sets G and H such that for any corresponding elements a, b, c of G and $\not a$, $\not b$, $\not c$ of H (with $a \leftrightarrow \not a$, etc.), $a \circ b = c$ whenever $\not a \not\circ \not b = \not c$.

You should note that Definition 6.17 calls for *a* one-to-one correspondence between the elements of G and H which yields corresponding results when the respective group operations are applied. It does not insist that *every* one-to-one correspondence between G and H produce such results. For instance, between $G = (\{D,E\}, +)$ and $H = (\{0,1\}, +)$ we could set up the correspondence

$$D \leftrightarrow 0$$

$$E \leftrightarrow 1$$

Since this correspondence fails to match E, the identity of G, with 0, the identity of H, we can already predict trouble. That it arrives we can show by taking $a = b = D \in G$. Then in G, $D + D = E$, according to Table 6.8. Under the above correspondence, this equation converts to $0 + 0 = 1$, which violates the (mod 2) table.

The point of Definition 6.17 is that the correspondence we first attempted *succeeds*. Although this is by now quite obvious, with these finite systems it is possible to double-check by comparing all four possible sums in G—$E + E$, $E + D$, $D + E$, $D + D$—with the corresponding sums in H. For larger finite systems such a task becomes very tedious, and for infinite systems, impossible. Fortunately, theoretical considerations and various short cuts combine to reduce the labor dramatically.

We have just seen that the groups $(\{D,E\}, +)$ and $(\{0,1\}, +)$ are isomorphic. An interesting question now is: Are there groups of *order two* (i.e., containing exactly two elements) whose structure is different from that of the preceding two groups? The answer is No. It turns out that *all* groups of order two have the same structure: in other words, any two *groups* of order two are isomorphic. (Of course, systems with two elements that are *not* groups might exist with quite different structures.) Moreover, all groups of order three are isomorphic (to each other). We have already produced two such groups: the arithmetic (mod 3) under addition[22] and the abstract group of Table 6.7.

For a larger example of isomorphic groups, we now compare in Table 6.9 three groups, or systems, of order 4. They are: (A), the additive group of the integers (mod 4); (B), the multiplicative group of the integers, *with* 0 *deleted*, (mod 5); and, (C), the group of symmetries of the rectangle. Note that each group has four elements. (Can two groups that do not have the same number of elements be isomorphic?)

TABLE 6.9

(A)

+	0	1	2	3
0	0	1	2	3
1	1	2	3	0
2	2	3	0	1
3	3	0	1	2

(B)

·	1	2	3	4
1	1	2	3	4
2	2	4	1	3
3	3	1	4	2
4	4	3	2	1

(C)

$\circ$	I	R	H	V
I	I	R	H	V
R	R	I	V	H
H	H	V	I	R
V	V	H	R	I

We need not be troubled that the operations for these groups differ: for Definition 6.17 permits G and H to have the operations $\circ$ and ϕ, which need not be the same.

Let us first compare (A) and (B). Can we tell at a glance whether they are isomorphic? Group (A) exhibits a cyclic character in that the order in which the numbers appear in each row (or column) is the same: 0, 1, 2, 3, 0, 1, 2 3, ... Group (B) does not seem to show this characteristic. However, our problem is deeper than surface appearances. There are, for sets with four elements each, $4\cdot 3\cdot 2\cdot 1$, or 24, possible one-to-one correspondences;[23] and the groups are isomorphic if only one of these 24 correspondences "works." To test these 24 possibilities is very tedious. Let us instead begin this way: clearly, if the two groups are isomorphic, their identities must correspond, for if a nonidentity in one group were made to correspond to the identity in

[22]See Exercises 6.2, Problem 1.

[23]See Exercises 2.7, Problem *7.

the second, the special property of the identity ($a \circ e = a$) in the second would not hold in the first. Thus we have: $0 \leftrightarrow 1$ (the identity 0 in (A) is matched with the identity 1 in (B)). Now let us examine inverses. In (A), $2' = 2$, while $1' = 3$, and $3' = 1$; thus 2 is the only nonidentity which is *its own* inverse. And in (B), $4' = 4$, while $2' = 3$, and $3' = 2$; thus 4 is the only nonidentity which is *its own* inverse. Obviously, we should make 2 in (A) correspond to 4 in (B). This narrows the choice of correspondences enormously. Let us try, then, the correspondence shown in Table 6.10.

TABLE 6.10

(A)		(B)
0	$\leftrightarrow$	1
1	$\leftrightarrow$	2
2	$\leftrightarrow$	4
3	$\leftrightarrow$	3

We now check a few equations: In (A), $2 + 3 = 1$; and the corresponding equation for (B), $4 \cdot 3 = 2$, is true. In (A), $3 + 1 = 0$; and the corresponding equation for (B), $3 \cdot 2 = 1$, is true. In fact, we shall find that any true equation for (A), when "translated" by means of Table 6.10 into an equation for (2), remains true there.

One method of checking whether all true equations in (A) yield corresponding true equations in (B) resorts to physical objects. Thus we may take twelve pieces of paper or cardboard whose two sides can be readily distinguished. On the first side of three of the pieces we place the symbol 0 from (A); on three more pieces, the symbol 1 from (A); etc. Then on the second side of each piece we write the corresponding element of (B). Now, beginning with all the pieces first side up, we select any two pieces, find a third piece which represents their sum according to the table for (A), and form a true equation in (A) — e.g., $2 + 3 = 1$. By turning each piece over in its place, we get the corresponding equation in (B) — e.g., $4 \cdot 3 = 2$ — whose truth we can verify from the table for (B). In this manner we can mechanically check all sixteen pairs of sums in the two groups. If we find that all equations are true in both groups, Definition 6.17 is satisfied, and the systems are isomorphic. By using the commutative and identity properties we may reduce this work substantially. Though trivial and time-consuming, this method does clarify the concepts of isomorphism.

As an alternative, briefer method, it is possible to start with the table for group (B) and rearrange its rows and columns so that the elements of this table occupy the same positions along its top and left side as do the corresponding elements of the table for group (A). If care is taken not to alter any products, then the rearranged table for (B) will appear as Table 6.11. The cyclic character of (B) is now fully evident, and its structure is seen to be identical to that of (A).

Now let us compare (A) and (C). Clearly, we make 0 correspond to I.

TABLE 6.11

(B)

·	1	2	4	3
1	1	2	4	3
2	2	4	3	1
4	4	3	1	2
3	3	1	2	4

As for inverses, in (C) $R' = R$, $H' = H$, $V' = V$. Thus in (C) *every* element is its own inverse; but in (A) only 0 and 2 have this property. Thus it is impossible to find any element of (C) that will behave as 1 does in (A): $1' = 3 \neq 1$. So (A) and (C) are *not* isomorphic, as is easily verified by finding an equation involving the element 1 which is true in (A) but whose analog in (C) is false. What about (B) and (C)?

As a result of the above, we see that there are *at least two* structurally different groups of order four: the *cyclic* group, represented by (A) and (B), and the so-called fours group, represented by (C). Note that both groups are commutative. It turns out that these two are the *only* abstractly different groups of order four: *any* group of order four is isomorphic either to (A) or to (C). Consequently, all groups of order four (and also, of order two and three) are necessarily commutative.

EXERCISES 6.6

*1. Use the successful correspondence between groups (A) and (B) to translate to (B) the following equations in (A), then check that the translated equations are true in (B): $0 + 3 = 3$, $1 + 1 = 2$, $3 + 3 = 2$, $3 - 1 = 2$ (How is subtraction in (A) translated to (B)?), $2 - 3 = 3$.

*2. (a) Consider the following correspondence between groups (A) and (B) of the text:

(A)		(B)
0	$\leftrightarrow$	1
1	$\leftrightarrow$	2
2	$\leftrightarrow$	3
3	$\leftrightarrow$	4

(i) Show that this correspondence is not successful by providing a pair of equations from (A) and (B) which demonstrate its failure.

(ii) Does the failure of this correspondence mean that (A) and (B) are not isomorphic after all? Explain.

(b) Consider the following correspondence between (A) and (B):

(A)		(B)
0	$\leftrightarrow$	1
1	$\leftrightarrow$	3
2	$\leftrightarrow$	4
3	$\leftrightarrow$	2

Is this a successful correspondence? Explain why it is, or else provide an example of its failure.

*3. (a) For comparison purposes, copy side-by-side the addition table (mod 3) and Table 6.7.

(b) Prove that the groups of part (a) are isomorphic by exhibiting an appropriate one-to-one correspondence between them and rearranging their tables (if necessary) to make clear that their structure is the same.

*4. Compare the system of the arithmetic (mod 4) under addition, as given in (A) of Table 6.9, with that of the arithmetic (mod 4) under multiplication. Are these two systems isomorphic groups? Why or why not?

*5. Construct from cardboard an isosceles triangle that is *not* equilateral. Work out the set of symmetries of such a triangle, interpreting "$\circ$" as "followed by." Does this set of symmetries form a group? Is this system isomorphic to any system discussed in this section? Explain your answer.

*6. It turns out that all groups of order 5 are isomorphic. Give an example of one such group.

*7. The additive group of the integers (mod 6) is one group of order 6. Are all groups of order 6 isomorphic to this group? (*Hint:* Consider the group of symmetries of an equilateral triangle found in Section 6.2.)

*8. It turns out that there are two structurally different (nonisomorphic) groups of order 6, and that all groups of order 7 are isomorphic. For order 8 there are actually five structurally different groups.

(a) Describe two of these groups. (*Hint:* Do not neglect the starred problems in Exercises 6.2.)

(b) For some values of n, the number of structurally different groups grows rapidly as n increases. Thus a book was published in 1964 listing all 267 groups of order 64. On the other hand, all groups of order 71 are isomorphic; and similar statements can be made for groups of orders 11 and 13, but not of order 9. If possible, conjecture a theorem.

Suggestions for Investigation

1. Two of the founders of group theory were Niels Abel (1802–1829) and Evariste Galois (1811–1834). Their lives were fascinating, mathematically speaking, but very brief and dogged with tragedy. Use the Bell reference and other histories to become acquainted with the details of their lives and mathematical contributions.

2. An interesting relationship between "clock" arithmetics and nuclear energy is suggested by the title of the very readable Scheid reference, which explains how arithmetics (mod n) are used to generate random numbers which, in turn, form part of a model applied to the study of protective shields from radioactive materials.

3. Arithmetics (mod n) form a basis for some methods of coding and decoding secret messages, especially when $n = 26$. One such code is the Caesar Cipher, so named because it was used by Julius Caesar. The Peck reference provides a lucid introduction to codes; the Sinkov reference is longer and more sophisticated. The Feistel reference provides an application.

4. If you wish to explore a geometric approach to groups, using cycles, the reference by McGee is beneficial. This method of portraying the structure of a group makes the concept of isomorphism visually evident.

5. A finite group considerably larger than the ones we have studied is formed by the symmetries of a cube, taking composition, as usual, as the operation. This group has 32 elements (motions). Try to find and describe each of them. While it is tedious to supply the entire operation table, try to work out enough of it to discover some characteristics of this group (e.g., whether it is commutative). A convenient physical cube to manipulate is a child's alphabet block with different letters on its six faces.

6. We have studied three properties of relations, and have found that some relations satisfy all three, and others satisfy none. Other possibilities are relations which are reflexive only, symmetric only, . . . , symmetric and transitive only, ... Altogether there are eight theoretically possible combinations of the presence or absence of these properties. For each of the eight, try to find and describe carefully a relation which illustrates it; or else try to explain why such an example does not exist.

Bibliography

Bell, E. T. *Men of Mathematics*. New York: Simon and Schuster, 1961. Especially Chapters 17 (Abel) and 20 (Galois).

Brumfiel, C. R. "Integers Modulo 3." *Twenty-Eighth Yearbook*. Washington: National Council of Teachers of Mathematics, 1963. Chapter 2.

Eggan, L. C. "Number Theory." *Twenty-Eighth Yearbook.* Washington: National Council of Teachers of Mathematics, 1963. Chapter 5.

Feistel, H. "Cryptography and Computer Privacy." *Scientific American,* 228:5 (May, 1973), 15–23.

Freund, J. E. *A Modern Introduction to Mathematics.* Englewood Cliffs: Prentice-Hall, Inc., 1956. Chapter 15.

Glicksman, A. M. "Why Must the Modulus Be Prime?" *New York State Mathematics Teachers' Journal,* 16:4 (1966), 158–61.

Hammel, Arnold. "Verifying the Associative Property for Finite Groups." *Mathematics Teacher,* XLI (1968), 136–39.

Ingraham, M. L. "A Permutation Group and Its Isomorphisms." *Twenty-Eighth Yearbook.* Washington: National Council of Teachers of Mathematics, 1963. Chapter 12.

Lieber, L. R. *Galois and the Theory of Groups.* Lancaster, Pa.: The Science Press Printing Co., 1932.

McGee, R. G. "Group Isomorphism: An Approach Using Cycle Graphs." *New York State Mathematics Teachers' Journal,*19:4 (1969), 147–53.

Mueller, F. J. "Modular Arithmetic." *Twenty-Seventh Yearbook.* Washington: National Council of Teachers of Mathematics, 1963. Chapter 4.

National Council of Teachers of Mathematics. *More Topics in Mathematics for Elementary School Teachers. Thirtieth Yearbook.* Washington: National Council of Teachers of Mathematics, 1968. "Relations" and "Equivalence Relations," 262–91.

Peck, L. C. *Secret Codes, Remainder Arithmetic and Matrices.* Washington: National Council of Teachers of Mathematics, 1961.

Richardson, Moses. *Fundamentals of Mathematics,* rev. ed. New York: The Macmillan Co., 1958. Chapter 17.

Scheid, Francis. "Clock Arithmetic and Nuclear Energy." *Mathematics Teacher,* LII (1959), 604–607. This article is reprinted in M. S. Bell, ed.,*Some Uses of Mathematics: A Source Book for Teachers and Students of School Mathematics, XVI.* Pasadena, Calif.: A. C. Vroman, Inc., 1967, 89–92.

Sinkov, Abraham. *Elementary Cryptanalysis: A Mathematical Approach.* New York: Random House, Inc., 1968.

Syer, H. W., ed. *Mathematical Systems,* SMSG *Supplementary and Enrichment Series,* 19. Pasadena, Calif.: A. C. Vroman, Inc., 1965.

7

The Integers and Further Extensions

How are integers related to whole numbers? What are negative numbers? How are subtraction, division, and taking roots closed? What roles do group and order properties play in extending the system of whole numbers? What kind of numbers are decimals which never end? How "real" are the real numbers?

Section 7.1 Giving Numbers a Direction

In using numbers to express size or quantity, we sometimes wish to describe their nature or direction as well. When we record temperatures, for example, we not only state the number of degrees, measured from zero, but also indicate whether the reading is *above* or *below* zero. For convenience we use the symbols $^{+}$ and $^{-}$, recording the temperature on a hot summer's day as $^{+}95°$ F and that during a winter cold wave as $^{-}25°$ F.

When signifying direction, the $^{+}$ and $^{-}$ symbols function as adjectives, not verbs: they denote two opposite *qualities* of the numbers to which they are affixed, rather than the *operations* of addition and subtraction. For this reason we print them in a slightly raised position. In practice, the $^+$ sign is often omitted when the context makes clear that the number has direction; but the $^-$ sign is always expressed. In interpreting a temperature, for example, we may assume that it has been measured upward from zero, unless the $^-$ sign is present. Another familiar use of directed numbers occurs in the financial pages of the newspaper, where changes in stock prices, up or down, are shown by $^+$ or $^-$ signs, respectively, followed by the amount of the change. Other examples of signed numbers will come easily to mind.

The directed numbers we have been illustrating are called *integers*. We may approach their study in several ways. One is to begin with applications, appealing primarily to intuition rather than to logic. You probably first met negative numbers in this fashion. It is worth noting that, historically, negative numbers were viewed as mysterious or even unacceptable (even by some mathematicians) as late as the seventeenth century.

Since we are stressing the logical approach to mathematics, however, we shall introduce the integers as a system with two binary operations, obeying certain laws. Here, too, there are several ways of proceeding, depending partly on how many assumptions we are willing to make and how many abstract concepts we utilize. We shall choose one of the quickest and easiest ways; but we will have to pay for it by making powerful assumptions. We can make it quite brief because the group concept is available to us.

We have already noted that the set N of natural numbers does not form a group under the operation of addition. (Why not?) If we include 0, we obtain the set W of whole numbers. Does W form a group under addition? What we will do now, in essence, is assume that the whole numbers can be supplemented by annexing just enough "new" numbers so that the extended system *does* form a group under addition. This means, of course, that N and W are subsets of the new system. Indeed, the very reason that the system $(W,+)$ fails to be a group shows what we must do: we must join to W enough new numbers to provide an additive inverse for each element of W, as well as for each new element.

EXERCISES 7.1

1. Give examples of the use of directed numbers in various fields of natural and social science and in everyday life.

2. Explain the meaning of the $^+$ and $^-$ symbols in the following situations:
   (a) If the latitude of Boston, Massachusetts is denoted by $^+42°$, by which symbol should we denote the latitude of Rome, Italy? Canberra, Australia?
   (b) If the longitude of Boston is denoted by $^-71°$, by which symbol should we denote the longitude of Rome? San Francisco, California? Peking, China?
   (c) If the highest land elevation is Mount Everest, 29,028 feet above sea level, and the deepest part of the ocean is the Philippine Trench in the

Pacific, 37,782 feet below sea level, represent these measurements by appropriate signed numbers.

(d) If the United States budget in fiscal 1943 showed a deficit of about 57.4 billion dollars and that in fiscal 1948 showed a surplus of about 8.4 billion dollars, represent these amounts by appropriate signed numbers.

(e) The distance required to stop a car, initially traveling 60 miles per hour on a dry road, is about 160 feet, which yields a deceleration of about 24 feet per second per second. The same car, starting from rest with the gas pedal pressed to the floor, accelerates at the rate of about $5\frac{1}{2}$ feet per second per second to reach a speed of 60 miles per hour in 16 seconds. Represent these accelerations by appropriate signed numbers.

3. (a) Explain why the set N fails to form a group under addition.

(b) Does the set W form a group under addition? Why or why not?

(c) Does either N or W form a group under multiplication? Explain your answers.

4. The arithmetic (mod 3) of Section 6.1 constitutes a system in which addition and multiplication are defined binary operations. Compare the group properties, under addition and under multiplication, of this system with those of Problem 3.

Section 7.2 Inventing the Integers

The exercises of the previous section confirm that addition in W is closed, associative, and even commutative (though commutativity is not required for a group); and 0 is the additive identity. But what about additive inverses? We would like to find, for each element $x \in W$, an element $x' \in W$ such that $x + x' = 0$.

To be specific, let us seek $1'$, the additive inverse of 1, among the whole numbers. We know that $1'$ must satisfy the equation $1 + 1' = 0$. Now clearly, $1' \neq 0$; for $1 + 0 = 1$ (Why?), rather than 0. So if $1' \in W$, then it must be a natural number. But if 1 and $1'$ are both natural numbers, and N is closed under addition, we must conclude that $1 + 1' \in N$ also and, therefore, cannot be 0. Similar difficulties arise when we try to find inverses for other natural numbers. Indeed, the only number $x \in W$ for which we can find an $x' \in W$ to satisfy $x + x' = 0$ is 0 itself: for by Theorem 4.4 with $a = 0$, $0 + 0 = 0$. So of all the whole numbers, only 0 has an inverse in W, and $0' = 0$. This last equation, of course, is an instance of Exercises 6.1, Problem *9.

In order, therefore, to satisfy the inverse requirement of the group definition (Definition 6.3), we must take the bold step of *creating* additional numbers, outside the set W, to act as additive inverses for the natural numbers. We could designate these new numbers by $1', 2', 3', \ldots$. In practice, however, we designate them by $^-1, ^-2, ^-3, \ldots$, and call them *negative integers*. For contrast, we may similarly designate the natural numbers as $^+1, ^+2, ^+3, \ldots$, and refer to them as *positive integers*. By our very method of creating the negative integers, we are insisting that

$$1 + {}^-1 = 0 \text{ (or } {}^+1 + {}^-1 = 0); \; 2 + {}^-2 = 0; \; 3 + {}^-3 = 0; \ldots$$

Before continuing with the theory, we pause to remind you of a standard geometric interpretation, or *model*, of the integers. Beginning with the number line described in Section 4.2, we may extend it to the left from the 0 point, using the same unit of measurement. See Figure 7.1. We then label the points to the left of 0 as $^-1, ^-2, ^-3, \ldots$. In other words, the natural numbers are depicted, as before, by equally spaced points to the right of 0, and their inverses by equally spaced points to the left. We note that the points of the line corresponding to $^-3$ and $^+3$ are the same distance from the zero point: 3 units away, but on opposite sides. This familiar model is a powerful intuitive aid in operating with integers and examining their order properties.

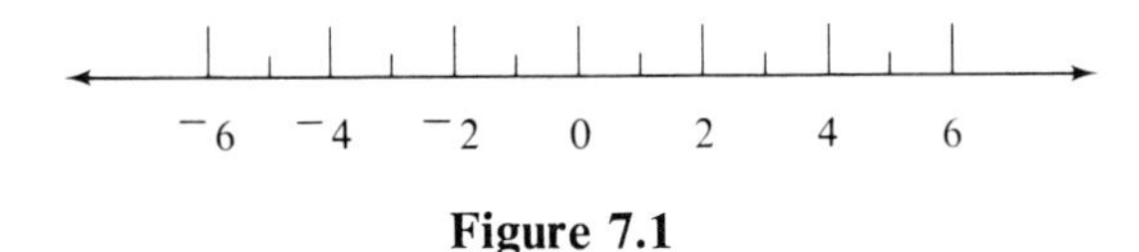

Figure 7.1

Since we shall often refer to the set of integers, we shall designate it, for convenience, by the letter Z. Then $(Z, +, \cdot)$ is the system of the integers, where $+$ and $\cdot$ here refer to the binary operations of addition and multiplication on Z. We shall develop these operations in detail in the following two sections. But if we define the system $(Z, +, \cdot)$ in terms of group properties, it will turn out that we have restricted the two fundamental operations on Z so that they behave just as our previous experience with signed numbers leads us to expect. We give, as Definition 7.1, the entire basis of the system of integers, and then trace out its consequences in more leisurely fashion.

Definition 7.1 System of Integers The system of integers, $(Z, +, \cdot)$, consists of the set W of whole numbers, together with the additive inverse of each element of W, designated as $^-1, ^-2, ^-3, \ldots$,[1] and two binary operations on Z, addition and multiplication, subject to the following requirements:

1. The system $(Z, +)$ forms a commutative group, with 0 as its identity.
2. The system $(Z, \cdot)$ is closed, associative, and commutative and has 1 as its identity.
3. Multiplication is distributive over addition.

We note that, according to Definition 7.1, W and N are subsets of Z, which

---

[1]Since $0' = {}^-0 = 0$, we do not list $^-0$.

therefore contains exactly the elements $0, 1, {}^-1, 2, {}^-2, \ldots$. Thus Z is an infinite set: an exercise suggests how to determine $n(Z)$, the cardinal number of Z. Also associated with Z are the usual two binary operations, which are subject to many of the same laws as are these operations on W.

For convenience we introduce the following obvious:

Definition 7.2 Positive, Zero, and Negative Integers If $x \in Z$, then

1. x is called a *positive* integer if $x \in N$.
2. x is called a *negative* integer if $x' \in N$.
3. 0 is called the *zero* integer.

You should check that every integer falls into exactly one of these three categories. Thus 4 (or ${}^+4$) is a positive integer, since $4 \in N$; ${}^-3$ is a negative integer, since $({}^-3)' = 3 \in N$; and 0, the zero integer, is neither positive nor negative.

One serious question, however, ought to be raised. Definition 7.1 assumes that the system it describes actually exists. This assumption is far from trivial; for it is quite easy to propose a system that is self-contradictory and so cannot logically exist. An example would be $(W,+,\cdot)$, with the additional requirement that $a \div 0 = 0$ for any $a \in W$. Since such a quotient would contradict Theorem 4.31, which is a consequence of other laws of $(W,+,\cdot)$, the proposed system is impossible. Does Definition 7.1 lead us into a similar trap? The answer, fortunately, is No; and a more sophisticated development of $(Z,+,\cdot)$ than ours would demonstrate this. But our "streamlined" approach forces us to accept the system $(Z,+,\cdot)$ on faith. The fact that we have worked with positive and negative integers since grade school at least gives us some psychological reassurance.

One important question related to existence, however, we can answer. Since we have defined ${}^-4$ as the additive inverse of 4, we know at once that $4 + {}^-4 = 0$; and similar statements are true of any natural number in Z. Moreover, 0 is its own inverse. But do the negative integers have inverses *within Z?* As an example we take ${}^-5$. According to Definition 7.1, ${}^-5$ is the additive inverse of 5, so that $5 + {}^-5 = 0$. Then by commutativity of addition (from Definition 7.1), ${}^-5 + 5 = 0$. But this equation implies that 5 is the additive inverse of ${}^-5$. So the additive inverse of ${}^-5$ is, in fact, a natural number, hence an integer. Can the preceding argument be generalized? If so, it means that the system $(Z,+,\cdot)$ is closed with respect to taking additive inverses. This question is settled by an exercise.

EXERCISES 7.2

1. (a) Draw a number line and locate on it points corresponding to the integers:

$$5,\ {}^-2,\ {}^+7,\ 0,\ {}^-6,\ 4' \text{ (the additive inverse of 4)},\ 0',\ ({}^-3)',\ ({}^+1)'$$

(b) Use Definition 7.2 to classify the integers of part (a) as positive, negative, or zero.

(c) Interpret Definition 7.2 geometrically on the number line by stating where the points corresponding to positive integers, negative integers, and zero lie.

2. (a) Give the additive inverse of each of the following integers:

$4, 0, {}^-3, {}^+6, {}^-6, ({}^+2)'$ (i.e., find the inverse of the inverse of $({}^+2)$), $({}^-5)'$

(b) For each number of part (a), write an equation expressing the relation of the given number to the corresponding answer. (*Hint:* See Definition 6.1.)

(c) (i) Represent each of the numbers of part (a), together with its inverse, on a number line.

(ii) State the distance between the given number and the zero point on the number line; also do this for the inverse of the given number and the zero point; then compare the two distances.

3. (a) If $n \in N$, is ${}^-n \in N$? Is ${}^-n \in Z$? Explain your answers, justifying them by reference to definitions and/or theorems.

(b) Since $0 \in W$, is it correct to assert that $0 \in Z$? that $0' \in Z$? Justify your answers as in part (a).

(c) Consider the following:

THEOREM 7.1 Closure of Additive Inverses The additive inverse of any integer is also an integer.

Illustrate Theorem 7.1 for the integers $3, {}^-2, 0, {}^+4$.

(d) An informal argument in the text shows that $({}^-5)' \in Z$. Generalize this argument and thus construct a formal proof of Theorem 7.1.

4. (a) Show that the set Z satisfies Definition 2.11 and so is an infinite set. (*Hint:* Set up a certain one-to-one matching between W and Z.)

(b) In view of Definition 2.10 and the results of part (a), determine $n(Z)$, the cardinal number of the set Z.

5. (a) (i) Given $3 \in Z$, find $3'$ and $(3')'$; then compare these answers with the given integer.

(ii) Repeat part (i) for ${}^-6 \in Z$ and $0 \in Z$.

(b) Write a statement generalizing the results of part (a), then compare it with Theorem 7.2:

THEOREM 7.2 Double Inverse If $x \in Z$, then $(x')' = x$.

(c) Prove Theorem 7.2. (*Hint:* Use Definition 6.1.)

*6. (a) Let Z^- be the set of negative integers. Determine $n(Z^-)$, justifying your answer logically.

(b) Use Z^- and other appropriate sets, together with Definition 4.4 extended to infinite sets, to show that $\aleph_0 + \aleph_0 = \aleph_0$. (*Hint:* See also Exercises *4.6, Problem *3.)

Section 7.3 Addition of Integers

Like whole numbers, integers were created historically through use; and operations on them ought to be defined so that they have the widest applications. What, then, would be the most useful decision concerning such sums as $^-4 + {}^-3$, $^-3 + {}^+5$, and $^+4 + {}^-7$?

A shopworn, but still intuitively appealing interpretation of positive and negative integers is to consider them as representing assets and debts, respectively. In that case, it is clear that, since a person who owes bills of \$4 and \$3 is altogether \$7 in debt, the sum of $^-4$ and $^-3$ should be $^-7$ in this interpretation. You can easily supply similar answers for the other two sums.

A second interpretation considers a motor launch going north (assumed to be the positive direction) or south, affected by an ocean current. Suppose the motor's controls are set so that the launch could travel south in still water at a speed of 4 knots ($^-4$); but that a current going south at 3 knots ($^-3$) is also present. Then the total speed of the launch, produced by the motor and the current, is represented by the sum, $^-4 + {}^-3$ knots. Again, you may interpret the remaining two sums according to this model.

Still another intuitive aid is the number line. We begin by locating the point for the first addend, $^-4$, on the line. Then we move along the line as many units as the second addend directs, in this case, 3 units *to the left* for $^-3$. Thus we arrive at the point $^-7$, which therefore represents the sum $^-4 + {}^-3$.

Since Definition 7.1 is so all-embracing, it might seem as though there is little freedom left for defining the sum of two integers. This is true, all the more so because we must not contradict the sums which we obtained for whole numbers in Chapter 4. Let us use these facts to try to determine a few sums logically.

We begin with the sum $^-4 + {}^-3 = x$. Intuitively, we would like x to be $^-7$. Since $^+7$ is the inverse[2] of $^-7$, we would expect that $x = {}^-7$ iff $x + {}^+7 = 0$. Replacing x by its equal, $^-4 + {}^-3$, and $^+7$ by its equal, $3 + 4$, we obtain the following sequence of equations:

$$\begin{aligned} x + {}^+7 &= ({}^-4 + {}^-3) + (3 + 4) \\ &= [{}^-4 + ({}^-3 + 3)] + 4 \\ &= ({}^-4 + 0) + 4 \\ &= {}^-4 + 4 \\ &= 0 \end{aligned}$$

[2]We are here using the fact that the additive inverse of an integer is unique. Since by Definition 7.1 $(Z,+)$ is a group, this uniqueness follows immediately from Exercises 6.2, Problem *9.

So, by transitivity,[3] we get $x + {}^{+}7 = 0$. This shows that we must have $x = {}^{-}7$ (Why?). You should justify each step of the above chain by Definition 7.1 or by previously proved properties. You should also reflect on why we expressed 7 as $3 + 4$ in this example, rather than as $5 + 2$ or even as $4 + 3$.

Passing now to the second sum, ${}^{-}3 + {}^{+}5$, we can similarly obtain the answer by using the debts-and-assets model, the motor launch model, or the number line. On the number line, for example, we start at the point for ${}^{-}3$ and move 5 units to the *right*, arriving at ${}^{+}2$.

Now it is easy to use Definition 7.1 and other properties to derive the sum of ${}^{-}3$ and ${}^{+}5$, logically and quite directly. We merely write ${}^{+}5$ as ${}^{+}3 + {}^{+}2$ (Why this breakdown?), and develop the following chain of equations:

$$\begin{aligned} {}^{-}3 + {}^{+}5 &= {}^{-}3 + ({}^{+}3 + {}^{+}2) \\ &= ({}^{-}3 + {}^{+}3) + {}^{+}2 \\ &= 0 + {}^{+}2 \\ &= {}^{+}2 \end{aligned}$$

so that ${}^{-}3 + {}^{+}5 = {}^{+}2$. You should supply a reason for each step. We note that the numerical value of the sum, 2, is this time the *difference* of the two addends $(5 - 3 = 2)$, whereas in the example ${}^{-}4 + {}^{-}3 = {}^{-}7$ it was their *sum*.

Since here we were forced to consider the numerical or *absolute* value of an integer, let us make this idea precise. Intuitively, if $x \in Z$, then the absolute value of x, denoted by $|x|$, is the distance (without regard to direction) between the point x on the number line and the zero point. Thus

$$|{}^{+}5| = 5, \quad |{}^{-}5| = 5, \quad |{}^{-}4| = 4, \quad |{}^{+}2| = 2, \quad |{}^{-}3| = 3$$

What is $|0|$? Formally, we have

Definition 7.3 Absolute Value If $x \in Z$, then the *absolute value* of x (in symbols, $|x|$) is given by

1. $|x| = x$ if x is positive.
2. $|0| = 0$.
3. $|x| = x'$ if x is negative.

You should check that the numerical examples above are consistent with Definition 7.3.

Returning to the sum ${}^{-}3 + {}^{+}5 = {}^{+}2$, we note that, of the two addends, the one whose absolute value is greater[4] is ${}^{+}5$, a positive integer, and that the sum,

---

[3]We shall assume that equality in Z has the same properties as that in W.

[4]Note that Definition 4.8 applies here, since, according to Definition 7.3, the absolute value of an integer is always a whole number.

$^{+}2$, is also positive. Consideration of more numerical examples and their models shows that this result is no accident.

We are now equipped to deal with the third sum, $^{+}4 + {}^{-}7$. You are invited to interpret this sum, also according to the three models. In particular, moving on the number line from $^{+}4$ seven steps to the left yields the point $^{-}3$.

Logically, we may devise the following chain of equations:

$$\begin{aligned} {}^{+}4 + {}^{-}7 &= {}^{+}4 + ({}^{-}4 + {}^{-}3) \\ &= ({}^{+}4 + {}^{-}4) + {}^{-}3 \\ &= 0 + {}^{-}3 \\ &= {}^{-}3 \end{aligned}$$

The first step uses the sum $^{-}4 + {}^{-}3 = {}^{-}7$, which we already worked out. You should supply a reason for each of the other steps. This time we should point out that the negative addend, $^{-}7$, has the larger absolute value, so the sum is accordingly negative. Moreover, the absolute value of this sum, $|{}^{-}3| = 3$, is again the *difference* of the absolute values of the two addends, namely, $7 - 4 = 3$.

Five other cases remain; they are illustrated by the sums $0 + 0$, $^{+}3 + {}^{+}4$, $^{+}4 + 0$, $^{-}3 + 0$, and $^{-}2 + {}^{+}2$. All cases can be calculated from Definition 7.1 and previously established properties for whole numbers. The first sum is too trivial to discuss. The second, $^{+}3 + {}^{+}4$, is an application of Definition 4.4, since $^{+}3$ and $^{+}4$ are whole numbers. Similarly, $^{+}4 + 0$ is an instance of Theorem 4.4; or it may be supported by the part of Definition 7.1 which names 0 as the additive identity. Since $^{-}3 \notin W$, the sum $^{-}3 + 0$ must be obtained only in the latter way, via Definition 7.1. As for $^{-}2 + {}^{+}2$, it is foreshadowed by Definition 6.1 and supported by the inverse property of the group $(Z, +)$ contained in Definition 7.1.

We need, therefore, only to generalize the examples of this section in order to produce a complete set of rules for adding integers. This important task is carried forward in the exercises.

EXERCISES 7.3

1. Explain intuitively, by use of the debt-asset and motor launch models, why

   (a) $^{-}3 + {}^{+}5 = {}^{+}2$ (b) $^{+}4 + {}^{-}7 = {}^{-}3$

2. Explain intuitively, by use of the models of Problem 1 and the number line, why

   (a) $^{+}3 + {}^{+}4 - {}^{+}7$ (b) $^{-}3 + 0 = {}^{-}3$ (c) $^{-}2 + {}^{+}2 = 0$

3. (a) In the text's chain of equations, justifying the sum $^{-}4 + {}^{-}3 = {}^{-}7$, $^{+}7$ was decomposed into $^{+}3 + {}^{+}4$. Why was this decomposition used rather than $^{+}2 + {}^{+}5$ or $^{+}4 + {}^{+}3$?

   (b) Answer similar questions regarding the chain of equations for

   (i) $^{-}3 + {}^{+}5 = {}^{+}2$ (ii) $^{+}4 + {}^{-}7 = {}^{-}3$

4. Supply a reason for each step in the text's logical argument showing that

   (a) $^{-}4 + ^{-}3 = ^{-}7$ (b) $^{-}3 + ^{+}5 = ^{+}2$ (c) $^{+}4 + ^{-}7 = ^{-}3$

5. (a) Explain how Definition 7.3 applies to yield $|^{+}5| = 5$, $|^{-}5| = 5$, $|^{+}2| = 2$, $|0| = 0$, $|^{-}3| = 3$.

   (b) Write the absolute value of each of the following integers:

   $$^{+}1,\quad ^{-}6,\quad 8,\quad ^{-}10,\quad ^{-}3 + ^{+}3,\quad ^{-}4 + ^{-}3,\quad ^{-}5 + ^{+}8,\quad ^{-}8 + ^{+}5$$

   (c) Sometimes it is carelessly stated that taking the absolute value of a number means simply "dropping its sign."

      (i) Does the statement appear to be true of the examples of part (b)? Explain your answer.

      (ii) To some persons the above statement might justify the equation $|^{-}b| = b$, where b is any integer. Is this equation correct? (*Hint:* Consider both positive and negative values for b, interpreting ^{-}b as "the additive inverse of b.")

6. (a) Given the indicated sum $^{-}2 + ^{-}6$, obtain the result

      (i) intuitively by use of some model.

      (ii) logically by a chain of equations, supplying a reason for each step.

   (b) Can the results of $^{-}4 + ^{-}3$ and $^{-}2 + ^{-}6$ be generalized to yield a rule for adding any two negative integers? If so, write such a statement formally as a theorem.

7. Answer Problem 6 for the following indicated sums:

   (a) $^{-}1 + ^{+}5$ (b) $^{+}9 + ^{-}6$ (c) $^{-}10 + ^{+}3$

8. State formally as theorems rules for adding integers that are generalizations of the equations

   (a) $^{+}7 + ^{+}11 = ^{+}18$ (b) $^{+}12 + 0 = ^{+}12$ and $0 + ^{-}6 = ^{-}6$

   (c) $^{+}9 + ^{-}9 = 0$

9. For each of the following theorems, provide a numerical example different from those discussed thus far, then prove it formally by generalizing the logical arguments in the text:

   (a)

THEOREM 7.3 Sum of Two Negative Integers If ^{-}a and ^{-}b are any two negative integers, then $^{-}a + ^{-}b = ^{-}(a + b)$.

   *(b)

THEOREM 7.4 Sum of Two Integers With Opposite Signs (Case I) If ^{+}a and ^{-}b are any pair of positive and negative integers, respectively, and if $|^{+}a| > |^{-}b|$, then $^{+}a + ^{-}b = ^{+}(a - b)$. (*Hint:* Care must be taken to show that $a - b \in N$.)

*(c)

THEOREM 7.5 Sum of Two Integers With Opposite Signs (Case II) If ^+a and ^-b are any pair of positive and negative integers, respectively, and if $|^-b| > |^+a|$, then $^+a + ^-b = ^-(b - a)$.

Section 7.4 Multiplication of Integers

Though the rules for multiplication of integers are simpler to state and apply than those for addition, they are harder to support intuitively by models. For instance, to extend the debts-assets model to cover multiplication, it is convenient to introduce a time variable. Thus, to investigate the product $(3)(^-2)$, we may assume that a person goes into debt at the rate of \$2 a day $(^-2)$ and we ask what his financial condition is after 3 days $(^+3)$, compared with his condition at the beginning. So this interpretation suggests that $(^+3)(^-2) = ^-6$.

The number line model for $(^+3)(^-2)$ yields a similar result if we interpret the indicated product as a command to perform 3 times the action of moving 2 steps to the left, starting at the zero point. This view is tantamount to considering multiplication as repeated addition:[5] $(^+3)(^-2)$ or $3(^-2) = ^-2 + ^-2 + ^-2 = (^-2 + ^-2) + ^-2$; then using Theorem 7.3 twice.

According to Definition 7.1, multiplication in Z is commutative. So we want also to have $(^-2)(^+3) = ^-6$. Interpreting this equation via the number line model or repeated addition, however, is less obvious, since it seems to call for committing an act a "negative-two" times. The debts-assets model is also a little forced, but can be rescued by assuming that "negative" time flows backward. Then $(^-2)(^+3) = ^-6$ may be considered as a comparison of a person's financial status two days ago $(^-2)$ with his present status, if he acquires assets at the rate of \$3 a day $(^+3)$.

Because these models are somewhat strained, it is a relief to be able to prove the two previous results by appeal to Definition 7.1. As in the first example of Section 7.3, we argue that $(^+3)(^-2) = x$ is actually $^-6$ iff $x + (^-6)' = 0$, or, equivalently, iff $x + 6 = 0$, because these equations show that x has 6 as its additive inverse. With a little luck we can produce a chain of equations which establishes that $x + 6 = 0$:

$$\begin{aligned} x + 6 &= (^+3)(^-2) + 6 \\ &= (^+3)(^-2) + (^+3)(^+2) \\ &= (^+3)(^-2 + ^+2) \\ &= (^+3)(0) \\ &= 0 \end{aligned}$$

Each step, of course, can be justified by Definition 7.1 or previously established properties of whole numbers. You should do this, and should also derive a similar chain for the product $(^-2)(^+3) = ^-6$.

[5] See Theorem 4.20, stated for natural numbers.

The remaining cases of products of integers are two, illustrated by $(^+3)(^+2)$ and $(^-3)(^-2)$. The first, the product of two natural numbers, is covered by Definition 4.6. The second is much more interesting and also difficult to handle intuitively. Since both factors are negative, there is no obvious interpretation in terms of repeated addition or motion on the number line. One could force the debts-assets model to yield a result by comparing a person's financial status with that of 3 days ago $(^-3)$ on the assumption that his debts were piling up at the rate of \$2 a day $(^-2)$. We would then describe his earlier status as \$6 better $(^+6)$ than his present one.

In these circumstances, the use of equations to produce a logical argument is particularly attractive. To prove that $(^-3)(^-2) = x$ is actually $^+6$, we show that $x + {^-6} = 0$, so that x is the additive inverse of $^-6$, or $^+6$, by Theorem 7.2. The chain of equations is

$$\begin{aligned} x + {^-6} &= (^-3)(^-2) + {^-6} \\ &= (^-3)(^-2) + (^+3)(^-2) \\ &= (^-3 + {^+3})(^-2) \\ &= (0)(^-2) \\ &= 0 \end{aligned}$$

You should supply reasons, noting that the substitution of $(^+3)(^-2)$ for $^-6$ utilizes a previous result of this section. As before, we pursue in the exercises the generalization of these examples to theorems for multiplying integers.

There is, however, a difficulty with the last step of this chain, $(0)(^-2) = 0$. Though we would like to use the zero product theorem, Theorem 4.16 does not apply here. Why not? But Definition 7.1 is so comprehensive and powerful that we can prove the zero product theorem all at once for Z. We announce the result as

THEOREM 7.6 Zero Product Theorem If $a \in Z$, then $a \cdot 0 = 0$.

The proof, though not hard, requires a bit of ingenuity. We seize upon the expression $a(a + 0)$, compute it in two different ways, and equate the results. First, $a(a + 0) = a \cdot a$, since $a + 0 = a$ (Why?). But also, employing the distributive law, $a(a + 0) = a \cdot a + a \cdot 0$. Equating the results, we get $a \cdot a = a \cdot a + a \cdot 0$. We note that this equation has the form $x = x + y$. Now, there is only one[6] integer y, which, when added to any integer x, produces just x: this is the additive identity, zero. So we must have $y = 0$, or $a \cdot 0 = 0$, as required.

There is one more theorem for whole numbers which we should like to extend to integers — Theorem 4.17. This theorem essentially states that $(W, +, \cdot)$

---

[6] It can be proved that the identity of a group is unique. See Exercises 6.2, Problem *7.

has no divisors of zero.[7] That Theorem 4.17 (hence also Theorem 4.18) can be extended to $(Z,+,\cdot)$ follows easily from Definition 4.6 and Theorems 7.8 and 7.9 in the exercises. For these theorems tell us that the products of different combinations of positive and negative integers are all positive or negative, never zero: so the only way to obtain a zero product is for at least one factor to be neither positive nor negative, hence zero. We announce our conclusion as

THEOREM 7.7 No Divisors of Zero If $a, b \in Z$ and $ab = 0$, then $a = 0$ or $b = 0$.

EXERCISES 7.4

1. Give interpretations, using the debts-assets model, for the products
   (a) $(^+3)(^+6)$ (b) $(^+4)(^-5)$ (c) $(^-2)(^+4)$ (d) $(^-5)(^-7)$
   (e) $(^-3)(0)$ (f) $(0)(^-4)$

2. Another model for the multiplication of signed numbers uses time (forwards and backwards) and change of temperature. Thus $(^+3)(^-1)$ may be interpreted as the change in temperature after 3 hours $(^+3)$ if the temperature is falling 1° $(^-1)$ each hour. Give interpretations, using this model, to the products of Problem 1.

3. (a) Supply a reason for each step of the text's chain of equations showing that $(^+3)(^-2) = {}^-6$.
   (b) Similarly analyze logically the product $(^-2)(^+3) = {}^-6$ *directly*, without commuting it to $(^+3)(^-2)$.
   (c) Supply a reason for each step of the text's chain of equations showing that $(^-3)(^-2) = {}^+6$, pointing out where Theorem 7.6 is used; also explain why Theorem 4.16 does not apply here.

4. Rewrite the proof of Theorem 7.6 in double-column form with numbered steps.

5. (a) Given the indicated product $(^+6)(^-5)$, obtain the result
      (i) intuitively by some model.
      (ii) logically by a chain of equations, explaining the argument step by step.
   (b) Can these results be generalized to yield a rule for multiplying any two integers, one positive and the other negative? Explain your answer.

6. Answer Problem 5 for the indicated product $(^-4)(^-6)$.

---

[7] See Definition 6.5 in Exercises 6.2, Problem 6.

7. For each of the following two theorems, provide a numerical illustration, then prove it by generalizing the logical arguments in the text:

   (a)

THEOREM 7.8 Product of Two Integers of Opposite Signs If ^{+}a and ^{-}b are any pair of positive and negative integers, respectively, then $(^{+}a)(^{-}b) = (^{-}b)(^{+}a) = {}^{-}(ab)$.

   (b)

THEOREM 7.9 Product of Two Negative Integers If ^{-}a and ^{-}b are any two negative integers, then $(^{-}a)(^{-}b) = {}^{+}(ab)$.

8. (a) We expect an identity to be unique; but is $^{+}1$ the only multiplicative identity in Z? Consider $^{-}1$: is it also an identity? Justify your answer.

   (b) Illustrate with several examples, then describe the effect of multiplying any integer x by $^{-}1$. Do not ignore the case where $x = 0$.

   (c) Generalize your discovery in part (b) by completing the statement of the following theorem:

THEOREM 7.10 If $x \in Z$, then $x \cdot (^{-}1) = \cdots$.

   (d) Supply a formal proof of Theorem 7.10.

9. Definition 7.1 includes the distributive property.

   (a) Illustrate this law by using each of the following triples of integers:

   (i) $^{+}2, {}^{+}5, {}^{-}1$ as $^{+}2(^{+}5 + {}^{-}1)$ (ii) $^{+}2, {}^{-}5, {}^{-}1$

   (iii) $^{-}3, {}^{+}5, {}^{-}1$ (iv) $^{-}4, {}^{-}5, {}^{+}2$

   (v) $0, {}^{-}2, {}^{+}3$ (vi) $^{-}2, 0, {}^{-}3$

   (vii) $^{-}3, {}^{+}2, {}^{-}2$ (viii) $^{-}4, {}^{-}2, {}^{-}3$

   (b) Does addition in Z distribute over multiplication? Justify your answer informally.

*10. (a) Compare the properties of the arithmetics (mod n), as developed in Sections 6.1 and 6.2, with those listed in Definition 7.1 for $(Z,+,\cdot)$. Which are common to both systems? Which hold only in $(Z,+,\cdot)$? Which hold only in the arithmetics (mod n)?

   (b) Use the results of part (a) to explain why the arithmetic (mod 6)[8] has divisors of zero, while, according to Theorem 7.7, $(Z,+,\cdot)$ does not.

---

[8]See Exercises 6.2, Problem (6)(c).

*11. In addition to the group, another important algebraic structure is the *ring*. The formal definition is

Definition 7.4 Ring Let R be a nonempty set, with $+$ and $\cdot$ two distinct binary operations defined on R. Then the system $(R,+,\cdot)$ is called a *ring* iff each of the following properties holds:

1. R is a commutative group under $+$.
2. R is closed and associative under $\cdot$.
3. In R, $\cdot$ distributes over $+$.

We commonly refer to the operation $+$ as *addition* and to the operation $\cdot$ as *multiplication.*

(a) According to Definition 7.4, is it necessary that multiplication in a ring be commutative? *Must* a ring contain a multiplicative identity? a multiplicative inverse for each element?

(b) Does $(Z,+,\cdot)$, as defined by Definition 7.1, form a ring? Explain.

(c) Does $(Z,+,\cdot)$ have the properties mentioned in part (a)? Explain.

(d) Answer parts (b) and (c) for the arithmetic (mod 2).

(e) Answer parts (b) and (c) for the arithmetic (mod 4); for the arithmetics (mod n), where $n \in W$, $n \geq 2$.

(f) May a ring contain divisors of zero? Explain your answer.

Section 7.5 Subtraction and Division of Integers

By now we have considered subtraction for the whole numbers and the modular arithmetics. It has always been defined as the operation inverse to addition. In Section 6.3 we generalized the concept to that of an inverse operation (*quifference*) in a group: the explicit statement is contained in Definition 6.6. In a completely analogous manner, we may introduce subtraction in Z by means of

Definition 7.5 Subtraction of Integers If $a, b \in Z$, then the *difference* $b - a$ is an integer x (if x exists) such that $a + x = b$.

Since this statement is an automatic extension of Definition 4.9 for the whole numbers, clearly whatever differences exist in $(W,+,\cdot)$ are the same in $(Z,+,\cdot)$. For instance, $13 - 5 = 8$ (or $^{+}13 - {}^{+}5 = {}^{+}8$) is true in both W and Z. But in the larger system, Z, we now have to consider also such proposed differences as $^{-}7 - {}^{-}2$, $^{-}9 - {}^{+}6$, $^{+}9 - {}^{-}6$, and $0 - {}^{-}2$; and we should also reconsider differences like $3 - 7$ (or $^{+}3 - {}^{+}7$) and $0 - 3$.

From a practical standpoint, having found models for adding integers, we should try to apply them to subtraction. Useful interpretations are debts and assets, elevators traveling between floors above and below ground level, and temperature differences. We may ask, for example, by how many degrees and in which direction the temperature has changed in going from $^{+}5°$ to $^{+}13°$. The change, $^{+}13 - {}^{+}5$, or $^{+}8$, has a positive sign, since the temperature *rose* during the interval. By contrast, a reading of $^{-}2°$ followed by one of $^{-}7°$, which is a drop of 5°, leads to the difference $^{-}7 - {}^{-}2$, which is $^{-}5$. You may similarly describe circumstances leading to the differences $^{-}9 - {}^{+}6$, $^{+}9 - {}^{-}6$, and $0 - {}^{-}2$.

From a theoretical point of view, however, instead of models, we must use Definition 7.5, focusing especially on the parenthetical caution, "if x exists" — all the more so because in W differences frequently do not exist. Thus we would consider the difference $^{+}13 - {}^{+}5$ to be an integer x if $x \in Z$ exists such that $^{+}5 + x = {}^{+}13$. Similarly, the difference $^{-}7 - {}^{-}2 = x$ would translate into the sum $^{-}2 + x = {}^{-}7$. It is easy to check, using the rules of signs for addition given in Section 7.3, that $x = {}^{+}8$ and $x = {}^{-}5$ satisfy the respective equations. These values, moreover, are exactly those suggested by our temperature model. You should likewise investigate the remaining differences of the preceding paragraph.

As further illustrations of Definition 7.5 we cite the following:

$$^{+}7 - {}^{+}3 = {}^{+}4 \quad \text{because} \quad {}^{+}3 + {}^{+}4 = {}^{+}7$$
$$^{+}3 - {}^{+}7 = {}^{-}4 \quad \text{because} \quad {}^{+}7 + {}^{-}4 = {}^{+}3$$
$$^{-}7 - {}^{-}3 = {}^{-}4 \quad \text{because} \quad {}^{-}3 + {}^{-}4 = {}^{-}7$$
$$^{-}3 - {}^{-}7 = {}^{+}4 \quad \text{because} \quad {}^{-}7 + {}^{+}4 = {}^{-}3$$
$$0 - {}^{-}5 = {}^{+}5 \quad \text{because} \quad {}^{-}5 + {}^{+}5 = 0$$
$$0 - {}^{+}5 = {}^{-}5 \quad \text{because} \quad {}^{+}5 + {}^{-}5 = 0$$

The definition of subtraction is obviously not a convenient tool for computation, as it does not directly yield the difference, but only provides a check for a proposed difference. But we can speedily invent an efficient algorithm for subtraction of integers. We need only make two preliminary remarks: first, Definition 7.1 establishes that $(Z,+)$ is a commutative group; second, by virtue of Definition 7.5, subtraction of integers is an operation inverse to addition in the sense of Definition 6.6. Thus we can interpret *quifference* ($\square$) in Definition 6.6 as *subtraction* ($-$) of integers and the *group operation* ($\circ$) as *addition* ($+$). This means that, by the above interpretation, Theorem 6.2 proves that subtraction is closed in Z; and Theorem 6.3 provides an algorithm for computing differences. In order to state this algorithm succinctly, we first recall that, in our notation, x' is the additive inverse of an element $x \in Z$. Then we can interpret Theorems 6.2 and 6.3 as the following two theorems for integers:

THEOREM 7.11 Closure of Subtraction If $a, b \in Z$, then $b - a$ is a unique element of Z.

THEOREM 7.12 Subtraction Algorithm If $a, b \in Z$, then $b - a = b + a'$.

The preceding two theorems are especially significant in view of the failure of subtraction to be closed in W. In fact, they provide the theoretical motivation for extending the whole-number system to that of the integers. The happy result, closure of subtraction, is thus an immediate consequence of Definition 7.1, which created the additive group of the integers.

At this point you may recall a rule in algebra for subtraction. This rule might be stated as: "To subtract integers, change the sign of the subtrahend, then proceed as in algebraic addition." In a modern text, the same rule might be expressed as: "The difference between two numbers can be found by adding the opposite of the second number to the first number."[9] Whatever the wording is, the meaning is essentially that of Theorem 7.12.

We now have a basis for bringing together three different concepts which have been introduced in this chapter and relating them to the ubiquitous, and perhaps confusing, *minus* sign in algebra. First, in Definition 7.2 we define a *negative integer:* $x = {}^-a$ if $x' \in N$. In particular, if $a = 4$, ${}^-4$ is a negative number, since $({}^-4)' = 4 \in N$. Here we have used a *raised* minus sign.

Second, in Definition 7.5 we define the *binary operation* of subtraction. When we write ${}^+7 - {}^+4$, with the minus sign at the usual level, we use the $-$ symbol as a verb, commanding the performance of a certain task. In algebra the symbol is used in the same way, as in $7 - 4$ or $x - y$.

Third, Definition 7.1 includes the additive inverse of any integer x, which we denote by x'. Thus we write $({}^+7)' = {}^-7$ and $({}^-7)' = {}^+7$. "Taking the additive inverse of" is a *unary operation*, since it is applied to a single number. Here the *prime* symbol (′) is used as a verb, but one with a different meaning from *subtract*. But in algebra the minus sign is commonly used instead of the prime sign. Thus we may write $-(7)$ or -7 to mean, "the additive inverse of 7." Similarly, $-(-7)$ may mean, "the additive inverse of ${}^-7$," which is ${}^+7$. Note that, according to this interpretation, the first minus sign in $-(-7)$ means *additive inverse of* and the second means *negative.*

But the appearance of the hedging words, *may* and *according to this interpretation,* in the last two sentences invites caution. Certainly *one* interpretation of $-(-7)$ is $({}^-7)'$, which in simplest form is ${}^+7$. But we could equally well regard both minus signs as commands to find the additive inverse. Then, in our more explicit notation, $-(-7)$ would become $[(7)']'$, or $[({}^+7)']'$, which looks different from the symbol for the first interpretation. But which number is $[({}^+7)']'$ in simplest form? Since $({}^+7)' = {}^-7$, we get $[({}^+7)']' = ({}^-7)' = {}^+7$. Despite appearances, then, both views of $-(-7)$ give the same result. Thus we have reconciled, for this example, the first and third concepts we described.

Since the second concept, that of subtraction, requires two elements, let us extend our example to the expression $x - (-7)$, where x is any integer. This expression we may regard as, "from x, subtract ${}^-7$." Theorem 7.12 then permits

---

[9]D. A. Johnson and J. J. Kinsella. *Algebra, Its Structure and Applications* (New York: The Macmillan Co., 1967), p. 164.

us to write $x - {}^-7 = x + ({}^-7)' = x + {}^+7$. We may, on the other hand, read $x - (-7)$ as a command to combine with or add to x the quantity $-(-7)$. But from the preceding paragraph, $-(-7)$ simplifies to ${}^+7$ according to either the first or the third concept. So we may write $x - (-7) = x + ({}^-7)' = x + {}^+7$, or $x - (-7) = x + [({}^+7)']' = x + ({}^-7)' = x + {}^+7$, in agreement with each other and the difference found near the beginning of this paragraph.

In summary, to express the three distinct concepts of *negative integer*, *subtraction*, and *additive inverse*, we have used three symbols: $^-$ (raised), $-$ (usual level), and $'$ (prime), respectively. How, then, can we get away with using only one symbol, $-$, in elementary algebra? The answer is that, since the three interpretations are perfectly consistent, it is immaterial which we choose at any given moment. Further examples will occur to you; some are included as exercises.

From this diversion we return to the second inverse operation for the integers, that of *division*. Clearly, the definition should be

Definition 7.6 Division of Integers If $a, b \in Z$, then the *quotient* $b \div a$ is an integer x (if x exists) such that $ax = b$.

As illustrations, we have

$$
\begin{aligned}
{}^+15 \div {}^+3 &= {}^+5 \quad \text{because} \quad ({}^+3)({}^+5) = {}^+15 \\
{}^-15 \div {}^+3 &= {}^-5 \quad \text{because} \quad ({}^+3)({}^-5) = {}^-15 \\
{}^+15 \div {}^-3 &= {}^-5 \quad \text{because} \quad ({}^-3)({}^-5) = {}^+15 \\
{}^-15 \div {}^-3 &= {}^+5 \quad \text{because} \quad ({}^-3)({}^+5) = {}^-15 \\
0 \div {}^+4 &= 0 \quad \text{because} \quad ({}^+4)(0) = 0
\end{aligned}
$$

For reasons similar to those advanced in Section 4.8, ${}^+4 \div 0$ and $0 \div 0$ are undefined.

So far, the development of division in Z parallels that for subtraction. Unfortunately, however, we cannot continue the process, for two related reasons. The first is that, while we could *define* the multiplicative inverse of an integer x as the integer x'' such that $x \cdot x'' = 1$, only two integers actually possess multiplicative inverses in Z (Which are they?). The second is that, because of the lack of inverses, we cannot claim that Z forms a group under multiplication. This means, in turn, that we cannot use group theorems to argue that division is closed in Z. The lack of closure is the second roadblock to further progress in this section. Counterexamples are ${}^+2 \div {}^+6$ and ${}^-5 \div {}^+3$; for the corresponding multiplicative equations are $({}^+6)x = {}^+2$ and $({}^+3)y = {}^-5$, and no integers x and y exist that will satisfy them.

In closing, it is worthwhile to ask whether we can, in turn, extend the system $(Z,+,\cdot)$ by introducing a multiplicative group of some kind and thus closing the operation of division. Definition 7.1, of course, does not do this; but an attempt to do so is the subject of Section 7.7.

EXERCISES 7.5

1. Use the temperature model to obtain the differences

$$^{-}9 - {}^{+}6, \quad {}^{+}9 - {}^{-}6, \quad 0 - {}^{-}2, \quad 3 - 7, \quad 0 - 3$$

2. Verify your answers to Problem 1 by using
   (a) Definition 7.5 (b) Theorem 7.12
3. The following cities have approximately the same latitude. Their longitudes are expressed in degrees measured from the meridian of Greenwich, taken as 0° longitude, positive if to the east of that meridian, negative if to the west. Compute the (signed) difference in longitude between New York City and Madrid, Spain, and similarly for each of the other five pairs of cities. In each case, explain the meaning of a positive difference; of a negative difference. The cities and their longitudes are:

New York City, $^{-}74°$; Naples, Italy, $^{+}14°$; Madrid, Spain, $^{-}4°$;
Tashkent, U.S.S.R., $^{+}69°$

4. Use the debts-assets model to obtain the differences

$$^{+}7 - {}^{+}3, \quad {}^{+}3 - {}^{+}7, \quad {}^{-}7 - {}^{-}3, \quad {}^{-}3 - {}^{-}7, \quad {}^{-}3 - {}^{+}7, \quad {}^{+}3 - {}^{-}7,$$
$$^{-}5 - 0, \quad 0 - {}^{-}5, \quad 0 - {}^{+}1$$

5. Verify your answers to Problem 4 by using the subtraction algorithm (Theorem 7.12).
6. Quote the rule for subtraction of integers which you learned in algebra and explain its relation to Theorem 7.12.
7. For each of the following indicated quotients, either find the answer, or else state that no quotient exists in Z:
   (a) $^{+}28 \div {}^{+}4$ (b) $^{-}28 \div {}^{+}4$ (c) $^{+}4 \div {}^{-}28$
   (d) $^{+}4 \div {}^{-}4$ (e) $^{-}28 \div {}^{-}1$ (f) $0 \div {}^{-}1$
   (g) $^{-}3 \div 0$ (h) $^{-}7 \div {}^{-}10$ (i) $^{+}12 \div {}^{-}16$
8. Simplify each of the following expressions:

$$(^{+}8)', \quad (^{-}10)', \quad (0)', \quad [(^{+}5)']', \quad [(^{-}4)']'$$

9. In the manner of the text's discussion, interpret each of the following expressions in three different ways, and show that the results are consistent:
   (a) $0 - (-5)$ (b) $x - 2$

10. (a) In the expression $-(-6)$, both minus signs may be interpreted as commands to "take the additive inverse of." Then $-(-6)$ means "the additive inverse of the additive inverse of $^+6$," or $[(^+6)']'$. Show that $[(^+6)']' = {}^+6$.

    (b) Similarly find $[(^-3)']'$, $[(0)']'$.

    (c) Generalize the examples in parts (a) and (b) by writing a formula for $[(a)']'$, and compare it with Theorem 7.2.

*11. Does multiplication in Z distribute over subtraction?

    (a) Investigate this question by means of numerical examples.

    (b) Compare this question with that in W. Then use this comparison, together with the results of part (a), if relevant, to prove or disprove distributivity of multiplication over subtraction in Z.

Section 7.6 Equality and Order Properties of Integers

The concept of equality for whole numbers can be extended without difficulty to the integers, despite the fact that Definition 4.3 does not apply to negative integers. Since each "new" integer is the additive inverse of a unique natural number, we can agree that two negative integers are equal iff their inverses are equal according to Definition 4.3. In symbols, $^-a = {}^-b$ iff $a = b$, where $a, b \in N$.

The consequence of this easy extension of the relation of equality to integers is that the same properties of equality continue to hold. Specifically, equality in Z is an equivalence relation,[10] hence transitive. In fact, we have already used this property in making chains of equations. Other equality properties analogous to those in Chapter 4 are mentioned in the exercises.

The order properties in Z require more careful study. The number line furnishes a natural model if we interpret *less than* as *to the left of* for the corresponding points on the line, as suggested in Exercises 4.7, Problem 6. Thus if we place points representing $^+3$, $^-1$, $^+4$, 0, and $^-6$ on the number line, as in Figure 7.2, we conclude, among other relations, that $0 < {}^+3$, $^+4 > {}^+3$, $^-1 < {}^+3$, $^-6 < {}^+4$, and $^-1 > {}^-6$. Which, if any, of these statements would be true of the absolute values of the above pairs of numbers?

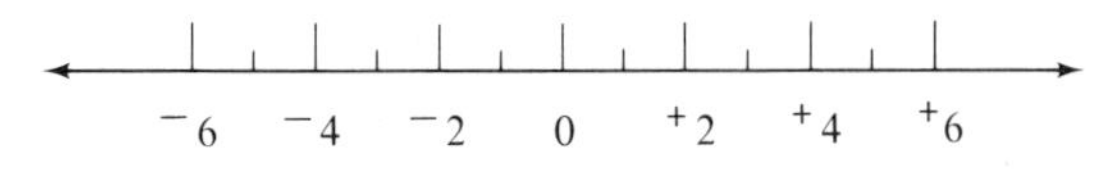

Figure 7.2

The formal definition of $<$ is a bit unexpected, turning, as it does, on the question of how to extend Definition 4.7 to Z. Before going farther, you should

---

[10]See Definition 6.12.

experiment with examples. Actually, only a very minor change is necessary, resulting in

Definition 7.7 Less Than If $a, b \in Z$, then $a < b$ iff there is a natural number (positive integer) x such that $a + x = b$.

The definition of $>$ is identical with Definition 4.8, except that it applies to all integers; so we need not state it here. Similar remarks hold for $\leq$ and $\geq$.

It is now intuitively clear that the trichotomy property (Theorem 4.23) can be extended to the integers. Indeed, an exercise deals with the special case where $b = 0$, a matter of some importance in connection with abstract algebraic systems.

The greatest surprise, however, comes in examining the order properties which relate to the operations of addition and multiplication. Now, and in the exercises, you are invited to experiment with numerical examples to see which, if any, of Theorems 4.25, 4.26, 4.27, and the *greater than* multiplication theorems can be automatically extended to Z and which must be modified or abandoned. We shall also investigate the proofs of these extended theorems.

EXERCISES 7.6

1. Equality in W is an equivalence relation. Using the fact that, for negative integers, $^-a = {}^-b$ iff $a = b$, explain intuitively why equality in Z is also an equivalence relation.

2. Besides the properties considered in Problem 1, the whole numbers possess several other equality properties, which are found in Chapter 4.
   (a) List each such property.
   (b) Provide a numerical example of each, using negative integers.
   (c) Can each of these properties be extended to Z? If so, state the extension; if not, explain why not.

3. (a) Insert the correct symbol ($<, =, >$) between each pair of integers in the following list:

   $^+8 \quad {}^+5, \qquad {}^+8 \quad {}^-5, \qquad {}^-8 \quad {}^+5, \qquad {}^-8 \quad {}^-5,$

   $6 \quad 0, \qquad {}^-6 \quad 0, \qquad {}^-4 \quad {}^-9, \qquad {}^-2 \quad {}^-2,$

   $^-2 \quad 2, \qquad 10 \quad {}^-10, \qquad 0 \quad {}^-8$

   (b) Check your answers to part (a) by locating the corresponding points on the number line.
   (c) Repeat part (a), using only the symbols $\leq$ and $\geq$.
   (d) Insert absolute value signs throughout the list in part (a); then put in the correct symbol ($<, =, >$).

4. At first glance it might seem as though Definition 4.7, when extended to the integers, should read, "If $a, b \in Z$, then $a < b$ iff there is an *integer* x such that $a + x = b$." Show by examples that such a definition would not only fail to accord with our intuitive notion of the number line but would also violate the trichotomy property.

5. Illustrate each of the following order definitions and theorems, extended to Z. In each case, use several examples involving one or more negative integers.
   (a) Definition 4.8 (greater than)
   (b) Theorem 4.23 (trichotomy)
   (c) Theorem 4.24 (transitivity for less than)
   (d) Theorem 4.25 (addition for less than)

6. (a) Write out the trichotomy law (Theorem 4.23), extended to integers, for the special case when $b = 0$.
   (b) Is the following statement true? "An integer a is positive iff $a > 0$." Explain your answer.
   (c) Write a similar statement for the case where $a < 0$ and state whether it is true.

7. (a) State the direct extension of the *multiplication-for-less-than* property (Theorem 4.26) to Z.
   (b) Test the truth of your answer to part (a) by means of several examples, choosing for a, b, and c several combinations of positive, negative, and zero integers. What do you conclude?
   (c) Does Theorem 4.26 appear to hold when extended to integers when $c > 0$? when $c < 0$? when $c = 0$? Explain your answers.

8. The results of Problem 7 may be summarized as

THEOREM 7.13 Multiplication for Less Than If $a, b, c \in Z$ and if $a < b$, then

1. $ac < bc$ if $c > 0$.
2. $ac > bc$ if $c < 0$.
3. $ac = bc$ if $c = 0$.

Now state and illustrate numerically a similar theorem for the *multiplication-for-greater-than* relation in Z.

9. (a) State the extension of the *transitive property for less than* (Theorem 4.24) to the integers.
   (b) Examine the proof sketched for Theorem 4.24 and determine which changes, if any, must be made in it to prove its extension to the integers.

*10. State and prove the extension to integers of the *addition-for-less-than* property (Theorem 4.25). (*Hint:* Note the changes made in Problem 9(b); also see Exercises 4.7, Problem *12(a).)

*11. Prove Theorem 7.13 in three parts:

(a) Prove Statement 1 in the manner of Exercises 4.7, Problem *12(b). In which respects, if at all, does this proof differ from that of Theorem 4.26?

(b) Prove Statement 2, so far as possible, by the method of part (a); but point out the modifications that must be made.

(c) Statement 3 is too trivial to write out. On which theorem does it primarily depend?

*12. Problem 6 leads to a formal definition of what is meant by order in a *ring*.[11] The system of integers and the arithmetics (mod n) are examples of ordered rings. The formal definition is

Definition 7.8 Ordered Ring A ring $(R,+,\cdot)$ is *ordered* when it contains a nonempty subset P, called the set of *positive* elements of R, such that

1. P is closed under addition and multiplication (i.e., sums and products of positive elements are positive).
2. For each element $a \in R$, exactly one of the following holds:

   (i) $a \in P$ (ii) $a = 0$

   (iii) $a' \in P$, where a' is the additive inverse of a. (This is known as the *trichotomy* property: compare Theorem 4.23.)

Answer the following questions with regard to Definition 7.8:

(a) How is Definition 7.8 related to Problem 6?

(b) Explain why the system $(Z,+,\cdot)$ is an ordered ring. (*Hint:* Use also Definition 7.4 in Exercises 7.4, Problem *11.)

(c) Show that the arithmetic (mod 3), though a ring, is not ordered. (*Hint:* Suppose $1 \in P$ and reach a contradiction; similarly, suppose $1' \in P$ and reach a contradiction. Now, since $1 \neq 0$, 1 cannot be assigned to any of the three categories mandated by Definition 7.8.)

Section 7.7 Inventing the Rational Numbers

In the history of mathematics, the next system to develop after the natural numbers was the rational numbers in the guise of *fractions*. These new numbers arose, of course, in response to the practical need of expressing parts of a unit

[11] See Definition 7.4 in Exercises 7.4, Problem *11.

when making measurements. As an early example, Egyptian symbols for fractions appeared in the Rhind papyrus, about 1650 B.C.; the scribe who wrote it said that he copied from an earlier document. The ancient Egyptians, however, used only fractions with unit numerators, like 1/2, 1/3, 1/4, . . . , except that they also had a special symbol for 2/3. Thus they wrote our fraction 2/5 as (1/3 + 1/15) and 2/7 as (1/4 + 1/28). Since they did not permit themselves to repeat fractions, (1/5 + 1/5) was not an acceptable representation for 2/5. Because they multiplied and divided by repeated doubling and halving, they often needed to express $2/n$, with n an odd integer, in terms of different unit fractions. A large part of the Rhind papyrus consists of a table for unit-fraction equivalents of $2/n$, where $n = 5, 7, 9, \ldots, 101$. With this restricted conception of fractions, it is no wonder that performing and applying arithmetic was an art, confined mostly to the priestly class.

In contrast to the Egyptians, who used a fixed numerator for their fractions, the Babylonians used a fixed denominator, as we do: but their denominator was 60, the same as the base for their *sexagesimal* numeral system. Thus they divided a unit into sixtieths and each sixtieth again into sixtieths, etc. We still use this system in dividing an hour, or a degree, into 60 *minutes* (*pars* ***minuta*** *prima* 'first small part') and a minute, in turn, into 60 *seconds* (*pars minuta* ***secunda*** 'second small part').

As in approaching other systems, we begin with models which appeal to the intuition. You will certainly remember the presentation of fractions by cutting up rectangles or circles. A more useful model for our purposes is a further development of the number line. Thus on the line in Figure 7.3, several integral points are marked for reference, as are other points corresponding to the fractions $^{-}11/6$ (or $^{-}1\frac{5}{6}$), $^{-}3/4$, $^{-}1/3$, 6/23, 1/2 (or 2/4), 2/3, and 5/4 (or $1\frac{1}{4}$).

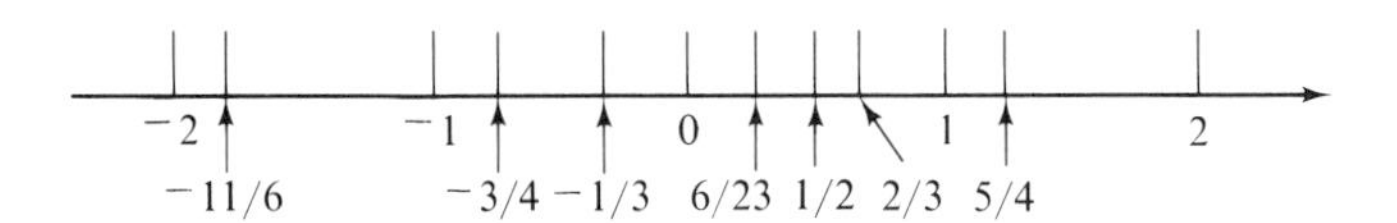

Figure 7.3

But for the rapid logical presentation of fractions — and, more generally, rational numbers — we shall borrow a leaf from the Egyptians, considering first unit fractions. We shall also introduce "new" numbers by means of a comprehensive definition analogous to Definition 7.1, assuming at once all the basic properties we want rationals to have. One of the desiderata is that we include *all* the fractions: 2/3, $^{-}2/3$, 4/1, $^{-}7/3$, and 0/4, as well as unit fractions, like 1/3 and $^{-}1/3$. Another is that, if possible, the operation of division be closed.

First, however, we shall define the operations of subtraction and division as inverse to addition and multiplication, respectively. The easy extension of Definitions 7.5 and 7.6 to the rationals is left for the exercises.

Just as subtraction is closed for the integers because each element of Z has an additive inverse in Z, so we can expect that if each element of the new system has a multiplicative inverse, then division will be closed. Even in Z, $^{+}1$ and $^{-}1$ have multiplicative inverses:[12] for example, $(^{-}1)'' = {}^{-}1$, because $(^{-}1)(^{-}1) = {}^{+}1$, the multiplicative identity. Then the inverse operation theorem (Theorem 6.3) suggests[13] how we might divide by $^{-}1$: e.g., $4 \div {}^{-}1 = 4 \cdot (^{-}1)'' = (4)(^{-}1) = {}^{-}4$. Can we similarly apply this suggestion to the quotient $4 \div 3$? Yes, if we can find $3''$. We can solve this problem by extending Z to include reciprocals of integers. Then $3'' = 1/3$, because $3 \cdot 1/3 = 1$; and our quotient becomes $4 \div 3 = 4 \cdot 3'' = 4 \cdot 1/3$. To give meaning to the expression $4 \cdot 1/3$, we invent a new number, $4/3$, and agree that $4 \cdot 1/3 = 4/3$. More generally, if $a, b \in Z$, with $b \neq 0$, we define $a \cdot 1/b = a/b$. So apparently we must deal with ordered pairs of integers,[14] a/b, with $b \neq 0$.

But we cannot stop here. If $4/3$ is a member of our new system, then $(4/3)''$ should be a member also. But what should $(4/3)''$ be? By definition of inverse, it must be an ordered pair of integers, x/y, with $y \neq 0$, such that $4/3 \cdot x/y = 1$. It turns out, as you know from your experience with arithmetic, that $x/y = (4/3)'' = 3/4$; for $4/3 \cdot 3/4 = (4 \cdot 1/3) \cdot (3 \cdot 1/4) = (4 \cdot 1/4) \cdot (1/3 \cdot 3) = 1 \cdot 1 = 1$, if we assume that the commutative and associative laws for multiplication still hold in our new system.

Now we can handle quotients like $5/7 \div 4/3$, for $5/7 \div 4/3 = 5/7 \cdot (4/3)'' = 5/7 \cdot 3/4 = 15/28$, provided we agree that the product of two ordered pairs, a/b and c/d, is ac/bd. Similarly, with $(^{-}2/5)'' = {}^{-}5/2$, the quotient $1/4 \div (^{-}2/5) = 1/4 \cdot (^{-}2/5)'' = 1/4 \cdot (^{-}5/2) = {}^{-}5/8$; so also $(^{-}3) \div (^{-}2/5) = (^{-}3/1)(^{-}5/2) = 15/2$.

Since the above discussion has been heuristic, based on elementary school knowledge of fractions, we cannot readily introduce a formal definition of rational numbers. But it may be helpful to write down an informal description, analogous to Definition 7.1 for Z. It is

Definition 7.9 System of Rationals (Description)[15] The system of rational numbers, $(Q,+,\cdot)$, consists of the set Z of integers (which may be written as $0/1$, $\pm 1/1$, $\pm 2/1, \ldots$) together with all other ordered pairs of integers a/b, with $b \neq 0$, and two binary operations on Q, subject to the following requirements:

1. The system $(Q,+)$ forms a commutative group, with 0 or $0/1$ as its identity.
2. The system $(Q^*,\cdot)$, where Q^* is the set of rationals with zero deleted, forms a commutative group, with 1 or $1/1$ as its identity.
3. Multiplication is distributive over addition.

---

[12]See Definition 6.2 for a notation for multiplicative inverse.

[13]Theorem 6.3 only *suggests* this procedure; for its hypothesis insists on group properties for multiplication, which we have not shown.

[14]The pair is *ordered* in the sense that the *order* of the letters makes a difference: e.g., 3/4 is not the same number as 4/3.

[15]This paragraph does not constitute a rigorous definition, but only a description.

First, we note that by Definition 7.9, Z (hence also W and N) is a subset of Q, and Q, therefore, contains an infinite number of elements. Surprisingly enough, the cardinal number of Q is also aleph-null, as an optional exercise shows. Second, we observe that Definition 7.9 is quite similar to Definition 7.1, but with one major property added (Which is it?). Third, the question of existence of the system described by Definition 7.9 can be answered here only with a quick and cavalier Yes, and a reassurance that a more time-consuming construction of the system $(Q,+,\cdot)$ would yield a proof.

Since our purpose is merely to sketch the number systems which extend beyond the integers, we shall not build them up logically by lists of definitions and theorems. But it is necessary to discuss briefly, for Q, the nature of the equality and inequality relations and the behavior of the two fundamental operations and their inverses. To begin with equality, this relation among elements of Q is more complex than in W or Z. Though the fractions $1/2$, $3/6$, and $^-4/^-8$ are not identical in appearance, we should like to be able to say that they are equal, or at least *equivalent*, in some sense. This we can accomplish by making

Definition 7.10 Equal Rational Numbers[16] If a/b, $c/d \in Q$, then $a/b = c/d$ iff $ad = bc$.

An exercise gives practice in applying Definition 7.10 and checking that it gives results in harmony with our experience with fractions.

As with our development of the integers, in order that addition and multiplication in Q behave as described in Definition 7.9, the calculation of sums and products must be done in specific ways. The familiar rule for multiplication, $a/b \cdot c/d = ac/bd$, has already been illustrated. Addition is accomplished either by the elementary school method of employing common denominators and then adding numerators, or by the equivalent formula, $a/b + c/d = (ad + bc)/bd$. Now it is easy to show that the commutative, associative, and distributive laws hold and that the identities and inverses are as expected. In particular, we have the multiplicative inverse formula, "If $a/b \in Q$ and $a/b \neq 0$, then $(a/b)'' = b/a$."

Continuing, we find that the special "sign rules" for operating with integers carry over to rationals: for instance, the product of two negative rationals is the expected positive rational, where the meanings of *positive* and *negative* accord with the number-line model. The laws of inequalities, which we examined for Z, also go over to Q without change. A consequence of the closure of division (*except by zero*) is that the quotient of any two integers, with the divisor not zero, always exists *as a rational number*. Formally, we have the following:

THEOREM 7.14 Rationals as Indicated Quotients of Integers If a, $b \in Z$ and if $b \neq 0$, then $a \div b = a/b \in Q$.

---

[16]A more sophisticated development would consider 1/2 and 3/6 as equivalent, but not equal, and both as members of the same equivalence class (see Definition 6.13).

For this reason, indeed, numbers of the form a/b are called *ratio*nal numbers: for they may be considered as *ratios* or indicated quotients of pairs of integers.

We close this section by mentioning a significant characteristic which Q possesses, but W and Z do not: between any two unequal rationals, no matter how close together, there is always another. Thus, for example, between 1/4 and 1/2 occurs the rational 1/3 (and many others also!); and between 3/216 and 3/215 occurs 6/431. We describe this property by saying that the rationals are *dense*. To make the concept of density more precise, we introduce

Definition 7.11 Dense Set An ordered set S, containing at least two elements, is said to be *dense* if between each pair of distinct elements of S there is another element of S.

Now Z has at least two elements and an order relation ($<$): but between the pairs x and $x + 1$ there is no integer y such that $x < y < x + 1$.[17] So Z and W are not dense sets. As for Q, it is also ordered and has at least two elements. By contrast to Z, if x and z are any pair of distinct elements of Q, we can find an element $y \in Q$ such that $x < y < z$. Intuitively, this is rather easy to see from the number-line model, since, if x and z are two distinct points on the line, the interval between them contains many points, specifically the midpoint. In fact, the standard method of proving that Q is dense actually chooses y to be this midpoint, the average of x and z: so $y = (x + z)/2$ and $x < (x + z)/2 < z$, as required. The details are left as an optional exercise.

EXERCISES 7.7

1. Draw a number line, choosing a large unit, and on it mark and label points corresponding to the rationals

$$1/3,\quad {}^-1/2,\quad 4/5,\quad 1\tfrac{1}{2},\quad {}^-5/3,\quad {}^-3/6,\quad 9/4,\quad 3/2,\quad {}^-2\tfrac{1}{4}$$

2. (a) Find, if possible, the multiplicative inverse of each of the following numbers; if none exists, so state:

$$5/1,\quad 6,\quad {}^-2/1,\quad 0,\quad {}^-10,\quad 4/5,\quad {}^-1/6,\quad 7/4,\quad 0/2$$

   (b) Similarly, determine the additive inverses of the rationals listed in part (a).

3. (a) State the definition of division in Q, obtained by extending Definition 7.6 for Z.

---

[17] The proof of this fact is not quite trivial and requires an assumption about positive integers which we have not discussed.

(b) Use this definition to show that $3/4 \div 2/3 = 9/8$.

4. (a) Use multiplicative inverses and Theorem 6.3 to compute the following quotients of rational numbers:

$$5 \div 4, \quad 5 \div {}^{-}4, \quad 2/3 \div 5, \quad 2 \div 3/5, \quad 1/4 \div 3/2,$$
$$({}^{-}3) \div 1/2, \quad ({}^{-}3/8) \div ({}^{-}2/3)$$

(b) Explain briefly how Definition 7.9 justifies the use of Theorem 6.3 in making the above computations.

(c) State the rule you learned in elementary school for dividing one fraction by another, then compare it with the method you used in part (a).

5. Give an informal argument justifying Theorem 7.14. (*Hint:* Change $a, b \in Z$ to the form $a/1, b/1 \in Q$.)

6. (a) Compare Definition 7.9 with Definition 7.1. In which respects do they differ?

(b) Explain how these differences relate to the question of closure of division in Z and in Q.

7. (a) Apply Definition 7.10 to select from the following rationals all pairs of equal rationals:

$$2/3, \quad 7/9, \quad 3/2, \quad {}^{-}4/6, \quad 0/{}^{-}5, \quad 3/{}^{-}2, \quad 8/2, \quad 3/3, \quad 10/13,$$
$${}^{-}8/{}^{-}12, \quad {}^{-}1/{}^{-}1, \quad 0/6, \quad 7/7, \quad {}^{-}4/{}^{-}1, \quad 1/4, \quad 40/52, \quad {}^{-}9/6$$

(b) (i) Draw a number line and locate on it a point for each rational listed in part (a).

(ii) How does this model represent equal rational numbers?

(c) Using the number line as a model and several specific pairs of rationals from the list of part (a), illustrate the use of the order symbols ($<, >$) for rationals.

8. (a) Compute each of the following sums by the method of finding common denominators, and also by use of the formula $a/b + c/d = (ad + bc)/bd$:

$$1/2 + 1/3, \quad 1/3 + 1/2, \quad 3/4 + 5/2, \quad ({}^{-}1/2) + ({}^{-}1/3),$$
$$({}^{-}3/4) + 1/2, \quad 2/3 + 0/1, \quad 3/4 + ({}^{-}1/2), \quad ({}^{-}2/3) + 2/3$$

(b) Which group and sign properties for addition in Q are illustrated by the sums of part (a)?

(c) Convert the sums of part (a) into products, then compute them.

(d) Which group and sign properties for multiplication in Q are illustrated by the products of part (c)?

(e) Using the rationals $^-3/5$, $1/2$, and $2/3$, illustrate the distributive law in Q.

9. (a) State the definition of subtraction in Q obtained by extending Definition 7.5 for integers.

(b) Use the definition to show that $2/3 - 1/2 = 1/6$.

(c) Explain why and how Theorem 6.3 can be used to provide a subtraction algorithm for Q.

(d) Apply the algorithm of part (c) to compute the following differences:

$$3/4 - 1/12, \quad (^-3/4) - (^+1/12), \quad 5/3 - 0/1, \quad 5/6 - 5/6,$$
$$(^+1/3) - (^-3/4), \quad 0 - (^-1/2)$$

10. (a) Between the integers 100 and 1000 there are many integers.

(i) How many are there?

(ii) Does this example imply that the integers are dense, according to Definition 7.11? Explain your answer.

(b) For each of the following pairs of rationals, find a rational between them:

$$1/13 \text{ and } 1/10, \quad 5/71 \text{ and } 5/70, \quad 0 \text{ and } 1/1{,}000{,}000$$

(c) Are the answers to part (b) unique? How many different rationals can be inserted between each pair? Explain your answers.

*(d) Prove that if x and z are rationals, then so is $(x + z)/2$.

*(e) Complete a proof that the set of rationals is dense by using the results of part (d) and also establishing that $x < (x + z)/2 < z$. (*Hint:* Since $x \neq z$, we may choose x and z so that $x < z$. Now prove that $x < (x + z)/2$, using the fact that this inequality is logically equivalent to $2x < x + z$ (Why?) and noting that $2x = x + x < x + z$. Then similarly prove that $x + z < 2z$.)

*11. A tempting definition of adding fractions, quite different in its effects from the formula $a/b + c/d = (ad + bc)/bd$, is given by the formula $a/b + c/d = (a + c)/(b + d)$. Thus, for example, $2/3 + 5/6 = 7/9$, according to this "new" formula.

(a) Compare this result for $2/3 + 5/6$ with that obtained by the earlier formula. Which seems more reasonable to you, and why?

(b) Which, if either, of the two formulas provides a true interpretation when applied to combining parts of a unit, such as placing a stick $5/6$ foot long at the end of another $2/3$ foot long? Explain.

(c) In sports, a team is often reported as winning, say, "2 out of 3 games" — a winning ratio of 2/3. Suppose a football team won 2 out of 3 games it played in October and 5 out of 6 games it played in November. To get the two-month winning ratio we may combine, or *add*, the two separate ratios, 2/3 and 5/6. Which, if either, of the two addition formulas provides a true interpretation for this situation? Explain.

(d) Investigate the properties of the "new" formula for addition of rationals, comparing them with the additive group properties of Definition 7.9. (*Hint:* Do not overlook the implications of our extended concept of equality of rationals, given in Definition 7.10: e.g., $1/2 + 1/3 = 2/5$, but $2/4 + 1/3 = \ldots$.)

*12. A *field* is an important, though more complicated, algebraic structure than a group or a ring. One definition of a field is

Definition 7.12 Field Let F be a nonempty set with two distinct binary operations, $+$ and $\cdot$, defined on F. Then the system $(F,+,\cdot)$ is called a *field* iff each of the following properties holds:

1. $(F,+)$ is a commutative group, with 0 as its identity.
2. $(F^*,\cdot)$ (i.e., F with 0 deleted) is a commutative group, with 1 as its identity.
3. $\cdot$ distributes over $+$.
4. $0 \neq 1$ (i.e., the two identities are distinct elements of F).

Using Definition 7.12, determine whether each of the following systems is a field:

(a) $(Q,+,\cdot)$ (b) $(Z,+,\cdot)$ (c) the arithmetic (mod 5)

(d) the arithmetic (mod 12) (e) the arithmetic (mod 2)

(f) $(W,+,\cdot)$ (g) the system $(S,+,\cdot)$, where $S = a$ and $+$ and $\cdot$ yield $a + a = a$ and $a \cdot a = a$.

(h) The system consisting of all the integers, together with their reciprocals (except for the reciprocal of 0), with $+$ and $\cdot$ taken as the usual arithmetic operations for integers and fractions.

*13. By definition, the cardinal number of the set of whole numbers is aleph-null: $n(W) = \aleph_0$. From Exercise 7.2, Problem 4(b), we may conclude that $n(Z) = \aleph_0$ also. What is $n(Q)$? Considering that Q is a dense set, while W and Z are not, we might expect that $n(Q) > \aleph_0$. Nevertheless, we can establish a one-to-one correspondence between W and Q by the following process:

(a) Consider the infinite array below, which contains positive rationals:

$$\begin{array}{ccccccccccccccc}
1/1 & \to & 2/1 & & 3/1 & \to & 4/1 & & 5/1 & \to & 6/1 & & \cdots & & \cdots \\
 & \swarrow & & \nearrow & & \swarrow & & \nearrow & & \swarrow & & & & & \\
1/2 & & 2/2 & & 3/2 & & 4/2 & & 5/2 & & \cdots & & \cdots & & \cdots \\
\downarrow & \nearrow & & \swarrow & & \nearrow & & \swarrow & & & & & & & \\
1/3 & & 2/3 & & 3/3 & & 4/3 & & \cdots & & \cdots & & \cdots & & \cdots \\
 & \swarrow & & \nearrow & & \swarrow & & & & & & & & & \\
1/4 & & 2/4 & & 3/4 & & \cdots & & \cdots & & \cdots & & \cdots & & \cdots \\
\downarrow & \nearrow & & \swarrow & & & & & & & & & & & \\
1/5 & & 2/5 & & \cdots & & \cdots & & \cdots & & \cdots & & \cdots & & \cdots \\
 & \swarrow & & & & & & & & & & & & & \\
1/6 & & \cdots & & \cdots & & \cdots & & \cdots & & \cdots & & \cdots & & \cdots \\
\downarrow & & & & & & & & & & & & & & \\
\cdots & & \cdots & & \cdots & & \cdots & & \cdots & & \cdots & & \cdots & & \cdots
\end{array}$$

(i) Verify that this array contains every positive rational at least once.

(ii) Does it contain any rationals more than once? If so, how many times?

(b) Write down the rationals of the array in the order indicated by the arrows, *omitting duplications*. Thus begin: 1/1, 2/1, 1/2, 1/3, ...

(c) It is possible to set up a one-to-one correspondence between the sequence obtained in part (b) and the sequence of natural numbers, 1, 2, 3, 4, ...

(i) Exhibit the first ten terms of this correspondence and verify that it is one-to-one.

(ii) According to the results of part (i), what is the cardinal number of the set Q^+, the *positive* rationals?

(d) Show that the one-to-one correspondence between N and Q^+ of part (c) can be extended to one which includes also 0/1 and the negative rationals. What does this imply about $n(Q)$?

(e) For an explicit formula giving a correspondence like that of part (c), study the Bailey and Musser reference. Note, however, that the arrangement of the array and the sequencing shown by the arrows in that article are different from those given here.

Section 7.8 Rationals in Decimal Form

The rationals, as we have seen, form a closed system with respect to the four fundamental operations (except for division by zero): briefly, they constitute a field.[18] Because they are dense, moreover, any physical measurement may be expressed approximately by rationals to any desired degree of accuracy. We may, for example, measure a physical length by means of appropriate instruments to the nearest mile, foot, inch, sixteenths of an inch, thousandths of an inch, etc.; and in every case we may record the result in the form a/b.

As a practical matter, however, this form is not very convenient, especially for performing arithmetic operations. It is not, to be sure, as cumbersome as the ancient Egyptian system with unit numerators, described in the previous section. But the Babylonians had a much better idea. They used a sexagesimal (base-sixty) notation throughout (both for integers and fractions), utilizing the same principle of position that we do. For example, the set of symbols $\| < < \| < < <$, which is equivalent to "two–ten–twelve–thirty,"[19] in an appropriate context stands for $2 \cdot 60 + 10 + 12 \cdot 60^{-1} + 30 \cdot 60^{-2}$, the last two places representing fractional values. As a more substantial example, an old clay tablet recorded the square root of two as $1 + 24 \cdot 60^{-1} + 51 \cdot 60^{-2} + 10 \cdot 60^{-3}$, a value which is in error by less than one millionth. Since fractional notation used the same symbols as those for whole numbers, changing only their position, it was as easy for the Babylonians to compute with one as with the other. And all this dates back to 2000 B.C.!

Our Hindu-Arabic system, using the principle of position with a base of ten, was developed much later. Apparently it began in India during the sixth century A.D., though the symbol for zero waited for three more centuries. But, strangely, this notation was applied only to whole numbers, so that half the advantage of the principle of position was lost. Not until the end of the sixteenth century did Simon Stevin, a Flemish mathematician, successfully urge the use of decimal *fractions*, and thus complete the development of the Hindu-Arabic system.

It is not our purpose to discuss computing with decimals. We merely point out, and illustrate in the exercises, that the same procedures can be applied in systems using bases other than 10, such as 12, 60, or even 2, as is done internally in electronic computers. Instead, we wish to examine decimal notation as an alternative method of expressing rational numbers. We shall also inquire about the meaning of decimal expressions which do not terminate, such as .3333 . . . , .323223222322223 . . . , and the expansions for such numbers as $\sqrt{2}$ and π.

You are aware that any rational can be converted into a decimal (base-ten) expression by division, i.e., by invoking Theorem 7.14. For instance, $1/2 = 1 \div 2 = .5$, $3/4 = .75$, $5/8 = .625$, but $1/3 = .3333\ldots$, a nonterminating decimal. We should like to probe this behavior in detail, encouraging you to experiment with converting a variety of fractions into decimals. By observing the results you may conjecture answers to the following questions, among others:

1. Which fractions convert to terminating decimals?

---

[18]See Definition 7.12 in Exercises 7.7, Problem *12.

[19]For these Babylonian symbols, see Exercises 5.1, Problem *6.

2. Which fractions convert to repeating decimals?
3. Do some fractions convert into decimals such as .323223222322223 . . . , which neither terminate nor repeat?
4. Is the behavior of the decimal equivalent of a fraction determined by the numerator alone, the denominator alone, or a combination of the two?
5. If a given fraction yields a terminating decimal, after how many places will it terminate?
6. If a given fraction yields a repeating decimal, after how many digits will it repeat?

The Babylonians were inveterate "table makers": they wrote tables of reciprocals, squares, cubes, square roots, cube roots, and even trigonometric functions. Where fractions were involved, they usually tabulated only *regular* numbers, namely, those numbers whose reciprocals terminate in base-sixty. Thus 2, 15, and 48 are regular, because $1/2 = 30 \cdot 60^{-1}$, $1/15 = 4 \cdot 60^{-1}$, $1/48 = 1 \cdot 60^{-1} + 15 \cdot 60^{-2}$. More generally, a regular number in the Babylonian numeration system is a positive integer whose reciprocal can be expressed exactly as a sum of multiples of negative powers of 60. This definition is equivalent to requiring that the number itself contain as factors only 2, 3, and 5, which are the factors of 60. Since $48 = 2^4 \cdot 3$, it is regular; but 7 and $99 = 3^2 \cdot 11$ are not. Are 450 and 525 regular?

The preceding discussion provides a partial answer to Question 1 for the sexagesimal system. Can we transfer this reasoning to our decimal system? Which of the following numbers are "regular" in base-ten: 2? 3? 4? 7? 40? 60? 800? You should compute the decimal equivalents of 1/2, 1/3, and the other reciprocals to ascertain which terminate. Careful observation may also yield an answer to Question 5. Moreover, consideration of fractions like 2/3, 4/7, 9/40, 7/60, 9/60, and 12/60 should suggest an answer to Question 4.

Since the term *regular* has proved so useful, we pause to give an official definition, generalized to the base b. It is

Definition 7.13 Regular Number A positive integer n is *regular* in the base b numeration system iff the rational $1/n$ is expressible as a terminating decimal in the form $a_1b^{-1} + a_2b^{-2} + \cdots + a_kb^{-k}$, where k is a positive integer and $a_i \in W$ with $a_i < b$.

The preceding examples show that some rationals convert to terminating decimals. We may ask whether, conversely, some terminating decimals may be expressed in the form a/b. This is easy! Just to read aloud the decimal .7 establishes that $.7 = 7/10$. Similarly, $.24 = 24/100 = 6/25$ and $3.209 = 3 + 209/1000 = 3209/1000$. Moreover, this tactic works for *all* terminating decimals. So we can assert that every terminating decimal is (or represents) a rational number.

Is the converse of the last sentence true? No, for the division process by which a/b is converted to a decimal need not terminate. Let us investigate what happens when it does not.

It is common knowledge that $1/3$ and $2/3$ yield the repeating, or *periodic*, decimals $.3333\ldots$ and $.6666\ldots$, respectively. Some rationals, however, repeat in blocks of two: thus $1/11 = .09090909\ldots$ and $5/33 = .15151515\ldots$ repeat with periods of two. A convenient way to write periodic decimals is to place a bar over the part which repeats. Thus we may write $.3333\ldots = .\bar{3}$, $.15151515\ldots = .\overline{15}$, $4.157575757\ldots = 4.1\overline{57}$. How are the following fractions expressed in the bar-notation: $4/9$? $1/6$? $6/11$? $2/7$? $1/17$?

Especially the last two examples raise the following question:

7. In converting a fraction to a decimal, how long must one continue the division in order to determine whether the decimal terminates, repeats, or does neither?

Actually, Question 7 is closely related to Question 3. To seek an answer, let us examine in detail the conversion of $1/7$. As shown in the following computation, the first nonzero quotient digit is 1, and the first remainder is 3; the second quotient digit is 4, and the second remainder is 2:[20]

```
    .1428571
 7)1.0000000
    7
   ---
    30
    28
    ---
     20
     14
     ---
      60
      56
      ---
       40
       35
       ---
        50
        49
        ---
         10
          7
         ---
          3 ← repetition starts
```

But how many *different* remainders can we get on division by 7? Since any remainder must be less than 7, there are only six possibilities: 1, 2, 3, 4, 5, and 6. (Why is 0 *not* a possibility for $1/7$, although it is for $1/4$?) Now if there are only six possibilities, then, after at most six division steps, one of the remainders

[20]The remainders are in boldface type for emphasis.

must recur. When it does, all the quotient digits and remainders from that point on must simply repeat the preceding steps. Thus 1/7 *must* repeat periodically after a maximum of six steps: i.e., it must convert to a repeating decimal whose period does not exceed 6. In this case, evidently the period *is* 6.

Similarly, 3/11, which has only ten *different* possible remainders (excluding 0), must convert to a repeating decimal with a period not exceeding 10.

$$\begin{array}{r} .272 \\ 11\overline{)3.000} \\ 2\,2 \\ \hline \mathbf{80} \\ 77 \\ \hline \mathbf{30} \\ 22 \\ \hline \mathbf{8} \end{array} \leftarrow \text{repetition starts}$$

As a matter of fact, only two different remainders occur: so 3/11 has a period of 2 rather than 10.

The chief significance, however, is that both of these rationals, 1/7 and 3/11, must convert to repeating decimals with definite periods. Extending the argument, it is easy to explain why 1/23 and 4/257 must convert to repeating decimals. Why, and what are their maximum possible periods? Although the preceding argument is not a formal proof, it can clearly be applied to any rational number to show that the decimal equivalent of the rational is periodic and that the period does not exceed a definite number.[21]

The work of the preceding paragraph shows that the answer to Question 3 is No. Equivalently, we assert that every rational can be expressed as a decimal which either terminates at some stage (with the remainder zero) or else repeats periodically. We now ask whether the converse is also true. Half of it certainly is; for we have already seen how easy it is to express a *terminating* decimal in the form a/b. Is there a method for dealing with a periodic decimal? Again we proceed by example.

Given the repeating decimal $.36363636\ldots = .\overline{36}$, which rational (if any) does it represent? Let us write $x = .36363636\ldots$; then $100x = 36.36363636\ldots$. The crux of the method is to subtract x from a suitable multiple of x so that the infinite "tails" drop off. Here the subtraction goes

$$\begin{aligned} 100x &= 36.36363636\ldots \\ x &= .36363636\ldots \\ \hline 99x &= 36.00000000\ldots \end{aligned}$$

[21]For an exact determination of the period, see the sixth Suggestion for Investigation at the end of this chapter.

or $99x = 36$. Solving this last equation for x, we get $x = 36/99 = 4/11$. We succeeded by choosing $100x$ as a "suitable multiple." To see how the method works, you should try subtracting x from $10x$, from $1000x$ and from $10000x$, recalling that the aim is to subtract off the infinite tails.

As a second example we try $4.16\overline{287} = 4.16287287287....$ Here the proper (smallest successful) multiple is $1000x$ (Why?), and we obtain

$$\begin{array}{rl} 1000x = & 4162.87287287\ldots \\ x = & \quad\; 4.16287287\ldots \\ \hline 999x = & 4158.71000000\ldots \end{array}$$

or $999x = 4158.71$; hence $x = 4158.71/999 = 415{,}871/99{,}900$, which is in the form a/b, though possibly not in lowest terms.

The preceding two examples had periods of 2 and 3, respectively. Can we extend the same technique to a repeating decimal whose period is 6, 13, or 2048? Certainly: we have only to multiply x by the appropriate power of 10, then subtract off the infinite tails and solve the resulting equation for x. Although this argument is not a formal proof, it obviously can be applied to any periodic decimal. Can it be applied to the decimal .323223222322223 . . . ? Is this decimal periodic?

So far we have illustrated general methods for converting any rational from the form a/b to a terminating or repeating decimal, and vice versa. This experience makes plausible the following:

THEOREM 7.15 Rationals as Decimals A decimal expression represents a rational number iff it terminates or is periodic.

We note that Theorem 7.15 asserts that the set of rationals (in the form a/b) and the set consisting of all terminating or repeating decimals are equal.

In Section 7.7 we showed that the set of rationals can be represented by points on the number line. Consequently, there is a point on the number line corresponding to every terminating or repeating decimal. But where do nonterminating, nonrepeating decimals like .323223222322223 . . . fit in? These numbers exist; yet by Theorem 7.15 they cannot be rationals. Can they be represented by points on the number line? If so, how can there be room for them, since the rationals are already dense on the line? And what kind of numbers are they? The next section deals with these questions.

EXERCISES 7.8

1. (a) Add the two numbers $2\cdot 60 + 10 + 12\cdot 60^{-1} + 30\cdot 60^{-2}$ and $5\cdot 60 + 30 + 19\cdot 60^{-1} + 45\cdot 60^{-2}$, expressing the sum in the same form.

(b) Express in the form a/b:

$$24\cdot 60^{-1},\quad 15\cdot 60^{-2},\quad 17\cdot 60^{-1} + 20\cdot 60^{-2},\quad 5 + 36\cdot 60^{-1},$$

$$2\cdot 60 + 10 + 12\cdot 60^{-1} + 30\cdot 60^{-2}$$

(c) Express the number $2\cdot 60 + 10 + 12\cdot 60^{-1} + 30\cdot 60^{-2}$ as a terminating or repeating decimal (in base-ten notation).

*(d) Express the Babylonian approximation for $\sqrt{2}$ (given in the text) in decimal form and compare it with the more nearly accurate value, 1.41421356+.

2. (a) Which of the following numbers are regular in the Babylonian system? Explain your answers. The numbers are 6, 24, 28, 75, 120, 143, 450, 525.

(b) Demonstrate that 6 and 24 *are* regular by giving the terminating sexagesimal expansions of 1/6 and 1/24, respectively.

*(c) For each of the remaining numbers of part (a) that is regular, express its reciprocal as a terminating sexagesimal.

3. (a) Which of the following numbers are regular in the decimal system? Explain your answers. The numbers are 2, 3, 4, 6, 7, 24, 28, 40, 60, 450, 800.

(b) For each of the numbers of part (a) that are regular, express its reciprocal as a terminating decimal.

(c) Explain how the above examples illustrate

THEOREM 7.16 Terminating Decimals The decimal expansion of the rational number $1/b$ terminates iff b is of the form $2^m\cdot 5^n$, where $m, n \in W$.

*(d) Examine the rationals 1/2, 1/5, 1/10, 1/4, 1/25, 1/40, and others; then conjecture a theorem which predicts the number of places the decimal expansion of $1/b$ requires for termination.

4. (a) Clearly 1/4, 2/4 = 1/2, and 3/4 all have decimal expansions which terminate. Is this also the case for

(i) 1/5, 2/5, 3/5, 4/5? (ii) 1/7, 2/7, 3/7, 4/7, 5/7, 6/7?

(iii) 1/6, 2/6, 3/6, 4/6, 5/6?

(b) After considering the results of part (a), conjecture an answer to Question 4 in the text.

(c) Illustrate the following theorem, which generalizes Theorem 7.16, and explain why your answer to part (b) is or is not consistent with it. The theorem is

THEOREM 7.17 Terminating Decimals (General Case) The decimal expansion of the rational number a/b, when a/b is expressed in lowest terms, terminates iff b is of the form $2^m \cdot 5^n$, where $m, n \in W$.

(d) Explain why the clause, "when a/b is expressed in lowest terms," may not be omitted from Theorem 7.17.

5. (a) Write each of the following repeating decimals in the bar notation:

$$.4444\ldots,\ .585858\ldots,\ 5.8888\ldots,\ .585555\ldots,$$

$$2.16666\ldots,\ 2.1066066066\ldots$$

(b) Express each of the following fractions in decimal form, using the bar notation: 4/9, 6/9, 1/6, 6/11, 3/7.

(c) Predict the maximum possible period of the decimal expansions of each of the following rationals; then find the expansions and compare them with your predictions. The rationals are 5/7, 5/6, 8/9, 8/7, 2/11, 5/13, 2/15, 1/17, 3/15, 7/24, 9/24.

(d) Generalize the preceding results by completing the following statement: "If the rational a/b is expressed in lowest terms and if b contains prime factors other than 2 or 5, then its decimal expansion repeats with a period not exceeding. . . ."

6. (a) Change each of the following decimals to the form a/b, expressing the results in lowest terms:

$$.9,\quad .73,\quad .08,\quad 2.5,\quad .625,\quad 3.05,\quad .175,\quad .000040$$

(b) Answer part (a) for .3, .33, .333, .3333, .33333 . . . = $.\bar{3}$.

(c) How does part (b) illustrate the fact that a rational can be approximated as closely as we please by a terminating decimal?

(d) Answer part (a) for $.\bar{5}$, $.\overline{54}$, $.5\bar{4}$, $.\bar{7}$, $.\overline{72}$, $1.0\bar{5}$, $1.\overline{05}$, $1.\overline{051}$, $.\overline{624}$, $.0\overline{624}$, $3.05\overline{161}$.

7. It is possible to consider all terminating decimals as repeating or periodic: e.g., 1/2 = .50000 . . . = $.5\bar{0}$.

(a) Write as periodic decimals 1/4, 3/8, 27/25, 68/100, 405/10,000.

(b) Convert to the form a/b the following repeating decimals: $.\bar{8}$, $.\bar{9}$, $.24\bar{9}$, $.374\bar{9}$, $.67\bar{9}$, $.00404\bar{9}$.

(c) In how many ways can the rational numbers 1/2 and 7/5 be expressed as periodic decimals? Write down all such expressions.

(d) Answer part (c) for the rationals 2/3 and 12/11.

(e) Is it true that every rational number a/b can be expressed *uniquely* in decimal notation? Explain your answer.

(f) Rewrite Theorem 7.15, using the enlarged concept of periodic decimals developed in this exercise.

8. Which, if any, of the following decimal expressions represent rational numbers? In each case, explain your answer.

   (a) 1.414 (b) $1.414141\ldots = 1.\overline{41}$

   (c) 1.414 1416 1418 14110 14112...

   (d) 1.414 2135 6327 30950 48801 ... $= \sqrt{2}$

*9. Give at least an informal proof of Theorem 7.15.

*10. (a) Which of the following numbers are regular in the duodecimal (base-twelve) system? For those which are, exhibit their reciprocals as terminating duodecimals. The numbers are 2, 3, 4, 5, 6, 7, 8, 9, T, E, 11.

   (b) Answer part (a) for the binary (base-two) system, considering the numbers 10, 11, 100, 101, 110, 111, 1000, 1001, 1010.

   (c) The duodecimal equivalent of 1/5 is periodic. Predict its period, then confirm your answer by computing the duodecimal form.

   (d) Answer part (c) for the binary equivalent of 1/101 (which is the same rational as that of part (c)).

*11. Apply Questions 1–7 of the text to the general base-b numeration system, with $b \in N$ and $b \geq 2$. How, if at all, would the answers differ from those for the special case $b = 10$?

*12. Solve the following cryptarithms, which involve repeating decimals:

   (a) EVE/DID = .TALKTALKTALK . . . (There are two solutions.)

   (b) In base-nine, SO/HE = .RANRANRAN . . .

      (i) Find the only solution with positive digits whose value is less than 1/2.

      (ii) How many other solutions with positive digits are there?[22]

   (c) ME/SS = .PIPIPI . . . , in base-ten, with the digits in arithmetic progression.[23]

   (d) In base-eleven, SI/HE = .CANCANCAN . . . , with a value less than 1/4.[24]

---

[22]Adapted from Problem 796, proposed by C. W. Trigg, in *Mathematics Magazine*, 44:3 (May, 1971), p. 165.

[23]Adapted from Problem 3284, proposed by C. W. Trigg, in *School Science and Mathematics*, 70:10 (October, 1970), p. 677.

[24]Adapted from Problem 233, proposed by C. W. Trigg, in *Pi Mu Epsilon Journal*, 5:2 (Spring, 1970), p. 87.

Section 7.9 Are the Rationals Really Closed?

In Section 7.7 we finished showing that the rationals are closed under the four fundamental operations of arithmetic, except for division by zero. But in Section 7.8 we left several questions hanging. Among them are the following:

1. What is the nature of nonrepeating, nonterminating decimals such as .323223222322223 . . .?
2. Do they fit on the number line, which is already dense in the rationals? If so, how?
3. What is the nature of numbers like $\sqrt{2}$, $\sqrt{3}$, $\sqrt[3]{2}$, and π? Where, if at all, do they fit on the number line?
4. Since raising numbers to powers and taking roots of numbers constitute another pair of operations, which, if any, of the systems W, Z, or Q are closed under these operations?

We shall show that these four questions are interrelated.

For convenience, we begin with powers and roots of whole numbers. Definition 5.1 gave meaning to the expression x^m, where $x \in W$ and $m \in N$. There is no problem in extending this definition to include $x \in Z$ and even $x \in Q$: for *raising to a power* is a special case of multiplication, which is well defined in all three number systems. It follows that the operation of raising to a positive integral power is closed in W, Z, and Q. In particular, we frequently have occasion to square a number, for example in computing the area of a square whose side is 3 units in length.

Attempts to perform the inverse operation, however, lead to difficulties. If a square has an area of 25 square units, then its side has a length of $\sqrt{25}$, or 5; but if its area is 26 square units, the precise length of its side is not obvious, even though we may write it formally as $\sqrt{26}$ units. We can, to be sure, find a rational number which is approximately correct: 5.1 is a possibility, since $5.1^2 = 26.01$, which is very close to the given area. Moreover, if we are unwilling to tolerate this discrepancy of .01 in the area, we may by trial and error produce a closer approximation, e.g., 5.099, since $5.099^2 = 25.999701$, which differs from 26 by less than .0003. Indeed, since the first estimate, 5.1, is too large and the second too small, we may conclude with certainty that $5.099 < \sqrt{26} < 5.100$. But though we have $\sqrt{26}$ "trapped" within a fairly narrow range of rationals, we shall soon be able to prove that $\sqrt{26}$ itself is not rational. In view of Theorem 7.15, this means that $\sqrt{26}$ cannot be expressed exactly as either a terminating or a periodic decimal.

Before proceeding, we should assign a definite meaning to the phrase "the (a) square root of 26." The above method of testing decimal approximations provides a clue: $\sqrt{26}$ is that number (if such a kind exists!) whose square is 26. In algebraic language, $x = \sqrt{26}$ iff $x^2 = 26$. Generalizing, we have the following:

Definition 7.14 Square Root If $r \in Q$, then *a square root of* r is a rational number x (if x exists) such that $x^2 = r$.

It is easy to check by Definition 7.14 that $\sqrt{25} = 5$, $\sqrt{1} = 1$, $\sqrt{1/4} = 1/2$, $\sqrt{4/9} = 2/3$, and $\sqrt{1.96} = 1.4$.

Similar reasoning and symbolism yield roots of higher order than two. Thus we write $\sqrt[3]{64} = 4$, since $4^3 = 64$; $\sqrt[4]{81} = 3$, since $3^4 = 81$; $\sqrt[3]{125/27} = 5/3$, since $(5/3)^3 = 125/27$; and $\sqrt[10]{1024} = 2$, since $2^{10} = 1024$. Formally we announce

Definition 7.15 ***n*th Root** If $r \in Q$, then *an nth root of r*, for $n \in N$, is a rational number x (if x exists) such that $x^n = r$.

At this point the question of closure becomes vexing. You should recall that closure has two parts, existence and uniqueness. The problem of uniqueness has already been foreshadowed by the careful phrasing of Definitions 7.14 and 7.15, which speak of *a* square root and *an* *n*th root. From the fact that, according to the laws of signs for multiplication in Z and Q, $(^+5)^2 = (^-5)^2 = 25$, it follows that 25 has *two* distinct square roots, $^+5$ and $^-5$. Is the same true of the other examples following Definition 7.14? Similarly, 16 has two fourth roots, since $(^+2)^4 = (^-2)^4 = 16$. For cube roots, on the other hand, there is only one possibility, since $(^+x)^3 \neq (^-x)^3$: thus $\sqrt[3]{^+8} = {}^+2$ only, while $\sqrt[3]{^-8} = {}^-2$ only. To avoid the sign ambiguity which occurs with even roots (Why then?), we shall henceforth agree that the radical *symbols* $\sqrt{r}$ and $\sqrt[n]{r}$ (for n even) represent positive roots only. So, though we must admit that "there are two square roots of 25," we restrict the symbol $\sqrt{25}$ to mean $^+5$ only. Similarly, $\sqrt{16/9} = {}^+4/3$, $\sqrt[4]{16} = {}^+2$, $^-\sqrt{49} = {}^-(^+7) = {}^-7$, $\sqrt[3]{8} = {}^+2$, but $\sqrt[3]{^-8} = {}^-2$, since here a negative root is the only one possible.

Turning now to the question of the *existence* of *n*th roots, we have already noted that $\sqrt{26}$ may not have an exact value that is rational. To see why, we shall investigate first the simpler example, $\sqrt{2}$. We begin by obtaining bounds for $\sqrt{2}$. Clearly $\sqrt{2} \neq 1$, since $1^2 = 1 < 2$; also, $\sqrt{2} \neq 2$, since $2^2 = 4 > 2$. We conclude, then, that $\sqrt{2}$ is some number lying between 1 and 2. We might guess $\sqrt{2} = 3/2 = 1.5$, but $(3/2)^2 = 9/4 > 2$; so apparently 3/2 is a little too large. Trying $7/5 = 1.4$, we get $(7/5)^2 = 49/25 < 2$; so 1.4 is too small. Hence $\sqrt{2}$ is between 1.4 and 1.5. Now we try, in turn, 1.41 and 1.42 and find that $\sqrt{2}$ is between these two rationals, etc. Each of these pairs of bounds is closer to $\sqrt{2}$ than the preceding ones, but none is exact. Thus we can get a sequence of inequalities:

$$1 < \sqrt{2} < 2$$

$$1.4 < \sqrt{2} < 1.5$$

$$1.41 < \sqrt{2} < 1.42$$

$$1.414 < \sqrt{2} < 1.415$$

$$1.4142 < \sqrt{2} < 1.4143$$

$$\cdots$$

Note that each bound is a rational number. So far, however, we have not succeeded in finding a rational whose square is *exactly* 2: i.e., we have not demonstrated that $\sqrt{2}$ is rational.

Since order relations can be easily visualized on the number line, let us call geometry to our aid. Each of the rationals occurring in the above sequence of inequalities corresponds to a unique point on the line; and $\sqrt{2}$ is "squeezed" within shorter and shorter intervals determined by pairs of these points. A magnified portion of the number line containing these intervals would look like Figure 7.4. Intuitively, we feel that $\sqrt{2}$ must correspond to a (or the?) point

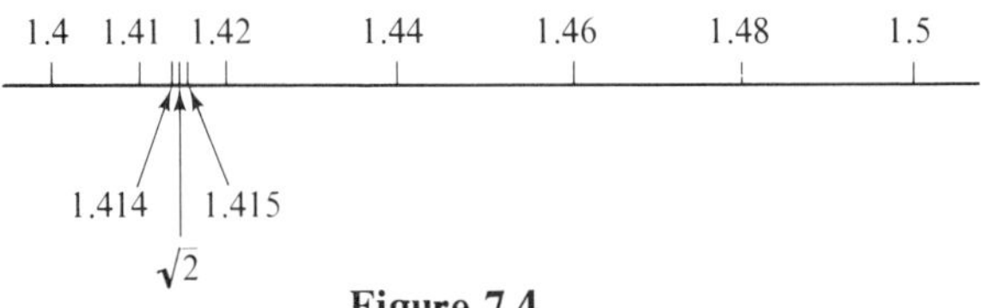

Figure 7.4

within each of these ever-narrowing intervals. Interestingly enough, this point may be located *exactly* by standard construction methods of elementary geometry, using only a straight edge and compass. We assume a given number line, with one point chosen to represent the number 0, and a given unit length laid off successively to the right and left of the 0 point with the division points corresponding to the positive and negative integers, as in Figure 7.5. We

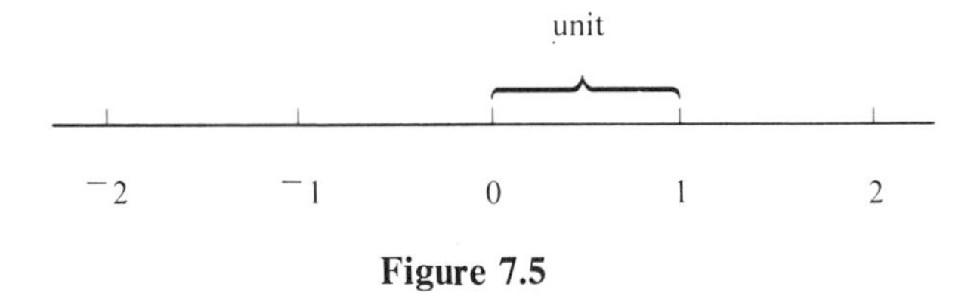

Figure 7.5

now construct above the line the square whose side is determined by the points corresponding to 0 and 1. Having drawn the diagonal of the square from the point 0, we establish from the Theorem of Pythagoras that its length is $\sqrt{2}$ units, the number we are seeking. If we now describe with the compass an arc with the point 0 as center and the length of the diagonal as radius, the arc meets the number line at the point which corresponds to $\sqrt{2}$. The completed figure is shown in Figure 7.6. Insofar as we can trust geometric intuition, $\sqrt{2}$ *is* representable as a point on the number line. Does this mean that $\sqrt{2}$ is rational?

Historically, this question is an ancient and explosive one; it is sometimes called "the scandal of Pythagoras." The Pythagorean mathematical brotherhood, dating from the sixth century B.C., held as part of its philosophy that "all is number." By this the Pythagoreans meant that the universe is explicable in terms of whole numbers. Since rationals are quotients of whole numbers,

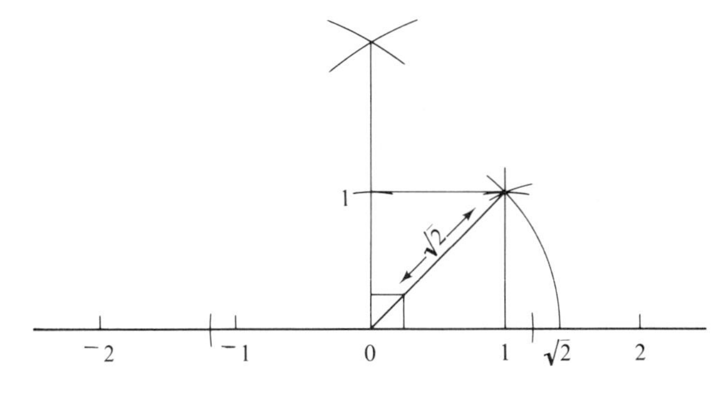

Figure 7.6

they also included these. In their mathematical researches, consideration of the square and its diagonal led them naturally to $\sqrt{2}$. Now, squares and their diagonals are certainly part of the universe which they were trying to explain. Imagine their consternation, therefore, when they discovered and actually proved that $\sqrt{2}$ is *not* rational and so could not be brought within their world scheme. Because the brotherhood was a secret society, they managed for a time to keep the damaging evidence to themselves; but eventually it came out. Legend has it that the brother who revealed this guilty secret was punished for his indiscretion by banishment or death.

In any case, the following proof that $\sqrt{2}$ is not rational is not only historically important but also constitutes a beautiful example of an indirect argument. You should compare its logical structure with that of Euclid's proof of the infinitude of primes (Theorem 4.36). Because of the significance of both method and result, the proof is given in considerable detail. The strategy, of course, is to assume that $\sqrt{2}$ *is* rational, reason thereafter with strict logic, and finally arrive at a contradiction, causing rejection of the assumption. So we present

THEOREM 7.18 Irrationality of $\sqrt{2}$ The (positive) square root of 2 is not rational. (*Rephrased:* If $x = \sqrt{2}$, then x is not rational.)

Proof:

1. Suppose $x = \sqrt{2}$ and x is rational. (This supposition is a possibility, but unsupported by any valid reason.)
2. Then $x = \sqrt{2} = a/b$, with $a, b \in Z$ and $b \neq 0$, by Definition 7.9.
3. Without loss of generality, we may assume that a/b is in lowest terms; for if it were not, we could reduce it.
4. Using Theorem 7.14 and Theorem 4.13, extended to integers, we may multiply the equation $a/b = \sqrt{2}$ by b or $b/1$, obtaining $a = \sqrt{2} \cdot b$.
5. By Definition 7.14 for a square root, $(\sqrt{2})^2 = 2$.
6. Using Theorem 4.14 extended to integers to multiply the equation in Step 4 by itself, and applying step 5, we get $a^2 = 2b^2$.

7. But $a^2 = 2b^2$ implies that a^2 is even, by Definition 4.17.
8. Theorem 4.35 now shows that *a is even*.
9. Since a is even, we may represent it by $a = 2c$, with $c \in Z$, by Definition 4.17 extended to integers.
10. Substituting from Step 9 into Step 7, we get $a^2 = (2c)^2 = 2b^2$, or $4c^2 = 2b^2$, using the definition of exponent and commutativity and associativity of multiplication for integers.
11. By Theorem 4.15 extended to integers, $4c^2 = 2b^2$ may be reduced to $2c^2 = b^2$, so that b^2 is even.
12. Again applying Theorem 4.35, *b is even*.
13. Since, from Steps 8 and 12, both a and b are even, they have 2 as a common factor, so that a/b is *not* in lowest terms.
14. But Step 13 *contradicts* Step 3.
15. In view of this contradiction, the consistency of the system of rational numbers (including the integers) and the valid reasoning in all steps except the first, Step 1 must be false, and the theorem is true.

By similar methods it is possible to prove that $\sqrt{3}$, $\sqrt[3]{2}$, and similar radicals are irrational; the exercises call for two such proofs. Many different proofs of the irrationality of $\sqrt{2}$ have been made. One which uses the Fundamental Theorem of Arithmetic (Theorem 4.32) is easy to generalize so as to prove

THEOREM 7.19 Irrationality of Radicals If $x \in Q$ and $n \in N$, with $n \geq 2$, then $\sqrt[n]{x}$ is irrational unless x is the nth power of some rational.

Let us recapitulate. We have located $\sqrt{2}$ on the number line and have also proved that it is not rational. The approximate decimal value of $\sqrt{2}$, as given in Exercises 7.8, Problem 8(d), does not appear either to terminate or to repeat. Yet looks can be deceiving; and because we are dealing with an infinite decimal expansion, we can never be sure of its character simply by carrying it out to a finite number of places, no matter how large. Here one solid proof is worth more than a million decimal places! In Theorems 7.15 and 7.18, we have the proof that the decimal expansion of $\sqrt{2}$ — if it has one! — can neither terminate nor repeat. That $\sqrt{2}$ *is* equivalent to some infinite decimal is a deep property which we are not in a position to prove; but the geometric model of the point corresponding to $\sqrt{2}$ makes this equivalence plausible. Another evidence is the arithmetic algorithm for computing square roots which, as you may recall, can be continued indefinitely.

But what about other radicals like $\sqrt{3}$, $\sqrt[3]{2}$, and $\sqrt[4]{2/7}$? Since it is possible to prove that they are not rational, their decimal expansions also must not terminate or repeat. Again, the existence of such expansions can be made plausible by making successive approximations and appealing to a geometric model to locate the corresponding points on the number line. Finally, it seems reasonable to suppose that "manufactured" decimals like $y = .323223222322223\ldots$, which must be irrational by Theorem 7.15, can also be associated with points on the number line. Indeed, it is easy to write down for y a sequence of inequalities similar to those for $\sqrt{2}$:

$$0 < y < 1$$
$$.3 < y < .4$$
$$.32 < y < .33$$
$$.323 < y < .324$$
$$\ldots$$

We now invite you to reconsider the four questions we raised at the beginning of this section. You should now be in a position to answer them, at least partially and with some confidence. But at least two mysteries may remain, the first a variant of Question 2 and the second an extension of Question 4. They are

5. How can the number line, already dense in the rationals, accommodate the plethora of irrationals implied by this section?
6. Since even our largest number system, Q, is not closed under the operation of extracting nth roots, what further extension of Q will be needed to close this operation (if this is possible at all!), and how can it be accomplished?

Though the complete answers to these questions are beyond the scope of this book, we shall conclude with some further intuitive remarks about them.

First, with regard to Question 5, we have already seen that a point representing $\sqrt{2}$ exists on the number line: it can even be found by a geometric construction. It is reasonable also to think that points for other roots and for numbers like π can be located by the *nested interval* method illustrated by the number y as performed in this section. This very procedure, indeed, is the basis of one way to introduce the *real* numbers, to which we shall now turn.

In common usage the adjective *real* means *genuine*, *actual*, *tangible*, *existing*, and the like. When, for example, we contrast the world of ideas or theory with the "real" world, we regard the latter as the totality of physical objects which we can see, feel, or otherwise apprehend with our senses, either directly or with the aid of instruments such as telescopes or microscopes. With this interpretation, numbers of all kinds are not part of the real world, for they are mental constructs, as we pointed out following Definition 4.1. For instance, we can physically observe the *numeral* 2, written on a piece of paper, and also sets of two physical objects; but we can perceive the cardinal *number* 2 only within the mind and not by any of the five senses.

Throughout the history of mathematics, however, scholars have been fond of assigning technical meanings to some of the most common words in our language. Some of these assignments are most apt and intuitively appealing, but other choices are not informative or are positively misleading. The phrase "real number" belongs in the last category: real numbers are just as abstract as any other kind of number!

So what is the meaning of the technical term *real number?* As examples we may take any rational number ($5/6$, $^{-}3/2$, $^{-}7$, $9, \ldots$) or any irrational number ($\sqrt{2}$, $\sqrt[3]{7/4}$, $^{-}\sqrt{5}$, $\pi, \ldots$). In fact, we may consider R, the set of real numbers, as an extension of the rationals, obtained by joining to them the irrationals, in this way forging another link in the chain of inclusion: $W \subset Z \subset Q \subset R$.

To define or logically create the real numbers is mathematically a very sophisticated task, which we shall have to forego. Instead, we shall resort to two rather intuitive descriptions, one algebraic and the other geometric. The algebraic description uses the fact that all rationals can be expressed as terminating or repeating decimals, and conversely, together with the reasonable assumption that all irrationals may be represented as nonterminating, nonperiodic decimals, and conversely. Integers are also included, of course, since 5, for example, can be written as $5.\bar{0}$ or $4.\bar{9}$. Thus we come to

Definition 7.16 Algebraic Description of the Reals (Description) A *real* number is any number which can be expressed as a decimal, either terminating or not.

The geometric description now evolves easily. In Section 7.7, we already showed how to locate on a number line points corresponding to rationals; in this section we have done the same for irrationals. So it is evident that, corresponding to every rational or irrational number, there is a unique point on the number line. It is reasonable to conclude, moreover, that every point on the line represents *some* rational or irrational number. This brings us to

Definition 7.17 Geometric Description of the Reals (Description) A *real* number is any number which can be represented by a point on the number line.

From the above description as well as from the informal use of the term *irrational* as a synonym for *not rational*, it appears that the set of reals is the union of two disjoint sets, Q and Q', the set of irrationals. Algebraically speaking, Q' consists of all the numbers representable as nonterminating, nonperiodic decimals. Geometrically, Q' corresponds to the set of points that are still missing from the number line after all the rationals have been located on it.

Now, in order to develop the system $(R, +, \cdot)$, we would have to consider the meanings and properties of the two binary operations $+$ and $\cdot$ on R. Without going into detail, we can say that the system of real numbers possesses the properties mentioned for the rationals in Definition 7.9, density as given in Definition 7.11, and *completeness* in the sense that the real numbers completely fill up the number line, leaving no gaps of any kind.

We might hope that the reals are also closed with respect to the operation of extracting nth roots. Unfortunately, this is true only for the positive reals and zero. To see the difficulty that arises from negative reals, let us consider the symbol $\sqrt{^-4}$. By Definition 7.14, extended to reals, we must seek a real number x such that $x^2 = {}^-4$. Now in R, as in Z and Q, a number x can be classified as positive, negative, or zero; moreover, the same laws of signs continue to hold, including those for multiplication. Consider, then, the equation $x^2 = {}^-4$. Clearly, $x = 0$ does not satisfy it. Moreover, $x = {}^+2$ and $x = {}^-2$ do not work either, because the squares of both are $^+4$, not $^-4$. Our reluctant conclusion must be that $\sqrt{^-4}$ is not a real number. It is, however, very definitely a number of some kind. Historically it has been called an *imaginary* number; and it is part of a still larger system, that of the *complex* numbers, $(C,+,\cdot)$, which includes the reals as a proper subset.

Although the complex numbers have many important applications, especially in physics and engineering, we shall not discuss them in this book. Instead, we shall summarize the relationships among the various kinds of numbers which we have mentioned. We do this by means of Figure 7.7:

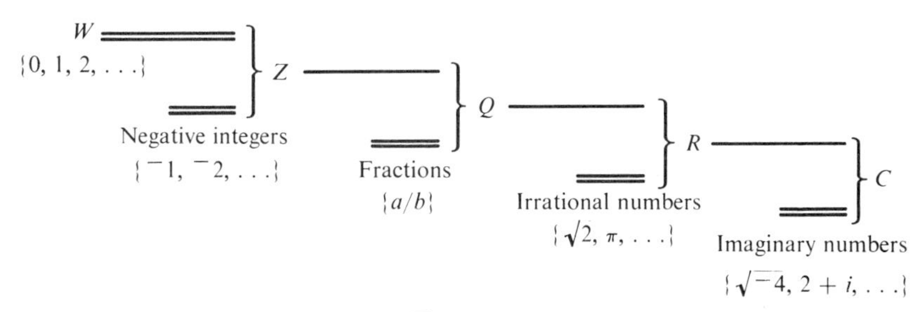

Figure 7.7

Note that each system in Figure 7.7 includes the system placed to its left as a proper subset: $W \subset Z \subset Q \subset R \subset C$. This means that the number 5, for example, is a whole number, an integer (5 or $^+5$), a rational number (5 or 5/1), a real number (5 or $5.\bar{0}$), and a complex number (5 or $5 + 0i$). Thus we may claim to have begun acquaintance with complex numbers in the kindergarten!

Let us end by viewing in retrospect our development of the five number systems in terms of gradually fulfilling the laws for groups, fields, and order relations, as well as closure of the various operations. We began with the whole numbers, which possess the properties of closure, commutativity, associativity, and identity for the two basic operations, addition and multiplication. But $(W,+,\cdot)$ fails to be a group for lack of inverses, both additive and multiplicative. The introduction of negative integers furnishes additive inverses and thus makes the system $(Z,+)$ a group and closes the operation of subtraction. When fractions are made available, the system $(Q^*,\cdot)$, with zero deleted, also becomes a group, and division (except by zero) is closed; and the entire system $(Q,+,\cdot)$ constitutes a field. In the meantime, we noted that the rationals, though dense, still do not completely fill the number line, and that they lack nth roots of most rationals. An extension to the reals, $(R,+,\cdot)$, makes the number line complete. It can be proved that the final extension, to $(C,+,\cdot)$, closes the operation of taking nth roots and preserves the field properties, though unfortunately the order properties no longer hold. During the course of this endeavor, moreover, we met, in one guise or another, most of the common laws of arithmetic and elementary algebra; and we found them to be identical.

But is this the end of the story, or are still further extensions possible? Certainly they are, and many have been attempted. Some of them employ the concepts of group and field; others sacrifice some of the laws, such as commutativity of multiplication or even the existence of a multiplicative identity, to gain a more general and flexible behavior. The study of such systems, though quite beyond the scope of this text, lies at the heart of modern algebra, one of the most extensive and active branches of mathematics.

EXERCISES 7.9

1. (a) State formally the extension of Definition 5.1, mentioned at the beginning of this section, to include powers of integers and of rationals.
   (b) Compute 13^2, 2^5, 0^7, $(^-13)^2$, $(^-2)^5$, $(^-4)^3$, $(^-4)^4$.
   (c) If x is a negative integer and m is a natural number, discuss the circumstances under which x^m is positive; under which x^m is negative.
   (d) If in part (c) x is a negative rational number, how, if at all, would your answer change?
   *(e) State part (d) formally as a theorem, and prove it.
2. (a) Using the extended definition of Problem 1(a), together with the multiplication formula for rationals, $a/b \cdot c/d = ac/bd$, compute $(1/3)^2$, $(4/3)^2$, $(^-4/3)^2$, $(3/4)^2$, $(.75)^2$.
   (b) Compare the last two results of part (a).
   (c) Generalize part (a) by computing $(a/b)^2$, where $a/b \in Q$.
   (d) Prove formally, but briefly

THEOREM 7.20 Square of a Rational If $a/b \in Q$, then $(a/b)^2 = a^2/b^2$.

   (e) Compute $(2/3)^3$, $(^-2/3)^3$, $(1/2)^4$, $(^-1/2)^4$, $(a/b)^3$, $(a/b)^m$ for $m \in N$. Note that the last part yields the following generalization of Theorem 7.20:

THEOREM 7.21 *m*th Power of a Rational If $a/b \in Q$ and $m \in N$, then $(a/b)^m = a^m/b^m$.

3. (a) Compute $\sqrt{49}$, $\sqrt{0}$, $\sqrt{1/9}$, $\sqrt{16/25}$, $\sqrt{9/4}$, $\sqrt{2.25}$.
   (b) Compare the last two results of part (a).
   (c) Test each of your answers to part (a), using Definition 7.14.

(d) Show that $\sqrt{21} \neq 5$ and $\sqrt{21} \neq 4.4$.

(e) By testing some of the rationals, 4.0, 4.1, 4.2, . . . , 4.9, find the best approximation (to tenths) of $\sqrt{21}$.

4. (a) Compute $\sqrt[3]{1}$, $\sqrt[3]{125}$, $\sqrt[3]{^-1}$, $\sqrt[3]{^-125}$, $\sqrt[3]{27/8}$, $\sqrt[4]{16}$, $\sqrt[5]{243}$.

(b) Test each of your answers to part (a) by using Definition 7.15.

5. (a) Which of the following statements are true?

(i) 6 is a square root of 36.	(ii) $\sqrt{36} = 6$
(iii) $^-6$ is a square root of 36.	(iv) $\sqrt{36} = {}^-6$
(v) $^-\sqrt{36} = {}^-6$	(vi) 6 is a cube root of 216.
(vii) $\sqrt[3]{216} = 6$	(viii) $^-6$ is a cube root of 216.
(ix) $^-6$ is a cube root of $^-216$.	(x) $\sqrt[3]{^-216} = {}^-6$

(b) Which of the following equations is (are) satisfied by $x = 4$?

(i) $5 + \sqrt{x} = 7$ (ii) $3 - \sqrt{x} = 1$ (iii) $3 + \sqrt{x} = 1$

*(c) (i) Does equation (i) of part (b) have any roots other than $x = 4$? Justify your answer algebraically.

(ii) Does equation (iii) of part (b) have any roots other than $x = 4$? Justify your answer algebraically.

6. (a) Show that $\sqrt{5}$ lies between 2 and 3, i.e., that $2 < \sqrt{5} < 3$.

(b) By testing the rationals 2.0, 2.1, 2.2, . . . , 3.0, find the tenths interval containing $\sqrt{5}$, i.e., the interval such that $2.? < \sqrt{5} < 2.?$

(c) Continue the sequence of rational bounds in part (b) to hundredths.

(d) Locate on an appropriate magnified portion of a number line a point corresponding to $\sqrt{5}$, as was done in the text for $\sqrt{2}$.

(e) Starting with a number line with unit lengths marked off from the 0 point right and left, show how to construct, with straight edge and compass, a point exactly corresponding to $\sqrt{5}$. (*Hint:* Construct on the number line a rectangle of length 2 and height 1, and consider its diagonal.)

(f) Is there a point on the number line of part (e) corresponding to $^-\sqrt{5}$? If so, locate it; if not, why not?

(g) Answer part (f) for $\sqrt{^-5}$.

7. Construct a sequence of 5 sets of inequalities giving narrower and narrower rational bounds for each of the following decimals:

(a) $y = .4141141114\ldots$ (b) $\sqrt{3} = 1.73205\ldots$

(c) $\pi = 3.14159\ldots$

8. An algorithm for computing $\sqrt{r}$, for r a positive rational in decimal form,

that is more efficient than the tedious guesswork of Problem 6(b) and (c) is the following, which we illustrate for $\sqrt{26}$:

1. Make a rough estimate: $\sqrt{26} \approx 5$
2. Divide 26 by this estimate: $\begin{array}{r} 5.2 \\ 5\overline{)26.0} \end{array}$
3. Average the divisor and the quotient of part 2 to get a new and closer approximation to $\sqrt{26}$: $(5 + 5.2)/2 = 5.1$
4. . . . Continue, dividing 26 by the approximation found in part 3, then averaging divisor and quotient to get a still closer approximation, etc. Here

$$\begin{array}{r} 5.098 \\ 5.1\overline{)26.0000} \\ 25\ 5 \\ \hline 500 \\ 459 \\ \hline 410 \\ 408 \\ \hline 2 \end{array}$$

and $(5.1 + 5.098)/2 = 5.099$; to get a still better approximation, compute $5.099\overline{)26.0000}$ and average again, etc.

Apply the above method to compute

(a) $\sqrt{34}$ to the nearest thousandth.

(b) $\sqrt{3.4}$ to the nearest ten-thousandth.

(c) $\sqrt{5}$ to the nearest ten-thousandth.

(d) $\sqrt{474}$ to the nearest hundredth.

(e) $\sqrt{34}$, again to the nearest thousandth, starting with the very poor estimate, $\sqrt{34} \approx 12$. Note that the method still succeeds, though it takes more steps.

9. In the text's discussion of $\sqrt{^{-}4}$ we actually tested (and rejected) only three possible values for x: 0, $^{+}2$, and $^{-}2$.

(a) Complete the argument by explaining why, if x is *any* positive real number (not necessarily $^{+}2$), x cannot be a square root of $^{-}4$.

(b) Similarly explain why no *negative* real number can be a square root of $^{-}4$.

(c) Can the preceding arguments be generalized to determine whether $\sqrt{r}$ is a real number, when r itself is a *negative* real number? Explain your answer.

(d) Summarize your conclusions regarding the expression $\sqrt{r}$ by completing the following sentence: "Given a real number r, the expression $\sqrt{r}$ represents a real number iff. . . ."

(e) Given that r is a real number, answer part (d) for the expressions

(i) $\sqrt[4]{r}$ (ii) $\sqrt[3]{r}$ (iii) $\sqrt[n]{r}$, where $n \in W$ and $n \geq 2$

10. (a) Determine whether nonterminating, nonperiodic decimals such as .323223222322223 . . . are rational numbers. Justify your answer logically.

(b) Can a point corresponding to a number like the one in part (a) be located on the number line? Justify your answer intuitively.

(c) Answer parts (a) and (b) for radicals such as $\sqrt{2}$, $\sqrt{3}$, and $\sqrt[3]{2}$.

(d) Which, if any, of the systems W, Z, or Q are closed under the operation of raising to mth powers? Justify your answer logically.

(e) Which, if any, of the systems W, Z, or Q are closed under the operation of extracting nth roots? Explain your answer, at least intuitively, treating separately the special case where the radicand is negative and the index is even.[25]

11. (a) According to Definition 7.16, which of the following are real numbers?

$3/5$, $5/3$, $^{-}2/7$, 1492, 0, $^{-}7{,}000{,}000$, 9.23×10^{-8}, 1.06×10^{15}, $\sqrt{36}$, $\sqrt{6}$, $\sqrt[3]{8/27}$, $\sqrt[3]{^{-}8/27}$, $\sqrt{^{-}4}$, $^{-}\sqrt{4}$, 2π, 13.725, $13.\overline{7249}$, $13.724\overline{9}$, $\sqrt[5]{\pi/2}$, $\sqrt[4]{^{-}\pi/3}$, $2.07\overline{61}$, $10{,}241_{\text{five}}$, $.014_{\text{twelve}}$, $.\overline{31}_{\text{eight}}$, $4.040040004\ldots$, $\sqrt[4]{4.040040004\ldots}$

(b) Illustrate Definition 7.17 by drawing a number line and locating on it, if possible, points corresponding to the following numbers:

$3/5$, $5/3$, $^{-}2/7$, $\sqrt{36}$, $\sqrt{6}$, $\sqrt[3]{^{-}8/27}$, $\sqrt{^{-}4}$, $^{-}\sqrt{4}$, 2π, $4.040040004\ldots$, $2.07\overline{61}$

(c) For each of the numbers of part (a) that is not real, explain why it is not.

*12. (a) (i) Compute $12^2 = 12 \cdot 12$, then $(4 \cdot 3)^2 = (4 \cdot 3)(4 \cdot 3)$.

(ii) Similarly compute, then compare, 6^3 and $(2 \cdot 3)^3$.

(iii) More generally, compare a^2b^2 and $(ab)^2 = (ab)(ab)$, where $a, b \in N$.

(b) Prove the following:

---

[25]In the symbol $\sqrt[n]{x}$, x is the *radicand* and n is the *index*.

THEOREM 7.22 Square of a Product If $a, b \in N$, then $(ab)^2 = a^2b^2$.

(c) Would it be possible to prove Theorem 7.22 in a system where multiplication is not commutative? Explain.

(d) Illustrate the following two generalizations of Theorem 7.22:

THEOREM 7.23 Power of a Product If $a, b, m \in N$, then $(ab)^m = a^m b^m$.

THEOREM 7.24 Power of a Product (General Case) If $a_1, a_2, \ldots, a_k$ and $m \in N$, then $(a_1 a_2 \ldots a_k)^m = a_1^m a_2^m \ldots a_k^m$.

*13. Let $c \in N$, with $c \geq 2$. Then by the Fundamental Theorem of Arithmetic (Theorem 4.32), we may represent c as $c = p_1^{n_1} p_2^{n_2} \ldots p_k^{n_k}$, where the p_i are primes and the $n_i \in N$.

(a) Use Theorem 7.24 for the case $m = 2$ to prove that $c^2 = (p_1^{n_1})^2 (p_2^{n_2})^2 \ldots (p_k^{n_k})^2$.

(b) Now apply Theorem 5.4 of Exercises 5.2, Problem *8(c), to part (a) to prove that $c^2 = p_1^{2n_1} p_2^{2n_2} \ldots p_k^{2n_k}$.

(c) Observe how part (b) leads to

THEOREM 7.25 Factorization of a Square If $c \in N$ and $c \geq 2$, then in the prime factorization of c^2 each prime factor of c occurs an even number of times.

*14. (a) Another indirect proof that $\sqrt{2}$ is not rational follows the text's proof of Theorem 7.18 as far as Step 6, which yields the equation $a^2 = 2b^2$, though Step 3 is unnecessary this time. Now, by Theorem 7.25, each of a^2 and b^2 has a prime factorization in which the factor 2 appears an even number of times, or else it does not appear at all. Devise a contradiction by counting the number of factors 2 which occur on each side of the equation $a^2 = 2b^2$.

(b) Explain how the argument of part (a) may be applied to prove that $\sqrt{3}$, $\sqrt{5}$, $\sqrt{6}$, and $\sqrt{8}$ are not rational.

(c) Adapt Problem *13 (including Theorem 7.25) so as to be able to prove that $\sqrt[3]{2}, \sqrt[4]{2}, \ldots, \sqrt[3]{3}, \ldots$, etc., are not rational.

*15. (a) If in Theorem 5.1 we permit m and n to be rational numbers, then we may obtain equations like

$$4^{1/2} \cdot 4^{1/2} = 4^{(1/2)+(1/2)} = 4^1$$

But $4^{1/2}$ is so far a meaningless symbol, since it is not covered by Definitions 5.1, 5.2, or 5.3. From the ends of the equation $4^{1/2} \cdot 4^{1/2} = 4$, however, it is clear from Definition 7.14 that $4^{1/2}$ *ought* to represent $\sqrt{4}$, or 2. Write similar equations to justify assigning meanings to $9^{1/2}$, $144^{1/2}$, $(4/9)^{1/2}$, $2^{1/2}$, $26^{1/2}$, $x^{1/2}$.

(b) Similarly develop meanings for $x^{1/3}$, $x^{1/4}$, . . . , $x^{1/n}$, with $x \geq 0$, leading to

Definition 7.18 Fractional Exponents If $x \in Q$, $x \geq 0$, and $n \in N$, then $x^{1/n} = \sqrt[n]{x}$.

(c) Why is it necessary to include the condition $x \geq 0$ in Definition 7.18?

(d) Consider the equation $8^{2/3} \cdot 8^{2/3} \cdot 8^{2/3} = 8^{(2/3)+(2/3)+(2/3)} = 8^{6/3} = 8^2 = 64$, so that $8^{2/3} = \sqrt[3]{64} = \sqrt[3]{8^2}$. Experiment also with the expressions $9^{3/2}$, $16^{3/4}$, $5^{2/3}$, and others, and show that they lead reasonably to

Definition 7.19 Fractional Exponents (General Case) If $x \in Q$, $x \geq 0$, and $m/n \in Q$ with $m, n > 0$, then $x^{m/n} = \sqrt[n]{x^m}$.

(e) Combine Definition 7.19 with Definition 5.3 in order to assign meanings to *negative* rational exponents, assuming that $x \neq 0$.

Remark: It can be proved that Definitions 7.18 and 7.19 are consistent both with Definitions 7.14 and 7.15 and with the "laws" of exponents embodied in Theorems 5.1, 5.3, 5.4, 7.21, and 7.24. Thus, from a simple "short-hand" beginning in Definition 5.1, we have extended the concept of exponent from N to Q. Still further extensions are possible and extremely important in advanced mathematics.

Suggestions for Investigation

1. Although fractions were used in Egypt before the seventeenth century B.C., negative numbers were a comparatively recent development. Investigate in books on the history of mathematics (use the index!) how the concept of negative numbers developed and gradually won acceptance among mathematicians and the general public.

2. A more sophisticated logical introduction of the system of integers, requiring fewer assumptions than are contained in Definition 7.1, uses ordered pairs of whole numbers and the concept of equivalence classes. Investigate this method, using references such as Campbell, Garstens and Jackson, Keedy, or Spreckelmeyer.

3. (a) The Babylonian numeration system, including its extensive use of sexagesimal fractions in tables, is worthy of much more study than the brief sketch in the text. More broadly, the entire field of Babylonian mathematics, which spanned the first two millenia B.C., is historically significant. The Boyer and Eves references, among others, are helpful.

(b) The much later invention of decimal fractions deserves study: this includes the role of Stevin and others in their popularization and in the development of our modern notation for them. This topic appears in Karpinski and other histories of mathematics.

4. Current elementary and junior high school mathematics texts cover the operations of subtraction of signed numbers and division of fractions. Examine how and at what grade levels these operations are discussed. Do the texts make any effort to show an analogy between the rules for the two operations or to establish a logical connection between them? It would also be interesting to discuss these topics with elementary or secondary mathematics teachers.

5. Which algebraic operations can be performed by geometric construction, using straight edge and compass only? Included among these is extracting a square root. Investigate these operations, particularly the construction of $\sqrt{a}$, given the segment a and a segment of unit length, by considering $\sqrt{a}$ as the mean proportional between 1 and a. Also develop a method of constructing $\sqrt{2}$, $\sqrt{3}$, $\sqrt{4}$, $\sqrt{5}, \ldots$ in sequence, beginning with the text's construction of $\sqrt{2}$, from it obtaining a rectangle of dimensions 1 by $\sqrt{2}$, whose diagonal is therefore $\sqrt{3}$, etc. Elementary geometry texts and Courant and Robbins (Chapter 3) are possible references.

6. We have seen that the maximum possible period of the decimal equivalent of a/b is $b - 1$, when a/b is in lowest terms and b contains prime factors other than 2 and 5. To predict the exact period, however, some concepts and theorems from the theory of numbers are necessary. Investigate these with the aid of a number theory text, such as Shockley (Chapter 5). Also study the Leavitt reference, which discusses another interesting property of repeating decimals.

7. (a) Since π has long been associated with the circle and problems of geometric measurement, it has an extensive history, marked by the discovery of increasingly accurate approximations and increasingly powerful methods for computing them. Histories of mathematics by Boyer, Eves (see especially pp. 89–95), and others are good references, as is the article by von Baravalle.

   (b) An interesting link between π and probability is furnished by Buffon's needle problem, discussed in the Thumm reference and elsewhere.

   (c) An amusing historical blunder was the attempt by some members of one state legislature to enact a legal value of π. The Greenblatt reference contains the story.

8. An interesting topic providing another method of representing both rational and irrational numbers is that of *continued fractions*. We can, for example, represent $\sqrt{2}$ as

$$\sqrt{2} = 1 + \cfrac{1}{2 + \cfrac{1}{2 + \cfrac{1}{2 + \cdots}}}$$

The Courant and Robbins (index) and Holmes references, among others, are helpful.

Bibliography

Bailey, W. T. and G. L. Musser. "A One-to-one Correspondence Formula." *Mathematics Teacher*, LXI (1968), 498f.

Bell, E. T. *The Development of Mathematics.* New York: McGraw-Hill Book Co., 1945.

Boyer, C. B. *A History of Mathematics.* New York: John Wiley & Sons, Inc., 1968.

Brown, S. I. "Signed Numbers: A 'Product' of Misconceptions." *Mathematics Teacher*, LXII (1969), 183–95.

Cajori, Florian. *A History of Mathematics.* New York: The Macmillan Co., 1929.

Campbell, H. E. *The Structure of Arithmetic.* New York: Appleton-Century-Crofts, 1970. Chapter 6.

Courant, Richard and Herbert Robbins. *What is Mathematics?* New York: Oxford University Press, 1941.

Eves, Howard. *An Introduction to the History of Mathematics*, 3rd ed. New York: Holt, Rinehart and Winston, 1969.

Garstens, H. L. and S. B. Jackson. *Mathematics for Elementary School Teachers.* New York: The Macmillan Co., 1967. Chapter 6.

Greenblatt, M. H. "The 'Legal' Value of π, and Some Related Mathematical Anomalies." *American Scientist*, 53 (1965), 427A–32A.

Gross, H. I. and S. F. McInroy. "Some Comments about Decimal Fractions." *New York State Mathematics Teachers' Journal*, 18:2 (1968), 82–85.

Holmes, J. E. "Continued Fractions." *Mathematics Teacher*, LXI (1968), 12–17.

Karpinski, L. C. *The History of Arithmetic.* New York: Rand McNally & Co., 1925.

Keedy, M. L. *A Modern Introduction to Basic Mathematics*, 2nd ed. Reading, Mass.: Addison-Wesley Publishing Co., 1969. Chapter 9.

Leavitt, W. G. "A Theorem on Repeating Decimals." *American Mathematical Monthly*, 74:6 (1967), 669–73.

Lieber, L. R. *Infinity*. New York: Rinehart & Co., Inc., 1953.

Ringenberg, L. A. "A Portrait of $\sqrt{2}$." *Mathematics Teacher*, LVIII (1965), 586–95.

Scandura, J. M. *Mathematics: Concrete Behavioral Foundations*. New York: Harper & Row, Publishers, 1971.

Schaaf, W. L., ed. *Nature and History of* π, SMSG *Reprint Series*, 6. Pasadena, Calif.: A. C. Vroman, Inc., 1967.

———. *Computation of* π, SMSG *Reprint Series*, 7. Pasadena, Calif.: A. C. Vroman, Inc., 1967.

———. *Infinity*, SMSG *Reprint Series*, 14. Pasadena, Calif.: A. C. Vroman, Inc., 1969.

Shockley, J. E. *Introduction to Number Theory*. New York: Holt, Rinehart and Winston, 1967.

Spreckelmeyer, R. L. *The Integers*. Boston: D. C. Heath and Co., 1964.

Struik, D. J. *A Concise History of Mathematics*, 2nd rev. ed. New York: Dover Publications, Inc., 1948.

Tammadge, Alan. "The Expression of a Fraction as a Sum of Unitary Fractions." *Mathematics Teacher*, LX (1967), 126–28.

Thumm, Walter. "Buffon's Needle: Stochastic Determination of π." *Mathematics Teacher*, LVIII (1965), 601–607.

von Baravalle, Herman. "The Number π." *Mathematics Teacher*, LX (1967), 479–87.

ANSWERS TO SELECTED EXERCISES

EXERCISES 1.3

1. (b)

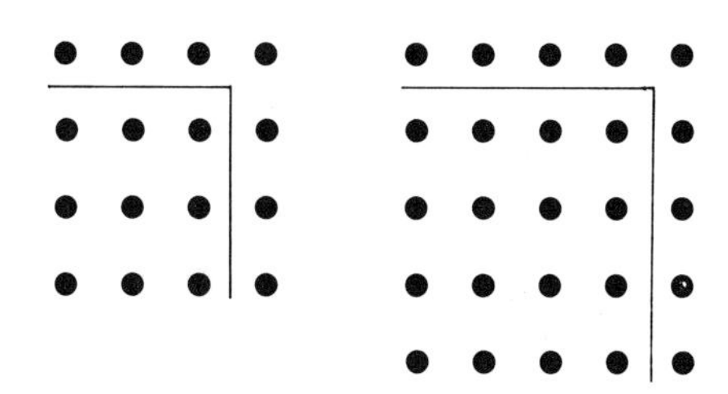

Each square array is obtained from the preceding one by annexing a border of dots. Since the border extends along two adjacent sides, with the corner dot included in both sides, the number of dots annexed to form a square of dimension n is one less than twice the dimension, or $2n - 1$, which is always an odd number.

3. To get the maximum number of regions, each new line must intersect each of the remaining lines in distinct points. Then the number of new regions created by inserting the new line is one more than the number of lines it crosses, since it separates each region it crosses into two parts. Thus the nth line creates n new regions. An algebraic formula for the maximum number of regions is $r = \frac{1}{2}(n^2 + n + 2)$.

4. (a) inductive, if you examine any number of specific squares of odd integers.

EXERCISES 2.1

1. (a) {Connecticut, Maine, Massachusetts, New Hampshire, Rhode Island, Vermont}
   (d) $\{1,3,5, \ldots\}$
   (e) {1892,1896,1904,1908,1912,1916}. The year 1900 was not a leap year!
2. (a) $\{x \mid x$ is a state of New England$\}$
   (d) $\{x \mid x$ is an odd whole number$\}$
   (e) $\{x \mid x$ is a leap year from 1892 to 1916, inclusive$\}$
3. (a) well defined.
   (c) not well defined: it is not clear whether *married women* includes widows or divorcees; the term *residing* is also ambiguous.
   (e) well defined, even though, as a matter of historical fact, this set contains no elements.
4. (a) $\in$ (c) $\notin$ (e) $\in$ (g) $\in$
5. No; many different sets exist which contain the elements 2, 4, and 6.

EXERCISES 2.2

1. (a) One answer is "American-made passenger cars."
   (c) One answer is "the set of planets of the solar system."
2. (a) All are subsets of U; all except H are proper subsets of U. Thus $A \subseteq U$, $B \subseteq U$, $\ldots$, $H \subseteq U$, $A \subset U$, $B \subset U$, $\ldots$, $G \subset U$, but $H \not\subset U$.
   (c) $A \subset H$, $B \subset H$, $C \subset A$, $C \subset F$, $C \subset H$, $D \subset H$, $E \subset H$, $F \subset H$, $G \subset H$

3. $T, T, F, T, F, T, F; T, F, F, F, T, F$

4. (a) (i) Yes. (iii) Yes. (iv) No.

   (b) (i) $M = \{1,2\}, T = \{1,2,3\}$

   (iii) same example.

   (iv) $M \subset T$ implies that T contains at least one element not in M; hence T cannot be a subset of M, so $T \subset M$ is false.

5. (a) always, since $M = M$.

6. (a) (i) cobras, snakes, reptiles, vertebrates.

   (ii) Yes, since any cobra is a snake and so belongs also to the more inclusive set of reptiles and, in turn, to the most general set, vertebrates.

7. (a) $M \subseteq T$ (When an example is requested, but may be easily supplied, generally no example will be given in the answers.)

8. (a) Let x be any element of $A : x \in A$. Then, by definition of subset, applied to $A \subseteq B$, $x \in B$ also. Since $x \in B$ and $B \subseteq C$ is given, $x \in C$ also. We have thus shown that any element of A must be in C, hence $A \subseteq C$ because the definition of subset is satisfied. (If A and/or B is a null set, as this term will be defined in Definition 2.5, the proof just given must be modified slightly, though the theorem is still true.)

EXERCISES 2.3

1. (a) $D' = \{0,2,4,6, \ldots\} = E$

   (c)

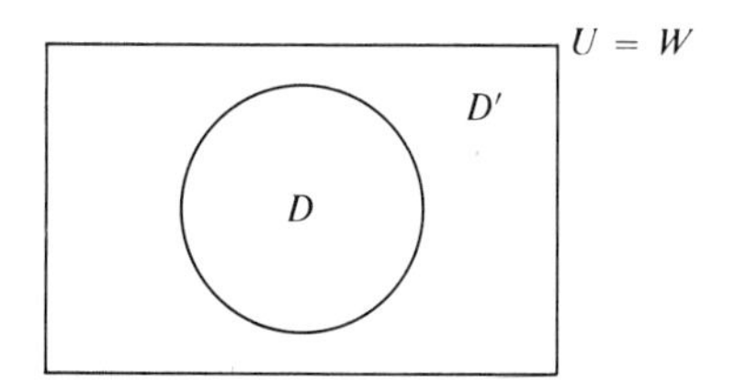

3. (a) $A' = \{b,e,h,i,j\}$; $F' = \{b,d,g,h,i,j\}$; $C' = \{a,b,c,d,e,g,h,i,j\}$; $H' = U' = \varnothing$ or $\{ \quad \}$.

   (c) $T, F, T, F, T, T; T, T, F, T, F, T; F, F, T, T$

4. (a) Since $F' = \{b,d,g,h,i,j\}$, $(F')' = \{b,d,g,h,i,j\}' = \{a,c,e,f,\} = F$.

(c) $(X')' = X$

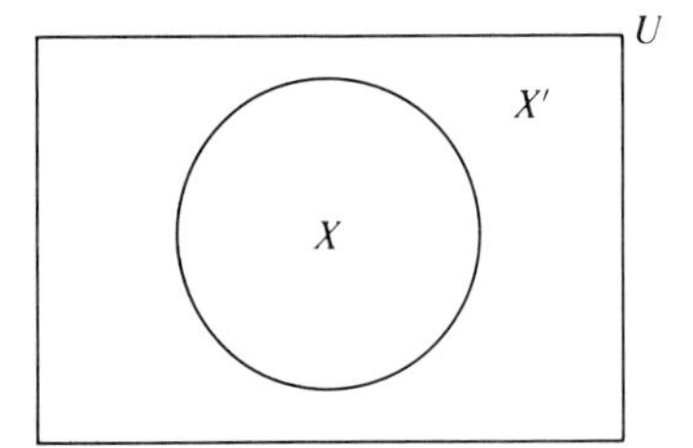

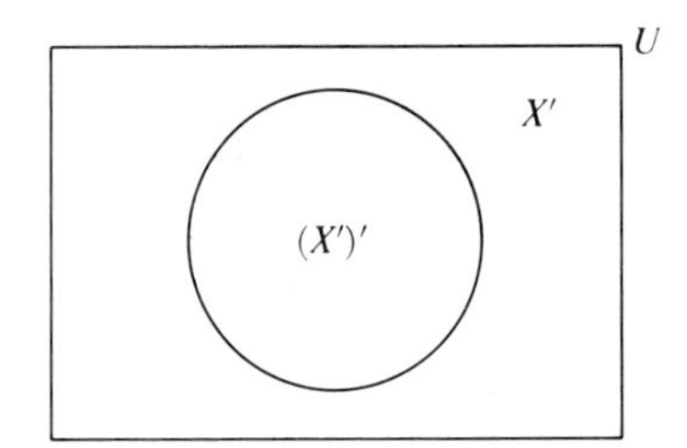

In the left-hand diagram, the set X is represented by the interior of the circle; so its complement, X', is represented by the remaining points of the rectangle (ignoring the circumference of the circle), namely, those lying outside the circle. In the right-hand diagram, X' is represented exactly as before; its complement, $(X')'$, is therefore represented by the points not lying outside the circle, namely, the points in its interior. But these points are precisely the same as those of X; hence $(X')' = X$.

5. (a) Yes; $\varnothing \subseteq \varnothing$ is true, since the subset relation includes equality of sets, and $\varnothing = \varnothing$.

   (c) No; $\varnothing \subset \varnothing$ is false.

6. (a) (ii) If $B = \varnothing$, necessarily $A = \varnothing$ in order to satisfy the hypothesis $A \subseteq B$. But then $A = \varnothing$ yields $A \subseteq C$, no matter which set C is. If $C = \varnothing$, we must have $A = B = \varnothing$ also, in order to satisfy the hypotheses $B \subseteq C$ and $A \subseteq B$. Again, $A = \varnothing$ yields $A \subseteq C$ for $C = \varnothing$.

EXERCISES 2.4

1. $S \cap X = \{\text{Venus}\}$; $L \cap S = \varnothing$; $M' \cap S = \{\text{Mercury,Venus}\}$; $M' \cap V' = \{\text{Mercury,Pluto}\}$; $(M \cap V)' = \{\text{Mercury,Venus,Uranus,Neptune,Pluto}\}$.

2. (b) (i) commutativity.

   (ii) associativity.

   (iii) identity.

   (d) The intersection of any set A with $\varnothing$ is $\varnothing$, since $\varnothing$ has no elements at all and so none which could be members of A.

3. (b) the set of nonboarders, or the set of commuters.

   (d) the set of freshman boarders who are English majors.

   (f) the set of nonfreshmen who are not boarders, or the set of upperclass commuters.

(g) the set of students who are not freshman boarders.

4. (a) (i) (iv)

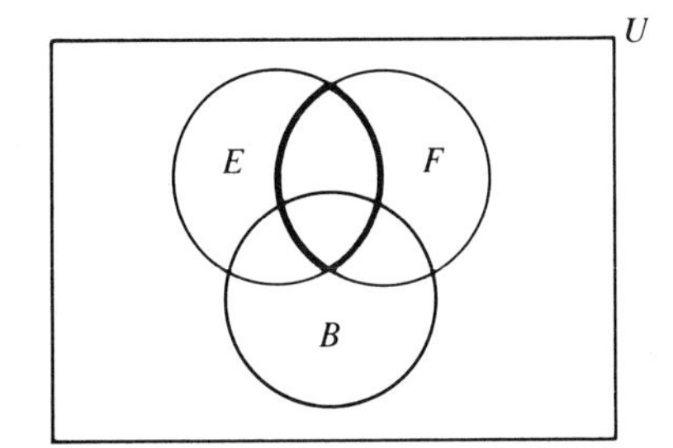

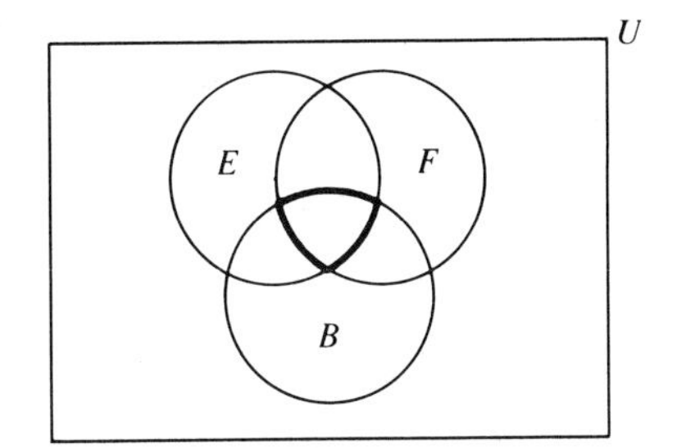

(v) the same result as (iv)

5. (a) $X \cap V = X$

(b) $A \cap B = A$

(c) Yes.

(f)

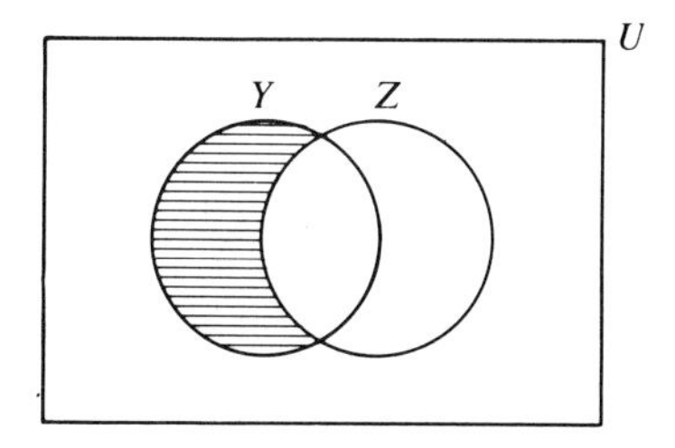

If $Y \cap Z = Y$, then the part of the Y circle which lies outside the Z circle must be empty: so it is shaded. But then all elements in Y are also in Z, hence $Y \subseteq Z$.

EXERCISES 2.5

1. $L \cup S =$
{Mercury, Venus, Earth, Mars, Jupiter, Saturn, Uranus, Neptune, Pluto} $= U$

$M' \cup S =$ {Mercury,Venus,Earth,Pluto}

$M' \cup V' =$ {Mercury,Venus,Uranus,Neptune,Pluto}

$(M \cup V)' =$ {Mercury,Pluto}

2. (b) (ii) associativity.

(iii) distributivity of intersection over union.

(iv) distributivity of union over intersection.

(c) $M \cup U = U$; $M \cup \varnothing = M$

(e) No. In Problem 1, $M' \cup V' \neq (M \cup V)'$, which is thus a counterexample.

3. (c) the set of students who are boarders and/or freshmen, and/or English majors.

   (e) the set of students who are not freshmen and/or not boarders; or the set of students who are upperclassmen and/or commuters.

   (f) the set of students who are not freshman boarders.

   Note that (e) and (f) describe the same set.

4. (a) (iii) (iv)

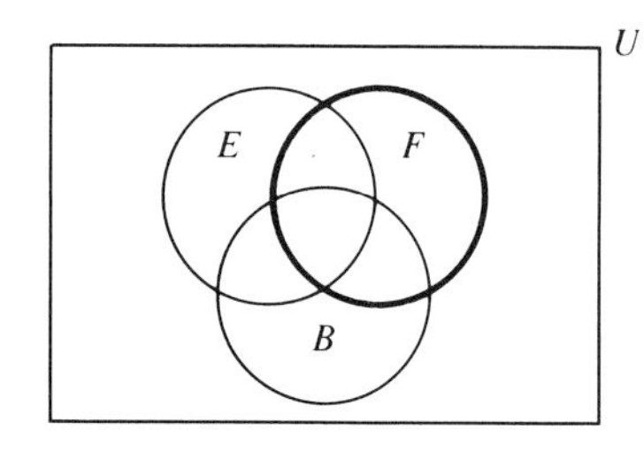

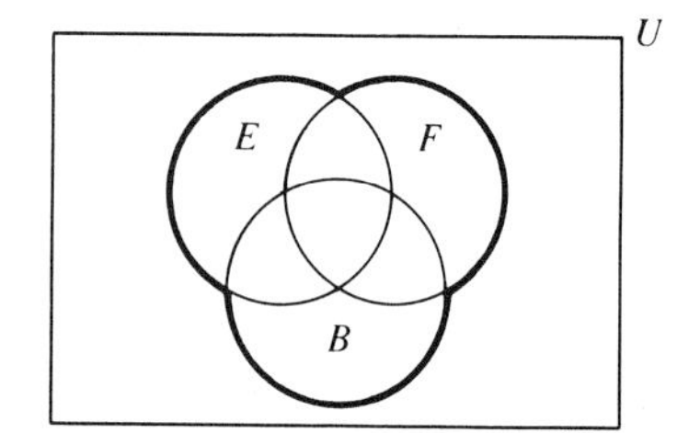

   (b) (iii) identity.

5. (a) One numbering is

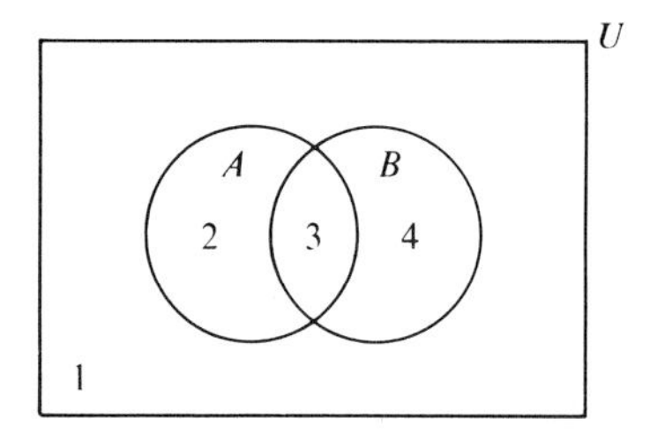

   (b) (i) $A \cup B = \{\#2,\#3,\#4\}$; so $(A \cup B)' = \{\#1\}$. $A' = \{\#1,\#4\}$; $B' = \{\#1,\#2\}$; so $A' \cap B' = \{\#1\}$ also.

EXERCISES 2.7

1. $A \sim E$; $B \sim C$; $D \sim F \sim G$; $H \sim U$. Yes: $D = G$.

2. (a) True, because two sets containing exactly the same elements can be matched one-to-one by pairing the identical elements.

   (b) False, because the two sets need not contain the same elements.

3. $n(S) = 3$, because of the one-to-one correspondence Mercury $\leftrightarrow$ 1, Venus $\leftrightarrow$ 2, Earth $\leftrightarrow$ 3, with $\{1,2,3\} \subseteq W$.

4. (a) Let each person correspond to the seat he is occupying. Then the set of people is equivalent to the set of seats, hence has the same cardinal number, 1500.

   (b) No: each book on the shelf corresponds to an author card, a title card, and (sometimes) one or more subject cards. Also, a set of several volumes with one title has only one author card, etc. Books with more than one author further complicate the correspondence.

5. (b) 0

   (f)

$$W = \{0,1,2,3, \ldots, k, \ldots\}$$

$$D = \{1,3,5,7, \ldots, 2k + 1, \ldots\}$$

   Thus $k \leftrightarrow (2k + 1)$.

   (g) $n(F) = \aleph_0$, using the one-to-one correspondence, $k \leftrightarrow 5k$.

6. (b) (i) Any attempt to set up a one-to-one correspondence between V and L fails because one element of L is left over: hence $V \nsim L$. But we can match V with $L^* = \{$Mars,Jupiter,Saturn,Uranus, Neptune$\}$ by matching the elements of V and L^* in order from left to right. So $V \sim L^*$ and $L^* \subset L$. Thus by Definition 2.13, $n(V) < n(L)$.

   (c) (ii) No, because $E \sim W$, contradicting the first sentence of Definition 2.13.

   (d) (ii) Yes, because, by Definitions 2.12 and 2.11, to have $A \sim B^* \subset B$ would automatically insure that $B^* \nsim B$, and so $A \nsim B$.

EXERCISES 2.8

1. (Albert,Dorothy), (Albert,Eve), (Bertrand,Dorothy), (Bertrand,Eve), (Charles,Dorothy), (Charles,Eve). $M \times F$.

2. (Albert,Bertrand), (Albert,Charles), (Albert,Dorothy), (Albert,Eve), (Bertrand,Albert), etc., a total of 20 slates. No, because the cartesian product $U \times U$, which is the only one large enough to contain all 20 of the slates, also includes the pairs (Albert,Albert), (Bertrand,Bertrand), etc., which are not slates.

3. 1/2, 3/2, 5/2, 1/7, 3/7, 5/7, 1/11, 3/11, 5/11. This *is* the set $N \times D$, except that cartesian products are written in the form (a,b), rather than a/b. The set $D \times N$ could be interpreted as the set of fractions whose numerators are chosen from the set D and whose denominators are chosen from the set N or as the set of reciprocals of the fractions in $N \times D$.

4. (b) Cartesian product is distributed over union.

(c) Yes: $E \times (F \cap G) = (E \times F) \cap (E \times G)$.

$$E \times (F \cap G) = E \times \{d,e\} = \{(a,d),(a,e),(b,d),(b,e)\}$$
$$E \times F = \{(a,d),(a,e),(a,f),(b,d),(b,e),(b,f)\}$$

and

$$E \times G = \{(a,c),(a,d),(a,e),(b,c),(b,d),(b,e)\}$$

so that

$$(E \times F) \cap (E \times G) = \{(a,d),(a,e),(b,d),(b,e)\}$$

5. (a) $n(N) = 3$; $n(D) = 3$; $n(N \times D) = 9$
   (c) $n(A)$ multiplied by $n(B)$ yields $n(A \times B)$.
   (d) Yes: suppose $B = \varnothing$; then $n(B) = 0$ and, since $A \times B = A \times \varnothing = \varnothing$, $n(A \times B) = 0$.
   (e) (i) If $A \neq B$, $A \times B \neq B \times A$, since the order of the elements in the ordered pairs would differ.
   (ii) $A \times B \sim B \times A$, since both sets would have the same number of ordered pairs: each element $(a,b) \in A \times B$ could be matched with an element $(b,a) \in B \times A$ and vice versa.

EXERCISES 3.1

1. (a) T (c) F (e) No, not a complete sentence. (g) F (i) propositional function; its truth value depends on which element x represents. (j) T (k) No, an exclamatory sentence does not have truth value. (m) T (n) No, a paradox.

2. (a) "All cows are not purple" could mean
   (i) "*All* cows are not purple but some are," which is equivalent to "Some cows are not purple," or
   (ii) "All cows are *not*-purple," which is equivalent to "No cows are purple."

EXERCISES 3.2

1. If "All trees are evergreens" is true, then "No trees are evergreens" is false, but if "All. . . ." is false, "No. . . ." is not necessarily true, for there might be some trees of each kind. So Definition 3.1 fails here.
   If "All trees are evergreens" is true, then "Some trees are not ever-

greens" is false, and if "All. . . ." is false, then "Some. . . ." is true. So Definition 3.1 is satisfied.

2. (a) All trees are evergreens.
   (c) The proposition r and its double negative, $\sim(\sim r)$, are logically equivalent — i.e., they always have the same truth value.

3. (a) New York City is not north of Philadelphia.
   (c) Augustus Caesar was not born in 1000 B.C.
   (d) Some tricycles do not have three wheels.
   (g) $2 + 9 \neq 29$
   (h) No giraffes have long necks.
   (j) All rectangles are squares.
   (l) Some mushrooms are edible.
   (m) In the United States census of 1970, Nevada did not have a greater population than Wyoming.

EXERCISES 3.3

2. (a) $9 + 4 = 14$ and $7 \in W$.
   (c) $9 + 4 \neq 14$
   (g) $A \cap \varnothing = A$ and $7 \notin W$.
   (k) It is not the case that $9 + 4 = 14$ and $\{1\} \subset \{3,1,2\}$.

3. (a) F (d) T (g) F (k) T

4. (b) (i) U (ii) $\varnothing$

5. (a) #3
   (b) Yes.

6. (a) It is raining and I forgot my umbrella.
   I forgot my umbrella and it is raining.
   Yes.
   (b)

p	q	$p \wedge q$	$q \wedge p$
T	T	T	T
T	F	F	F
F	T	F	F
F	F	F	F

7.

p	q	$p \wedge q$	$\sim(p \wedge q)$
T	T	T	F
T	F	F	T
F	T	F	T
F	F	F	T

8. (a) $\sim(p \wedge q)$ means "It is not the case that it is raining and I forgot my umbrella."

$(\sim p) \wedge (\sim q)$ means "It is not raining and I did not forget my umbrella."

They do not have the same meaning.

(b)

p	q	$\sim p$	$\sim q$	$(\sim p) \wedge (\sim q)$
T	T	F	F	F
T	F	F	T	F
F	T	T	F	F
F	F	T	T	T

The two statements, $\sim(p \wedge q)$ and $(\sim p) \wedge (\sim q)$, do not have the same truth table, so they are not logically equivalent.

9.

p	q	r	$p \wedge q$	(b) $(p \wedge q) \wedge r$	$q \wedge r$	(c) $p \wedge (q \wedge r)$
T	T	T	T	T	T	T
T	T	F	T	F	F	F
T	F	T	F	F	F	F
T	F	F	F	F	F	F
F	T	T	F	F	T	F
F	T	F	F	F	F	F
F	F	T	F	F	F	F
F	F	F	F	F	F	F

EXERCISES 3.4

1. (b) inclusive *or*.

3. (a) $9 + 4 = 14$ or $7 \in W$.

(c) $9 + 4 \neq 14$ or $\{1\} \subset \{3,1,2\}$.

(f) $A \cap \varnothing = A$ or $7 \notin W$.

(h) It is not the case that $9 + 4 = 14$ or $\{1\} \subset \{3,1,2\}$.

(j) $9 + 4 = 14$, or $7 \in W$ and $\{1\} \subset \{3,1,2\}$.

4. (a) T (c) T (f) F (h) F (j) T

5. (a) #2, #3, #4. (b) Yes. (c) Yes. Yes.

7. (a) (i) $\sim (p \vee q)$ means "It is not the case that it is raining or that I forgot my umbrella."

$(\sim p) \vee (\sim q)$ means "It is not raining or I did not forget my umbrella."

(ii)

p	q	$p \vee q$	$\sim(p \vee q)$	$(\sim p)$	$(\sim q)$	$(\sim p) \vee (\sim q)$
T	T	T	F	F	F	F
T	F	T	F	F	T	T
F	T	T	F	T	F	T
F	F	F	T	T	T	T

(b) Yes: $\sim(p \vee q)$ is logically equivalent to $(\sim p) \wedge (\sim q)$ and $\sim(p \wedge q)$ is logically equivalent to $(\sim p) \vee (\sim q)$.

9. (a)

p	q	r	$q \vee r$	$p \wedge (q \vee r)$	$p \wedge q$	$p \wedge r$	$(p \wedge q) \vee (p \wedge r)$
T	T	T	T	T	T	T	T
T	T	F	T	T	T	F	T
T	F	T	T	T	F	T	T
T	F	F	F	F	F	F	F
F	T	T	T	F	F	F	F
F	T	F	T	F	F	F	F
F	F	T	T	F	F	F	F
F	F	F	F	F	F	F	F

EXERCISES 3.5

2. (a) hypothesis: A person is a resident of Pennsylvania.

conclusion: He is a resident of the United States.

(c) hypothesis: The barometer falls rapidly.

conclusion: It will rain.

(e) hypothesis: The sun comes out.

conclusion: I'll go swimming.

(g) hypothesis: A figure is a square.
conclusion: It is a rectangle.

(i) hypothesis: It is a rat.
conclusion: It is a rodent.

(k) hypothesis: 4 is greater than 5.
conclusion: A triangle has three sides.

(m) hypothesis: A snake is poisonous.
conclusion: It is a rattlesnake.

3. The conditionals which must be true are: a, f, g, h, i, j, k, and l.

4. (a) (a) If a person is a resident of the United States, then he is a resident of Pennsylvania.
(c) If it is going to rain, then the barometer falls rapidly.
(e) If I'll go swimming, then the sun comes out.
(g) If a figure is a rectangle, then it is a square; or, every rectangle is a square.
(i) If an animal is a rodent, then it is a rat; or, all rodents are rats.
(k) If a triangle has three sides, then 4 is greater than 5.
(m) If a snake is a rattlesnake, then it is poisonous.
(b) The converses which must be true are b, f, h, and m.

6. (a) (a) If a person is not a resident of Pennsylvania, then he is not a resident of the United States.
(c) If the barometer does not fall rapidly, then it will not rain.
(e) If the sun does not come out, then I will not go swimming.
(g) If a figure is not a square, then it is not a rectangle.
(i) If an animal is not a rat, then it is not a rodent.
(k) If 4 is not greater than 5, then a triangle does not have three sides.
(m) If a snake is not poisonous, then it is not a rattlesnake.
(b) The inverses which must be true are b, f, h, and m.
(d)

p	q	$\sim p$	$\sim q$	$(\sim p) \rightarrow (\sim q)$
T	T	F	F	T
T	F	F	T	T
F	T	T	F	F
F	F	T	T	T

(e) The converse of a conditional is logically equivalent to its inverse.

7. (a) (a) If a person is not a resident of the United States, then he is not a resident of Pennsylvania.

(c) If it is not going to rain, then the barometer does not fall rapidly.

(e) If I won't go swimming, then the sun does not come out.

(g) If a figure is not a rectangle, then it is not a square.

(i) If an animal is not a rodent, then it is not a rat.

(k) If a triangle does not have three sides, then 4 is not greater than 5.

(m) If a snake is not a rattlesnake, then it is not poisonous.

(c) A conditional of the form $p \rightarrow q$ has for its converse, $q \rightarrow p$, its inverse, $(\sim p) \rightarrow (\sim q)$, and its contrapositive, $(\sim q) \rightarrow (\sim p)$. A given conditional and its contrapositive are logically equivalent; so also are the converse and the inverse; but the first pair of conditionals is not logically equivalent to the second pair.

8. (b)

p	q	$p \rightarrow q$	$(\sim p)$	$(\sim p) \vee q$
T	T	T	F	T
T	F	F	F	F
F	T	T	T	T
F	F	T	T	T

9. (a)

p	q	$p \wedge q$	$(p \wedge q) \rightarrow p$
T	T	T	T
T	F	F	T
F	T	F	T
F	F	F	T

(c)

p	q	$p \vee q$	$p \rightarrow (p \vee q)$
T	T	T	T
T	F	T	T
F	T	T	T
F	F	F	T

(e)

p	q	$p \to q$	$q \wedge (p \to q)$	$[q \wedge (p \to q)] \to p$
T	T	T	T	T
T	F	F	F	T
F	T	T	T	F
F	F	T	F	T

(g)

p	q	$(\backsim q)$	$p \wedge (\backsim q)$	$p \to q$	$\backsim (p \to q)$	$[p \wedge (\backsim q)] \to [\backsim (p \to q)]$
T	T	F	F	T	F	T
T	F	T	T	F	T	T
F	T	F	F	T	F	T
F	F	T	F	T	F	T

(i)

p	q	r	$p \to q$	$q \to r$	$(p \to q) \wedge (q \to r)$	$p \to r$	$[(p \to q) \wedge (q \to r)] \to (p \to r)$
T	T	T	T	T	T	T	T
T	T	F	T	F	F	F	T
T	F	T	F	T	F	T	T
T	F	F	F	T	F	F	T
F	T	T	T	T	T	T	T
F	T	F	T	F	F	T	T
F	F	T	T	T	T	T	T
F	F	F	T	T	T	T	T

EXERCISES 3.6

1. (a) The conditionals of Exercises 3.5, Problem 2, which are implications are a, f, g, h, i, j, k, and l.

3. (a) For the truth table of $(p \wedge q) \to q$, see the answer to Exercises 3.5, Problem 9(a). Since this statement is always true, it is an implication.

   (c) The truth table is given in the answer to Exercises 3.5, Problem 9(c). This statement is also an implication.

   (e) The truth table is given in the answer to Exercises 3.5, Problem 9(e). This statement is not always true. Therefore, it is not an implication.

4. (a) Yes, $B \subseteq I$.

(b)

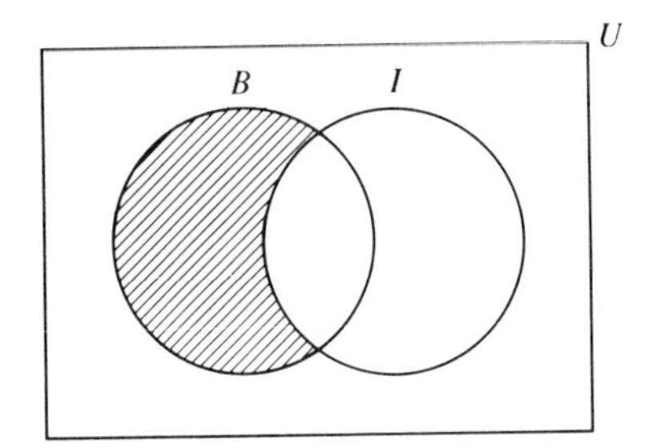

(d) Yes, because the conditional asserts that every element for which p is true (i.e., every butterfly) is an element for which q is true (i.e., an insect).

EXERCISES 3.7

1. (a) If $x = 6$, then $x - 5 = 1$. If $x - 5 = 1$, then $x = 6$ and if $x = 6$, then $x - 5 = 1$. $x - 5 = 1$ if and only if $x = 6$.

   (c) If a set is a null set, then it contains no elements. If a set contains no elements, then it is a null set and if it is a null set, then it contains no elements. A set contains no elements if and only if it is a null set.

   (e) If $8 - 2 = 5$, then $2 + 3 = 7$. If $2 + 3 = 7$, then $8 - 2 = 5$ and if $8 - 2 = 5$, then $2 + 3 = 7$. $8 - 2 = 5$ if and only if $2 + 3 = 7$.

2. Those biconditionals of Problem 1 which are logical equivalences are a, c, and e; b and d are not, because $p \rightarrow q$ or $q \rightarrow p$ is not always true.

3. (a)

p	q	$\sim p$	$(\sim p) \rightarrow q$	$p \vee q$	$[(\sim p) \rightarrow q] \leftrightarrow (p \vee q)$
T	T	F	T	T	T
T	F	F	T	T	T
F	T	T	T	T	T
F	F	T	F	F	T

   (b) Yes.

4. (a)

p	$(\sim p)$	$\sim(\sim p)$	$[\sim(\sim p)] \leftrightarrow p$
T	F	T	T
F	T	F	T

   Correct.

(c)

p	q	$\backsim p$	$\backsim q$	$p \to q$	$(\backsim p) \to (\backsim q)$	$(p \to q) \leftrightarrow [(\backsim p) \to (\backsim q)]$
T	T	F	F	T	T	T
T	F	F	T	F	T	F
F	T	T	F	T	F	F
F	F	T	T	T	T	T

Not correct.

(f)

p	q	r	$p \to q$	$r \to q$	$(p \to q) \wedge (r \to q)$	$p \vee r$	$(p \vee r) \to q$	$[(p \to q) \wedge (r \to q)] \leftrightarrow [(p \vee r) \to q]$
T	T	T	T	T	T	T	T	T
T	T	F	T	T	T	T	T	T
T	F	T	F	F	F	T	F	T
T	F	F	F	T	F	T	F	T
F	T	T	T	T	T	T	T	T
F	T	F	T	T	T	F	T	T
F	F	T	T	F	F	T	F	T
F	F	F	T	T	T	F	T	T

Correct.

5. (c)

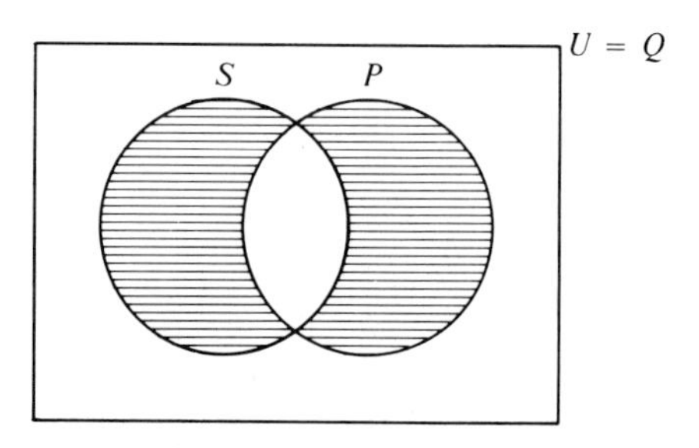

(d) This example illustrates the relation: if $S \subseteq P$ and $P \subseteq S$, then $S = P$. So the logical equivalence of two statements corresponds to the equality of their truth sets.

6. (a) (1) If x is a member of the club, then x may eat in the private dining room.

(2) x may eat in the private dining room if x is a member of the club.

(3) x may eat in the private dining room only if x is a member of the club.

(4) x may eat in the private dining room if and only if x is a member of thc club.

(c) No, this sign would admit to the dining room only those people who who are members of the club, but it does not guarantee admittance even to them.

8. (a) "If I go swimming, then the weather is sunny and the temperature is above 80° F." The conditional is not logically equivalent to its converse.

(c) (i) $(p \wedge r) \rightarrow q$ means "if the weather is sunny and I go swimming, then the temperature is above 80° F." $(r \wedge q) \rightarrow p$ means "if I go swimming and the temperature is above 80° F, then the weather is sunny."

(ii) The truth tables for the three compound conditionals, $(p \wedge q) \rightarrow r$, $(p \wedge r) \rightarrow q$, and $(r \wedge q) \rightarrow p$ follow. By comparing them or working out the expressions $[(p \wedge q) \rightarrow r] \leftrightarrow [(p \wedge r) \rightarrow q]$, etc., you may determine whether $\Leftrightarrow$ is correct.

p	q	r	$p \wedge q$	$p \wedge r$	$r \wedge q$	$(p \wedge q) \rightarrow r$	$(p \wedge r) \rightarrow q$	$(r \wedge q) \rightarrow p$
T	T	T	T	T	T	T	T	T
T	T	F	T	F	F	F	T	T
T	F	T	F	T	F	T	F	T
T	F	F	F	F	F	T	T	T
F	T	T	F	F	T	T	T	F
F	T	F	F	F	F	T	T	T
F	F	T	F	F	F	T	T	T
F	F	F	F	F	F	T	T	T

No two of the three compound conditionals are logically equivalent.

(e) (i) $[p \wedge (\sim r)] \rightarrow (\sim q)$ means "if the weather is sunny and I do not go swimming, then the temperature is not above 80 F°." $[(\sim r) \wedge q] \rightarrow (\sim p)$ means "if I do not go swimming and the temperature is above 80° F, then the weather is not sunny."

(ii) As in the answer to (c) (ii), the truth tables of the three compound conditionals in question are given, and you are asked to compare them.

p	q	r	$\sim p$	$\sim q$	$\sim r$	$p \wedge q$	$p \wedge (\sim r)$	$(\sim r) \wedge q$	$[(p \wedge q)] \rightarrow r]$	$[p \wedge (\sim r)] \rightarrow (\sim q)$	$[(\sim r) \wedge q] \rightarrow (\sim p)$
T	T	T	F	F	F	T	F	F	T	T	T
T	T	F	F	F	T	T	T	T	F	F	F
T	F	T	F	T	F	F	F	F	T	T	T
T	F	F	F	T	T	F	T	F	T	T	T
F	T	T	T	F	F	F	F	F	T	T	T
F	T	F	T	F	T	F	F	T	T	T	T
F	F	T	T	T	F	F	F	F	T	T	T
F	F	F	T	T	T	F	F	F	T	T	T

All three compound conditions are logically equivalent.

9. (a) The Rule of Detachment applies to (i), (ii), and (vi).

EXERCISES 3.9

1. By noting the reactions of many different patients to their first and later self-administered injections, the nurse has assembled many *particular* facts, on which she bases a *general* statement that the first self-administered injection is the hardest. But this generalization is not inescapable or forced on her from the data: it has only a certain degree of probability of being true. So the next patient could possibly be an exception to the general statement.

EXERCISES 3.10

1. (b) valid.

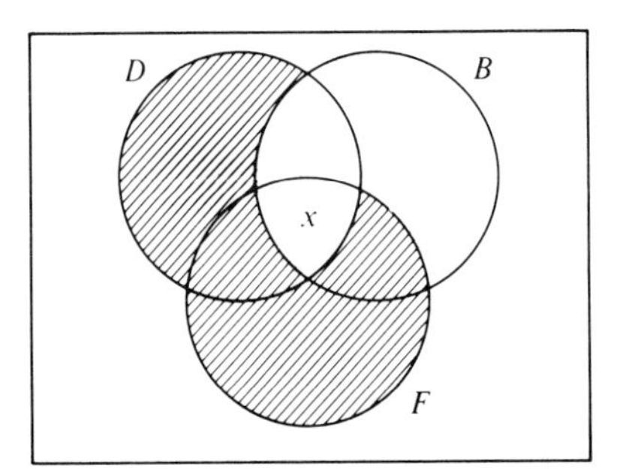

2. (a) No. Both parts of #1 illustrate deductive reasoning because they are both valid. Part (a) proceeds from particular to general (terriers—dogs—animals) while part (b) proceeds from general to specific (barkers—dogs—Fido).

3. invalid.

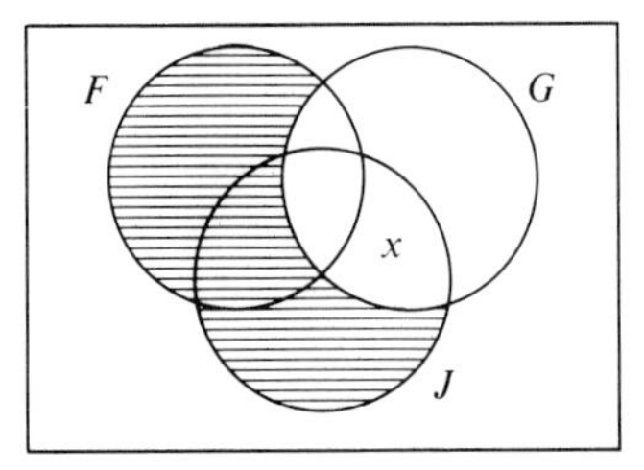

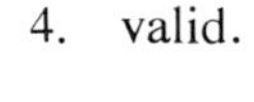
4. valid.

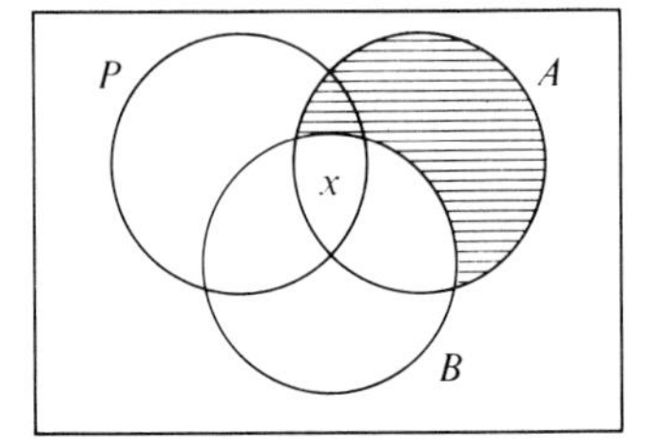

5. invalid.

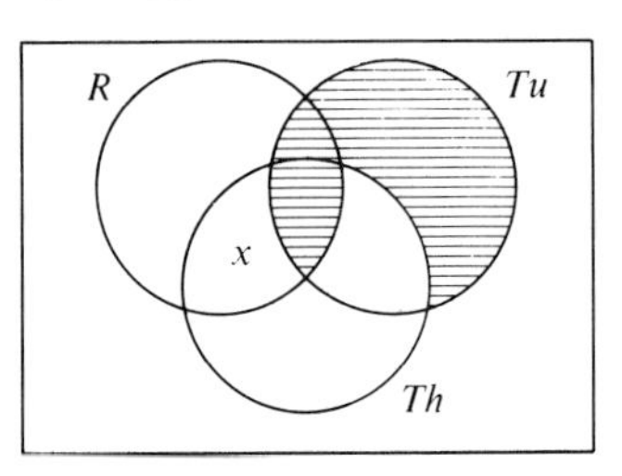

6. valid.

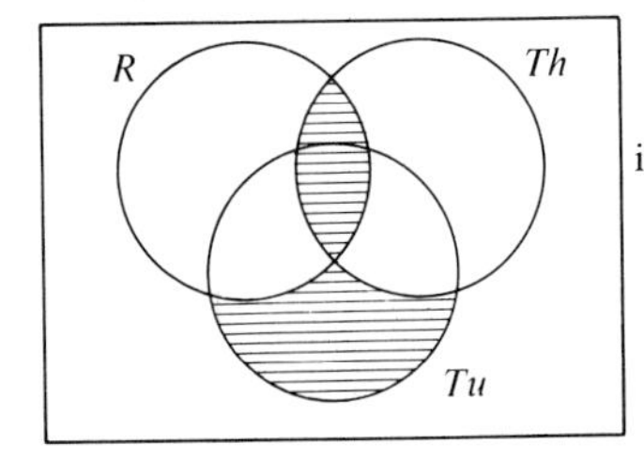

$R \cap Tu$ is empty.

7. invalid.

8. invalid.

9. valid.

10. invalid.

11. valid.

12. valid. Use two diagrams, one showing that the first two hypotheses imply "All A is C" and a second showing that this result, together with the third hypothesis, imply the conclusion.

13. valid. From the first two hypotheses we obtain the accompanying diagram. We look for an inescapable conclusion relating sets A and C: the strongest one is "No A is C," obtained by noting that the shading requires that the intersection of A with C be empty. Now use "No A is C" and "All D is C" to make a second diagram and determine whether A and D are related as the conclusion claims.

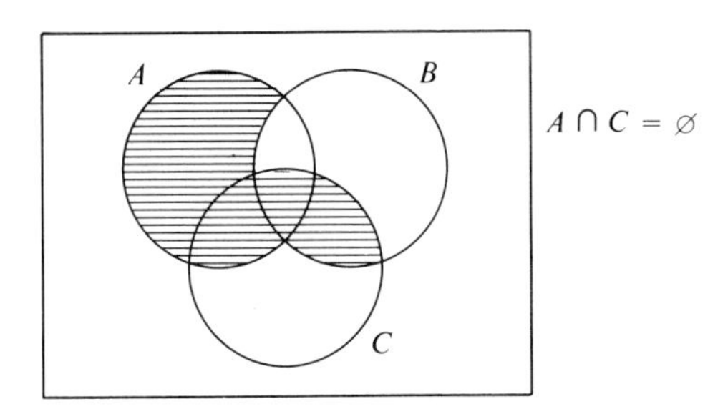

14. From the accompanying diagram we seek a way to relate the sets Q and C. Since their intersection is empty, we may derive the (inescapable) conclusion: "No residents of Quebec are Canadians."

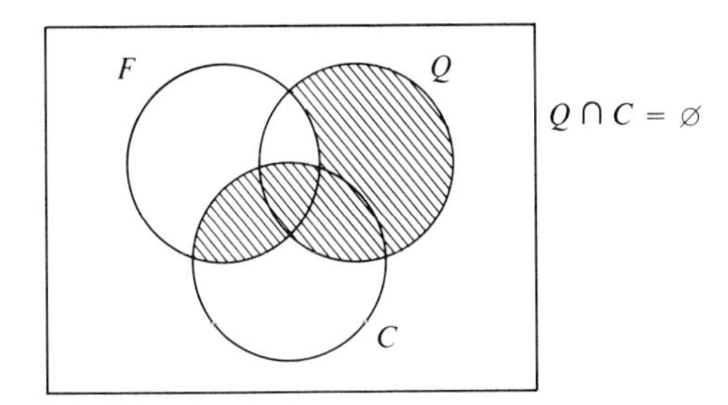

15. In the accompanying diagram the first hypothesis claims that there is an element x and/or an element y, as indicated. The second hypothesis claims that there is an element y and/or z, as indicated. We seek a conclusion relating sets C and M (milk drinkers). We cannot be sure whether y exists, because x and z would satisfy the hypotheses, with or without y; similarly, we cannot be sure whether x exists. Also, we have no clue as to

whether the remaining parts of C and M are empty or not. Thus, we are not forced to any conclusion regarding the milk-drinking habits of cats.

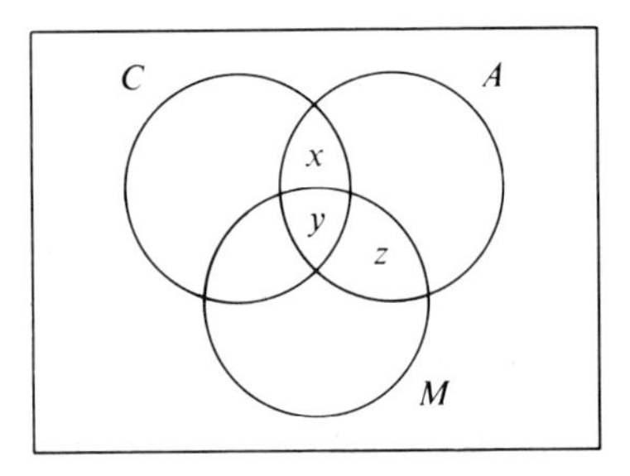

16. Some orchestra conductors write music.

17. No conclusion. 18. Some fractions are not whole numbers.

22. If it is a terrier, then it is a dog.
    If it is a dog, then it is an animal.

    ∴ If it is a terrier, then it is an animal.

    Yes, since the answer to Problem 1(a) was "valid."

24. (b) No. $p \Rightarrow q$

    $$\frac{r \Rightarrow q}{p \Rightarrow r}$$

    (c) The implication pattern is invalid; it does not yield correct reasoning. This fact can be shown by means of the truth table for $[(p \rightarrow q) \wedge (r \rightarrow q)] \rightarrow (p \rightarrow r)$, which shows that the expression is not a tautology.

28. Hypotheses: Pompous people are very talkative people.
    My neighbor is a very talkative person.

    Conclusion: My neighbor is a pompous person.

    invalid. Replace the first hypothesis by its converse to achieve validity.

30. invalid. Additional hypotheses are needed, such as: "Leading a blind man across the street is helping the handicapped."

EXERCISES 3.11

2. A dictionary uses circular definition.

3. (a) "This framework is given some kind of structure" refers to the undefined terms and axioms. The "rigorous conclusions" are the theorems.

(b) Inductive reasoning is used primarily in the choice of the framework and structure; it is also used in gathering data to "test" the model. Deductive reasoning is employed in studying and drawing conclusions from the model.

EXERCISES 4.1

1. (a) The cardinals are 150, 16, and 13; the ordinals are nineteenth, 23, 3, and twelfth.
2. (a) The following are true: $E \subseteq W$, $N \subseteq W$, $N \subset W$, $D \subseteq N$, $E \subseteq N$.
3. (a) 366, yes.
   (c) $\aleph_0$, because of the one-to-one correspondence between N and W: $n = w + 1 \leftrightarrow w$; no, $\aleph_0$ is a transfinite number.
   (e) No; though each element of W is a finite cardinal number, W itself contains an infinite number of elements: it is not a cardinal number, but a *set* of cardinal numbers.

EXERCISES 4.2

1. $3 + 4$:

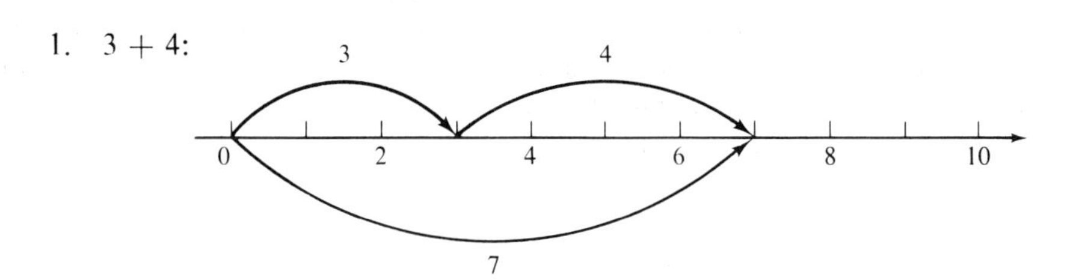

2. (a) (i) True, $4 + 2 = 6$. (iii) False, $3 + 6 \neq 7$.
3. (a) Let $A = \{a,b,c\}$, $B = \{d,e,f,g\}$; then $A \cap B = \varnothing$ and $A \cup B = \{a,b,c,d,e,f,g\}$, so $3 + 4 = n(A) + n(B) = n(A \cup B) = 7$.
4. (a) Commutativity of addition.
   (c) Closure of addition.
   (d) Associativity of addition.
   (e) Commutativity of addition and equality for addition.
   (g) Closure of addition and commutativity of addition.
   (i) Equality for addition.
5. (a) $5 + (12 + 8)$ (c) $(64 + 36) + 18$
6. (a) Yes: the sum of two even numbers is always even.

(c) Yes: final digits $0 + 0$ or $5 + 5$ yield 0; final digits $0 + 5$ or $5 + 0$ yield 5.

(d) No: $1 + 1 = 2$ is not in this set.

(f) Yes: the whole numbers are closed, and the sum of any two nonzero whole numbers is nonzero (i.e., a natural number).

7.

THEOREM 4.3 If $a, b, c \in W$, then $(a + b) + c = a + (b + c)$.

Proof: By definition of addition in W and footnote 5, we select disjoint sets, A, B, C such that $n(A) = a$, $n(B) = b$, $n(C) = c$; then $a + b = n(A \cup B)$ and $(a + b) + c = n(A \cup B) + n(C) = n[(A \cup B) \cup C]$. On the other hand, $b + c = n(B \cup C)$, and $a + (b + c) = n(A) + n(B \cup C) = n[A \cup (B \cup C)]$. But from set theory, union is associative, and, therefore, $(A \cup B) \cup C = A \cup (B \cup C)$. But since equal sets have the same cardinal number, $n[(A \cup B) \cup C] = n[A \cup (B \cup C)]$, whence, by definition of equality of whole numbers, $(a + b) + c = a + (b + c)$.

8.

THEOREM 4.6 If $a, b, c, d \in W$ and if $a = b$ and $c = d$, then $a + c = b + d$.

Proof: By Theorem 4.5, $a + c = b + c$, since $a = b$ was given. Also, by Theorem 4.5, $c + b = d + b$, since $c = d$ was given. But by the commutative law of addition, $c + b = b + c$ and $d + b = b + d$. However, by the definition of equality for whole numbers, all of $a + c$, $b + c$, $c + b$, $d + b$, $b + d$ represent the same cardinal number: in particular, $a + c = b + d$.

EXERCISES 4.3

1.

THEOREM 4.8 If $a, b, c \in W$ and if $a = b$ and $b = c$, then $a = c$.

Proof: $a = b$ means that a and b represent the same cardinal number, by Definition 4.3; similarly, $b = c$ means that b and c represent the same cardinal number. Therefore, a and c represent this same cardinal number and so by Definition 4.3 again, $a = c$.

2. (a)

THEOREM If $a, b, c \in W$, then $a + (b + c)$ is a whole number.

Proof:

1. $a, b, c \in W$	1. Hypothesis
2. $(b + c)$ is a single whole number.	2. Closure of addition
3. $a + (b + c)$ is a single whole number.	3. Closure of addition

3. (a) An appropriate chain is: $(x + y) + z = z + (x + y) = z + (y + x) = (z + y) + x$.

(b)

THEOREM If $x, y, z \in W$, then $(x + y) + z = (z + y) + x$.

Proof:

1. $x, y, z \in W$	1. Hyp.
2. $(x + y)$ is a single whole number.	2. Clos. ad.
3. $(x + y) + z = z + (x + y)$	3. Comm. ad.
4. $x + y = y + x$	4. Comm. ad.
5. $(y + x)$ is a single whole number.	5. Clos. ad.
6. $z + (x + y) = z + (y + x)$	6. Equal. ad. (Theorem 4.5)
7. $z + (y + x) = (z + y) + x$	7. Assoc. ad.
8. $(z + y), (x + y) + z, z + (x + y), z + (y + x), (z + y) + x$ are single whole numbers.	8. Clos. ad.
9. $(x + y) + z = (z + y) + x$	9. Trans. equal. (Steps 3, 6, 7)

4. (a) Numerical example: $(3 + 2) + 4 = (2 + 3) + 4, (5) + 4 = (5) + 4$.

An appropriate chain is $(x + y) + z = (y + x) + z$.

THEOREM If $x, y, z \in W$, then $(x + y) + z = (y + x) + z$.

Proof:

1. $x, y, z \in W$	1. Hyp.
2. $x + y = y + x$	2. Comm. ad.
3. $(x + y)$ and $(y + x)$ are single whole numbers.	3. Clos. ad.
4. $(x + y) + z = (y + x) + z$	4. Equal. ad. (Theorem 4.5)

(c) Numerical example: $(1 + 2) + 3 = 1 + (3 + 2)$, $3 + 3 = 1 + 5$, $6 = 6$.

An appropriate chain is $(x + y) + z = x + (y + z) = x + (z + y)$.

THEOREM If $x, y, z \in W$, then $(x + y) + z = x + (z + y)$.

Proof:

1. $x, y, z \in W$	1. Hyp.
2. $(x + y) + z = x + (y + z)$	2. Assoc. ad.
3. $y + z = z + y$	3. Comm. ad.
4. $(y + z)$ and $(z + y)$ are single whole numbers.	4. Clos. ad.

5. $x + (y + z) = x + (z + y)$	5. Equal. ad. (Theorem 4.5)
6. $(x + y) + z$, $x + (y + z)$, $x + (z + y)$ are single whole numbers.	6. Clos. ad.
7. $(x + y) + z = x + (z + y)$	7. Trans. equal. (Steps 2, 5)

(d)

THEOREM If $x, y, z, w \in W$, then $[(x + y) + z] + w = (x + y) + (z + w)$.

Proof:

1. $x, y, z, w \in W$	1. Hyp.
2. $(x + y)$ is a single whole number.	2. Clos. ad.
3. $[(x + y) + z] + w = (x + y) + (z + w)$	3. Assoc. ad.

5.

THEOREM If $a, b, c \in W$, then $a + b + c = a + (b + c)$.

Proof:

1. $a, b, c \in W$	1. Hyp.
2. $a + b + c = (a + b) + c$	2. Def. $a + b + c$
3. $(a + b) + c = a + (b + c)$	3. Assoc. ad.
4. $(a + b)$, $(b + c)$, $a + b + c$, $(a + b) + c$, $a + (b + c)$ are single whole numbers.	4. Clos. ad.
5. $a + b + c = a + (b + c)$	5. Trans. equal.

6. (b) (ii)

THEOREM. If $a, b, c, d \in W$, then $a + b + c + d = a + [(b + c) + d]$.

Proof:

1. $a, b, c, d \in W$	1. Hyp.
2. $a + b + c + d = [(a + b) + c] + d$	2. Def. part (a)
3. $(a + b) + c = a + (b + c)$	3. Assoc. ad.
4. $[(a + b) + c] + d = [a + (b + c)] + d$	4. Equal. ad. (Theorem 4.5)
5. $(b + c)$ is a single whole number.	5. Clos. ad.
6. $[a + (b + c)] + d = a + [(b + c) + d]$	6. Assoc. ad.

7. $(a + b), (a + b) + c,$ $[(a + b) + c] + d, a + b + c + d,$ $a + (b + c), [a + (b + c)] + d,$ $(b + c) + d$ and $a + [(b + c) + d]$ are single whole numbers. — 7. Clos. ad.

8. $a + b + c + d = a + [(b + c) + d]$ — 8. Trans. equal.

EXERCISES 4.4

1. (a) $A = \{a,b\}$, $B = \{c,d,e,f\}$, $n(A) = 2$, $n(B) = 4$, and $A \times B = \{(a,c), (a,d), (a,e), (a,f), (b,c), (b,d), (b,e), (b,f)\}$. Thus, $n(A \times B) = 8$ and $ab = 8$.

2. (a) (i)

THEOREM If $a, b, c \in W$, then $a(bc) \in W$.

Proof:

1. $a, b, c \in W$	1. Hyp.
2. (bc) is a single whole number.	2. Clos. mult.
3. $a(bc)$ is a single whole number.	3. Clos. mult.

(b) They are identical, except for replacing addition by multiplication.

3.

THEOREM 4.12 If $a \in W$, then $a \cdot 1 = 1 \cdot a = a$.

Proof: Since $a \cdot 1 = 1 \cdot a$ follows from the commutative property for multiplication (Theorem 4.10), we have only to prove that $a \cdot 1 = a$. By Definition 4.1 and footnote 5, we select a set A such that $n(A) = a$. Now 1 is the cardinal number of a unit set, a set with a single element: we name this set B, with $B = \{z\}$, so that $n(B) = 1$. Applying the definition for multiplication in W (Definition 4.6), we get $a \cdot 1 = n(A) \cdot n(B) = n(A \times B)$. But, by definition of cartesian product, we have $A \times B = \{(x,z) \mid x \in A\}$, since B contains only the single element, z. Since there is an obvious one-to-one correspondence between the elements $(x,z) \in (A \times B)$ and the elements $x \in A$ — $(x,z) \leftrightarrow x$ — we have $(A \times B) \sim A$, whence $n(A \times B) = n(A)$. It follows that $a \cdot 1 = n(A \times B) = n(A) = a$, or $a \cdot 1 = a$.

4. (a) This proof is analogous to that of Exercises 4.3, Problem 4(a).

(b) This proof is analogous to that of Exercises 4.3, Problem 4(d).

(d)

THEOREM If $a, b, c, d \in W$, then $ab + cd = dc + ba$.

Proof:

1. $a, b, c, d \in W$	1. Hyp.
2. $ab = ba, cd = dc$	2. Comm. mult.
3. ab, ba, cd, dc are single whole numbers.	3. Clos. mult.
4. $ab + cd = ba + dc$	4. Equal. ad. (Theorem 4.6)
5. $ba + dc = dc + ba$	5. Comm. ad.
6. $ab + cd, ba + dc, dc + ba$ are single whole numbers.	6. Clos. ad.
7. $ad + cd = dc + ba$	7. Trans. equal. (Steps 4, 5)

7. (b) Though $1 \cdot 0 = 2 \cdot 0$, yet $1 \neq 2$.

8. (a) Yes; if $ab = 0$, then $a = b = 0$ would be possible, and conversely.

EXERCISES 4.5

1. (b) The left array of dots has 4 rows and 5 columns: so there are $4(2 + 3) = 4 \cdot 5$ dots. The second array is separated into 4 rows of 2 dots each $(4 \cdot 2)$ and 4 rows of 3 dots each $(4 \cdot 3)$: this makes a total of $(4 \cdot 2) + (4 \cdot 3)$, or $8 + 12$ dots.

2. (b) $9 \cdot 42 = 9(40 + 2) = (9 \cdot 40) + (9 \cdot 2) = 360 + 18 = 378$

   (e) $41 \cdot 61 = (40 + 1)(60 + 1) = (40 \cdot 60) + (40 \cdot 1) + (1 \cdot 60) + (1 \cdot 1) = 2400 + (40 + 60) + 1 = 2400 + 100 + 1 = 2501$

3. (a) By the distributive law (Theorem 4.21), $4m + 5m = (4 + 5)m = 9m$.

4. (b) Theorem 4.20.

6. (c) distributivity: multiplication over addition.

   (d) closure of addition.

   (f) additive identity.

   (i) multiplicative identity.

   (k) commutativity of addition and multiplication.

7. (b) (i) \$.03 + \$.06 = \$.09; \$.08

   (ii) No; because, according to the tax table, the sum of the amounts of tax on the individual prices does not equal the amount of tax on the sum of the prices.

(c) It reveals a concrete situation in which the abstract system, W, does not strictly apply.

9. (a)

THEOREM If $x, y, z \in W$, then $x(y + z) = zx + yx$.

Proof:

1. $x, y, z \in W$	1. Hyp.
2. $(y + z)$ is a single whole number.	2. Clos. ad.
3. $x(y + z) = (y + z)x$	3. Comm. mult.
4. $(y + z)x = yx + zx$	4. Distr.
5. yx, zx are single whole numbers.	5. Clos. mult.
6. $yx + zx = zx + yx$	6. Comm. ad.
7. $x(y + z) = zx + yx$	7. Trans. equal.

(c)

THEOREM If $a, b, c, d \in W$, then $a(b + c + d) = ab + ac + ad$.

Proof:

1. $a, b, c, d \in W$	1. Hyp.
2. $b + c + d = (b + c) + d$	2. Def. triple sum
3. $(b + c), (b + c) + d, b + c + d$ are single whole numbers.	3. Clos. ad.
4. $a(b + c + d) = a[(b + c) + d]$	4. Equal. mult. (Theorem 4.13)
5. $a[(b + c) + d] = a(b + c) + ad$	5. Distr.
6. $a(b + c) + ad = (ab + ac) + ad$	6. Distr.
7. $(ab + ac) + ad = ab + ac + ad$	7. Def. triple sum
8. $a(b + c + d) = ab + ac + ad$	8. Trans. equal.

EXERCISES 4.7

1. (a) $3 < 7$ is true because $x = 4 \in N$ satisfies $3 + x = 7$.

   $6 > 0$ is true because $0 < 6$ (here $x = 6 \in N$ with $0 + 6 = 6$).

   $8 < 6$ is false because there is no natural number x such that $8 + x = 6$.

   $4 < 4$ is false because $x = 0$ satisfies $4 + x = 4$, but $0 \notin N$.

   (c) The fact that $6 - 8$ does not exist in W proves that subtraction is not closed in W.

3. (a) The "or" is exclusive, by the trichotomy property (Theorem 4.23).

4. (b) (ii) $n(C) < n(D)$, or $1 < 3$, according to Definition 4.7, because $1 + 2 = 3$ and $2 \in N$.

   (c) Yes, since the text asserts that, for whole numbers, the two "less than" definitions are logically equivalent, and Definitions 4.8 and 2.13 (last part) each define $a > b$ as logically equivalent to $b < a$.

5. All. A numerical example for $>$: $4 > 3$ and $3 > 2$ imply $4 > 2$.

6. (a) The point representing a is to the left of that representing b.

   (c) ... then exactly one of the following holds: a coincides with b, a is to the left of b, or a is to the right of b.

7. (c) No. If $c = 0$, by the zero product theorem, $ac = bc = 0$, which makes the conclusion, $0 < 0$, a false statement.

8. (a) No. $3 - 2 \neq 2 - 3$

10. (a)

THEOREM 4.24 If $a, b, c \in W$ and if $a < b$ and $b < c$, then $a < c$.

Proof:

1. $a < b, b < c$	1. Hyp.
2. There exists $x \in N$ so that $a + x = b$. There exists $y \in N$ so that $b + y = c$.	2. Def. $<$ applied to $a < b, b < c$.
3. $(a + x) + y = b + y$	3. Equal. ad. (Theorem 4.5)
4. $(a + x) + y = a + (x + y)$	4. Assoc. ad.
5. $a + (x + y) = b + y$	5. Trans. equal. (Steps 3, 4)
6. $a + (x + y) = c$	6. Trans. equal (Steps 2, 5)
7. $(x + y) \in N$	7. Clos. ad.
8. $a < c$	8. Def. $<$ applied to Steps 6, 7.

11. (a)

THEOREM If $a, b \in W$, then $(a + b) - b = a$.

Proof:

1. $a, b \in W$	1. Hyp.
2. $(a + b) \in W$	2. Clos. ad.
3. Taking $(a + b)$ as minuend, $(a + b) - b = a$ iff $b + a = (a + b)$.	3. Def. sub. (Definition 4.9)
4. But $b + a = a + b$.	4. Comm. ad.

5. Hence $(a + b) - b = a$. | 5. Def. logical equivalence, applied to Step 3.

EXERCISES 4.8

1. (a) $4 \mid 8$ because $4x = 8$ has a solution in W: $x = 2$; 35 is a multiple of 7 because $7 \mid 35$ ($7x = 35$, with $x = 5$); $4 \mid 5$ is false because $4x = 5$ has no solution in W.

   (b) In the first three cases, the quotients exist, because the corresponding equations in Definition 4.11 have solutions in W and these equations define the quotients.

2. (a) Both. A numerical example concerning "is a factor of" is: if $4 \mid 8$ and $8 \mid 16$, then $4 \mid 16$.

   (b)

THEOREM If $a, b, c \in W$ and $a \mid b$ and $b \mid c$, then $a \mid c$. We give a chain of statements which can be expanded into a proof: $a \mid b$ means there exists $x \in W$ so that $ax = b$; $b \mid c$ means there exists $y \in W$ so that $by = c$; multiply the first equation by y to get $(ax)y = by$ or $a(xy) = by = c$; but $(xy) \in W$; hence $a(xy) = c$ implies that $a \mid c$.

5.

THEOREM 4.30 If $a \in N$, then $0 \div a = 0$.

Proof: By the definition of division, the quotient $0 \div a = x$ iff $ax = 0$. But by the zero product theorem, if $x = 0$, then $ax = 0$ is satisfied. (*Note:* So far we have *not* established that 0 is the *only* answer!) Now $a \in N$ implies that $a \neq 0$. So if $x \in W$ were not zero, the sets in the definition of multiplication which represent a and x — i.e., for which $n(A) = a$ and $n(X) = x$ — would not be empty; hence the cartesian product, $A \times X$, would contain at least one ordered pair of the form (a,x), so that $(A \times X) \neq \varnothing$, and $ax = n(A \times X) \neq 0$, contradicting $ax = 0$. So the answer $x = 0$ is unique, and $0 \div a = 0$.

6. Use the same argument as for $1 \div 0$, merely replacing 1 by a and noting that $a \in N$ implies that $a \neq 0$.

7. (a)

THEOREM If $a, b \in W$, then $(a \cdot b) \div a = b$.

Since the definitions of subtraction and division are completely analogous, the proof of Theorem 4.28 can be carried over, with the substitution of "$\cdot$" for "$+$" and "$\div$" for "$-$."

EXERCISES 4.9

1. (a) 2.
   (c) Neither part is true: "prime" does not imply "odd" (compare 2); "odd" does not imply "prime" (compare 9).

3. (a) The primes above 50 are: 53, 59, 61, 67, 71, 73, 79, 83, 89, 97.
   (b) 7. Since the *square* of the next prime, 11, is larger than 100, the only multiples of 11 in the array (up to 100) are also multiples of lesser primes and so have already been crossed out (except 11 itself).
   (c) $k = 17$, since $17^2 < 300$ and $19^2 > 300$.

4. (a) The primes between 100 and 150 are: 101, 103, 107, 109, 113, 127, 131, 137, 139, 149.
   (b) $k = 11$, since the square of the next prime, 13, exceeds 150.

5. (a)

Interval	*Number of Primes*	*Interval Density*	*Cumulative Density*
1–25	9	.36	.360
26–50	6	.24	.300
51–75	6	.24	.280
76–100	4	.16	.250
101–125	5	.20	.240
126–150	5	.20	.233

6. $255 = 3 \cdot 5 \cdot 17$; $468 = 2^2 \cdot 3^2 \cdot 13$; $387 = 3^2 \cdot 43$; 197 and 953 are primes; the remainder factor.

8. (a) They all are primes.
   (c) No: the formula might fail for $n \geq 41$. In fact, it does fail for $n = 41$. Why?

EXERCISES 4.10

1. (a) 0 is even, since $0 = 2k$, where $k = 0 \in W$.
   (c) For 27, $k = 13$, since $2 \cdot 13 + 1 = 27$.

4.

THEOREM 4.37 If the square of a whole number is odd, then the number itself is odd.

Proof: The contrapositive of Theorem 4.37 is "If a whole number is *not odd*, then its square is *not odd*." But by Definition 4.18, "not odd" means "even." Thus, the contrapositive of Theorem 4.37 is equivalent to Theorem 4.33, which has already been proved, and this establishes the contrapositive. Since the contrapositive is logically equivalent to the original theorem, Theorem 4.37 is itself established.

5.

p	q	r	$p \to q$	$\backsim q$	$p \wedge (\backsim q)$	$\backsim r$	$r \wedge (\backsim r)$	$[p \wedge (\backsim q)] \to [r \wedge (\backsim r)]$	$(p \to q) \leftrightarrow \{[p \wedge (\backsim q)] \to [r \wedge (\backsim r)]\}$
T	T	T	T	F	F	F	F	T	T
T	T	F	T	F	F	T	F	T	T
T	F	T	F	T	T	F	F	F	T
T	F	F	F	T	T	T	F	F	T
F	T	T	T	F	F	F	F	T	T
F	T	F	T	F	F	T	F	T	T
F	F	T	T	T	F	F	F	T	T
F	F	F	T	T	F	T	F	T	T

EXERCISES 5.1

1. (a) "13," "3," 13, sixth (c) \$2000.00, "2000.00"

2. (a) 4125 = four thousand one hundred twenty-five; 1,000,240 = one million two hundred forty.

   (b) twenty thousand sixty-one = 20,061; one less than two million fifty thousand = 2,050,000 − 1 = 2,049,999.

3. (a) (i) 142 (iii) 1268

   (c) 250 = 𓍢𓍢 𓎆𓎆𓎆𓎆𓎆; 1,020,304 = 𓁨 𓂭𓂭 𓍢𓍢𓍢 𓏺𓏺𓏺𓏺

4. (a) LXVI = 66; MDLIV = 1554

   (c) 52 = LII; 1991 = MCMXCI

EXERCISES 5.2

1. $2^5 = 32$; $4^3 = 64$; $0^5 = 0$; $1^{100} = 1$; $3^{-2} = 1/9$; $100^{-2} = 1/10{,}000$

2. $3 \cdot 3 \cdot 3 \cdot 3 \cdot 3 \cdot 3 = 3^6$; $2^6 \cdot 2^5 = 2^{11}$; $7^4 \div 7^6 = 7^{-2} = 1/7^2$; $10^{16} \div 10^5 = 10^{11}$; $5^4 \cdot 5^0 = 5^4$; $9 \cdot 27 = 3^2 \cdot 3^3 = 3^5$

4. (a) $359 = 3 \cdot 10^2 + 5 \cdot 10 + 9$;
   $1{,}005{,}632 = 1 \cdot 10^6 + 5 \cdot 10^3 + 6 \cdot 10^2 + 3 \cdot 10 + 2$;
   $705.309 = 7 \cdot 10^2 + 5 + 3 \cdot 10^{-1} + 9 \cdot 10^{-3}$

   (b) $1 \cdot 10^3 + 9 \cdot 10^2 + 6 \cdot 10^1 + 9 \cdot 10^0 = 1969$;
   $4 \cdot 10^4 + 1 \cdot 10 + 6 + 3 \cdot 10^{-1} + 2 \cdot 10^{-2} = 40{,}016.32$;
   $1 + 2 \cdot 10^{-1} + 5 \cdot 10^{-4} = 1.2005$

5. (a) $2400 = 2.4 \times 10^3$; $.00166 = 1.66 \times 10^{-3}$

   (b) $3.7 \times 10^4 = 37{,}000$; $1.28 \times 10^{-4} = .000128$

   (c) $2{,}100{,}000 \times 15{,}000 = (2.1 \times 10^6) \times (1.5 \times 10^4) = (2.1 \times 1.5) \times (10^{6+4}) = 3.15 \times 10^{10}$

6. (a) .00 000 006 670 cm^3/g-sec^2;

   (c) 29,977,600,000 cm/sec; 983,514,000 ft/sec, or 1.86272×10^5 mi/sec

   (e) 5.7×10^3 years; $(5.7 \times 10^3) \times (3.65 \times 10^2) = (5.7 \times 3.65) \times 10^{3+2} = 21 \times 10^5 = 2.1 \times 10^6$ days (approximately)

   (g) .00 000 000 048 0 e.s.u.

   (i) The proton's mass is greater. $(1.67248 \times 10^{-24}) \div (9.1066 \times 10^{-28})$, which is about $.184 \times 10^{-24+28} = .184 \times 10^4 = 1.84 \times 10^3 = 1840$. So the proton has nearly two thousand times the mass of the electron.

EXERCISES 5.3

1. "5" in second partial product comes from multiplying 1 by 5 tens, yielding "5 tens," so that "5" belongs in the tens' column. "1" in third partial product comes from multiplying 1 by 1 hundred, yielding "1 hundred," so that "1" belongs in the hundreds' column.

3. (b)

THEOREM $27 + 49 = 76$.

Proof: $27 + 49 = (2\cdot 10 + 7) + (4\cdot 10 + 9)$ by Hindu-Arabic principles of numeration
$= (2\cdot 10 + 4\cdot 10) + (7 + 9)$ by comm., assoc. ad.
$= (2 + 4)\cdot 10 + (7 + 9)$ by distr.
$= 6\cdot 10 + 16$ by ad. table
$= 6\cdot 10 + (1\cdot 10 + 6)$ by H-A prin. of num.
$= (6\cdot 10 + 1\cdot 10) + 6$ by assoc. ad.
$= (6 + 1)\cdot 10 + 6$ by distr.
$= 7\cdot 10 + 6$ by ad. table
$= 76$ by H-A prin. of num.

(d)

THEOREM $43\cdot 25 = 1075$.

Proof: $43\cdot 25 = (4\cdot 10 + 3)(2\cdot 10 + 5)$ by H-A prin. of num.
$= (4\cdot 10)(2\cdot 10) + (4\cdot 10)\cdot 5 + 3(2\cdot 10) + 3\cdot 5$ by binomial expansion theorem (Theorem 4.22)
$= (4\cdot 2)\cdot 10^2 + (4\cdot 5)\cdot 10 + (3\cdot 2)\cdot 10 + 3\cdot 5$ by comm., assoc. mult., def. exp.
$= 8\cdot 10^2 + 20\cdot 10 + 6\cdot 10 + 15$ by mult. table
$= 8\cdot 10^2 + (2\cdot 10)\cdot 10 + 6\cdot 10 + (1\cdot 10 + 5)$ by H-A prin. of num.
$= 8\cdot 10^2 + 2\cdot 10^2 + 6\cdot 10 + (1\cdot 10 + 5)$ by assoc. mult. def. exp.
$= (8\cdot 10^2 + 2\cdot 10^2) + (6\cdot 10 + 1\cdot 10) + 5$ by gen. assoc. ad.
$= (8 + 2)\cdot 10^2 + (6 + 1)\cdot 10 + 5$ by distr.
$= (10)\cdot 10^2 + 7\cdot 10 + 5$ by ad. table
$= 1\cdot 10^3 + 7\cdot 10 + 5$ by def. exp., ident. mult.
$= 1075$ by H-A prin. of num.

EXERCISES 5.4

1. eight: 1, 2, 3, 4, 5, 6, 7, 10, 11, 12, 13, 14, 15, 16, 17, 20

   two: 1, 10, 11, 100, 101, 110, 111, 1000, 1001, 1010, 1011, 1100, 1101, 1110, 1111, 10000

2. five: 44, 100, 25; three: 22, 100, 9

3. (a) sixteen, counting 0. (b) 1, 2, 3, 4, 5, 6, 7, 8, 9, A, B, C, D, E, F, 10, 11, 12, 13, 14

4. $631_{\text{eight}} = 6 \cdot 8^2 + 3 \cdot 8 + 1$

   $15TE_{\text{twelve}} = 1 \cdot 12^3 + 5 \cdot 12^2 + 10 \cdot 12 + 11$

   $1101011_{\text{two}} = 1 \cdot 2^6 + 1 \cdot 2^5 + 1 \cdot 2^3 + 1 \cdot 2 + 1$

5. $631_{\text{eight}} = 6 \cdot 8^2 + 3 \cdot 8 + 1 = 6 \cdot 64 + 24 + 1 = 409$

   $15TE_{\text{twelve}} = 1728 + 720 + 120 + 11 = 2579$

   $1101011_{\text{two}} = 64 + 32 + 8 + 2 + 1 = 107$

6. 209_{ten} to base-8:

   8)209 R 1
   8)26 R 2
   3
   $= 321_{\text{eight}}$

   209_{ten} to base-12:

   12)209 R 5
   12)17 R 5
   1
   $= 155_{\text{twelve}}$

   209_{ten} to base-2:

   2)209 R 1
   2)104 R 0
   2)52 R 0
   2)26 R 0
   2)13 R 1
   2)6 R 0
   2)3 R 1
   1
   $= 11010001_{\text{two}}$

7. (a)

+	0	1	2	3	4
0	0	1	2	3	4
1	1	2	3	4	10
2	2	3	4	10	11
3	3	4	10	11	12
4	4	10	11	12	13

(b)
$$\begin{array}{r} 123 \\ +341 \\ \hline 1014_{\text{five}} \end{array}$$

(c)
$$\begin{array}{r} 38 \\ +96 \\ \hline 134_{\text{ten}} \end{array}$$

Checks!

9. (a) 4344_{five} (c) 1142_{seven} (e) 10111110_{two} (g) 104_{five}
   (i) 1185_{twelve}

10. (b) (ii) $101011011110_{\text{two}}$

11. (a) 4341_{five} (c)
$$\begin{array}{r} 3725 \\ \times\ 364 \\ \hline 17524 \\ 27376 \\ 13577 \\ \hline 1673404_{\text{eight}} \end{array}$$

12. (a)
$$\begin{array}{r} 124 \\ 3\overline{)432} \\ 3 \\ \hline 13 \\ 11 \\ \hline 22 \\ 22 \\ \hline \end{array}$$
Answer: 124_{five}

(d)
$$\begin{array}{r} 2032 \\ 14\overline{)34103} \\ 33 \\ \hline 110 \\ 102 \\ \hline 33 \\ 33 \\ \hline \end{array}$$
Answer: 2032_{five}

EXERCISES 6.1

1.

+	1	2	3	4	5	6	7	8	9	10	11	12
1	2	3	4	5	6	7	8	9	10	11	12	1
2	3	4	5	6	7	8	9	10	11	12	1	2
3	4	5	6	7	8	9	10	11	12	1	2	3
4	5	6	7	8	9	10	11	12	1	2	3	4
5	6	7	8	9	10	11	12	1	2	3	4	5
6	7	8	9	10	11	12	1	2	3	4	5	6
7	8	9	10	11	12	1	2	3	4	5	6	7
8	9	10	11	12	1	2	3	4	5	6	7	8
9	10	11	12	1	2	3	4	5	6	7	8	9
10	11	12	1	2	3	4	5	6	7	8	9	10
11	12	1	2	3	4	5	6	7	8	9	10	11
12	1	2	3	4	5	6	7	8	9	10	11	12

·	1	2	3	4	5	6	7	8	9	10	11	12
1	1	2	3	4	5	6	7	8	9	10	11	12
2	2	4	6	8	10	12	2	4	6	8	10	12
3	3	6	9	12	3	6	9	12	3	6	9	12
4	4	8	12	4	8	12	4	8	12	4	8	12
5	5	10	3	8	1	6	11	4	9	2	7	12
6	6	12	6	12	6	12	6	12	6	12	6	12
7	7	2	9	4	11	6	1	8	3	10	5	12
8	8	4	12	8	4	12	8	4	12	8	4	12
9	9	6	3	12	9	6	3	12	9	6	3	12
10	10	8	6	4	2	12	10	8	6	4	2	12
11	11	10	9	8	7	6	5	4	3	2	1	12
12	12	12	12	12	12	12	12	12	12	12	12	12

2. (a) Yes; yes; because each location in the tables contains exactly one element of the set $\{1,2,\ldots,12\}$. (b) Yes.
   (d) Each table is symmetric about its main diagonal.
4. $1' = 11$, $2' = 10$, $3' = 9, \ldots, 11' = 1$, $12' = 12$. All exist.
5. $1'' = 1$, $5'' = 5$ (since $5 \cdot 5 = 1$), $7'' = 7$, $11'' = 11$; the others do not exist.
6. In each case, the identity is its own inverse.
7. (a)

+	0	1	2	3
0	0	1	2	3
1	1	2	3	0
2	2	3	0	1
3	3	0	1	2

·	0	1	2	3
0	0	0	0	0
1	0	1	2	3
2	0	2	0	2
3	0	3	2	1

   (b) Yes, yes; because each location in the tables contains exactly one of the elements of the set $\{0,1,2,3\}$.
   (e) Additive inverses: $0' = 0$, $1' = 3$, $2' = 2$, $3' = 1$
   (f) Multiplicative inverses: $1'' = 1$, $3'' = 3$; $0''$ and $2''$ do not exist.

EXERCISES 6.2

1. (b) (i) Yes. (ii) No, because it lacks the multiplicative inverse, $0''$.
   (iii) Yes.
2. (b) It lacks two multiplicative inverses, $0''$ and $2''$.
   (c) With 0 deleted, $2 \cdot 2$ has no answer (closure violated); also, $2''$ does not exist.

4. It is commutative.

$\circ$	I	R	H	V
I	I	R	H	V
R	R	I	V	H
H	H	V	I	R
V	V	H	R	I

5.

$\circ$	I	R_1	R_2	F_1	F_2	F_3
I	I	R_1	R_2	F_1	F_2	F_3
R_1	R_1	R_2	I	F_3	F_1	F_2
R_2	R_2	I	R_1	F_2	F_3	F_1
F_1	F_1	F_2	F_3	I	R_1	R_2
F_2	F_2	F_3	F_1	R_2	I	R_1
F_3	F_3	F_1	F_2	R_1	R_2	I

This is a group, since the table shows closure, we accept associativity, I is present, and each element has an inverse: $R_1^* = R_2$, $R_2^* = R_1$, and the remaining elements are their own inverses. But the operation is not commutative, since the table is not symmetric about the main diagonal.

6. (a) No.

(b) Yes; three examples of divisors of zero for the arithmetic (mod 12) are (2,6), (3,8), and (4,3).

EXERCISES 6.3

1. (a) First, $4 - 1 = x$ iff $1 + x = 4$ by the definition of subtraction; from the addition table (mod 5), $x = 3$. Second, using Theorem 6.3, $4 - 1 = 4 + 1' = 4 + 4 = 3$.

(c) First, $1 - 4 = x$ iff $4 + x = 1$; from the table, $x = 2$. Second, $1 - 4 = 1 + 4' = 1 + 1 = 2$.

(e) Similarly, $2 - 0 = 2$.

(g) First, $3 \div 1 = x$ iff $1x = 3$ by the definition of division; from the multiplication table (mod 5), $x = 3$. Second, using Theorem 6.3, $3 \div 1 = 3 \cdot 1'' = 3 \cdot 1 = 3$.

(i) First, $3 \div 2 = x$ iff $2x = 3$; from the table, $x = 4$. Second, $3 \div 2 = 3 \cdot 2'' = 3 \cdot 3 = 4$.

(k) Similarly, $0 \div 2 = 0$.

(l) The quotient $2 \div 0$ does not exist. Both methods fail, since $0x = 2$ has no solution, and $0''$ does not exist.

(m) The quotient $0 \div 0$ is not unique.

2. (a) 2 (b) 8 (e) 5 (g) $2 \div 6 = x$ iff $6x = 2$; but no such x exists, since 6 times any number (mod 12) yields 6 or 12 only. (i) The quotient does not exist. (k) $12 \div 6 = x$ iff $6x = 12$; but the quotient is not unique, as $x = 2, 4, 6, 8, 10, 12$ will satisfy the equation. (m) The quotient is not unique.

3. (a) (i) 2 (iii) 1 (v) 2

   (b) (i) 3 (iii) 0 (iv) 3 (v) The quotient does not exist.

5. (b) (ii) $4^2 - 3 \cdot 4 + 2 = 4 - 0 + 2 = 0$ (iii) 5 is another root.

6. (a) Yes (all parts), because, since each of these are groups for the operation of addition, by Theorem 6.2 the inverse operation is closed.

   (b) No (all parts), because the systems are not groups *under multiplication.* But (i) *with* 0 *deleted* is a group under multiplication and so is closed under division (except by zero).

EXERCISES 6.4

2. Yes; the equivalence classes are the 180 keyed regions, and two cities are equivalent iff they are in the same region.

3. (a) T (only), no. (c) T, no. (e) S, no.

   (g) R, S, T, yes. (i) T, no. (k) R, S, T, yes.

   (m) S, no. (o) R, T, no. (p) R, S, T, yes.

   (q) R, T, no.

4. (a) R: Any set can be put into one-to-one correspondence with itself, by matching each element with itself.

   S: If the elements of Set A can be matched one-to-one with Set B, the matching can also be done from Set B to Set A.

   T: Each element (say x) of Set A can be matched with a distinct element (say y) of Set B; y, in turn, can be matched with a distinct element (say z) of Set C. This follows from the given facts, $A \sim B$ and $B \sim C$. By matching x with z, we set up a one-to-one correspondence which shows that $A \sim C$.

EXERCISES 6.5

1. (a) $9 - 5 = 4 \in W$; $17 - 11 = 6 \in W$

   (c) By Theorem 4.23, one of the following must be true: $a = b$, $a < b$, $a > b$. If $a = b$, then $a - b = 0 \in W$. If $a < b$, then $b - a$ will be an element of W, and if $a > b$, then $a - b$ will be an element of W.

2. (a) $8 \equiv 3$, because $5 \mid (8 - 3)$; $9 \equiv 24$, because $5 \mid (24 - 9)$.
   (b) [3], [4], [4], [2], [1], [0]

3. (a) [4] (c) [2]

5.

Definition 6.16 Multiplication (mod n) If $[a]$ and $[b]$ are any two equivalence classes (mod n), then $[a] \cdot [b] = [ab]$, where $[ab]$ signifies the equivalence class containing ab.

6. (a) [3] (c) [4] (e) [1]

8. (a) $[0] = \{0,4,8,12,16, \ldots\}$; $[1] = \{1,5,9,13,17, \ldots\}$;
   $[2] = \{2,6,10,14,18, \ldots\}$; $[3] = \{3,7,11,15,19, \ldots\}$

9. (a)

THEOREM If $a \equiv b \pmod n$, then $b \equiv a \pmod n$.

Proof: By Definition 6.14, the given statement, $a \equiv b \pmod n$, implies that $n \mid (a - b)$ or $n \mid (b - a)$, whichever difference exists. Now, if $b \geq a$, then $b - a$ exists and $n \mid (b - a)$. But $n \mid (b - a)$ is also a sufficient requirement for $b \equiv a \pmod n$, by Definition 6.14. The case $b < a$ is similar.

10. (c) Yes, the system of equivalence classes (mod n) is a group under addition since it satisfies all the requirements of Definition 6.3. Then, by Theorem 6.2, the inverse operation is closed.

EXERCISES 7.1

2. (b) $+, -, +$
   (e) $^{-}24$ feet per second per second. $^{+}5\frac{1}{2}$ feet per second per second.

3. (b) No, because no whole number, except 0, has an additive inverse in the set W.
   (c) No, because the elements of these sets, except 1, lack multiplicative inverses in these sets.

EXERCISES 7.2

1. (b) Positive: 5, $^{+}7$, $(^{-}3)'$; negative: $^{-}2$, $^{-}6$, $4'$, $(^{+}1)'$; zero: 0, $0'$.

2. (a) (last three numbers): $^{+}6$ (or 6), $^{+}2$ (or 2), $^{-}5$.
   (b) (last three numbers): $^{-}6 + {}^{+}6 = 0$, $(^{+}2)' + {}^{+}2 = 0$ or $^{-}2 + {}^{+}2 = 0$, $(^{-}5)' + {}^{-}5 = 0$ or $^{+}5 + {}^{-}5 = 0$.

(c) (i) (last three numbers):

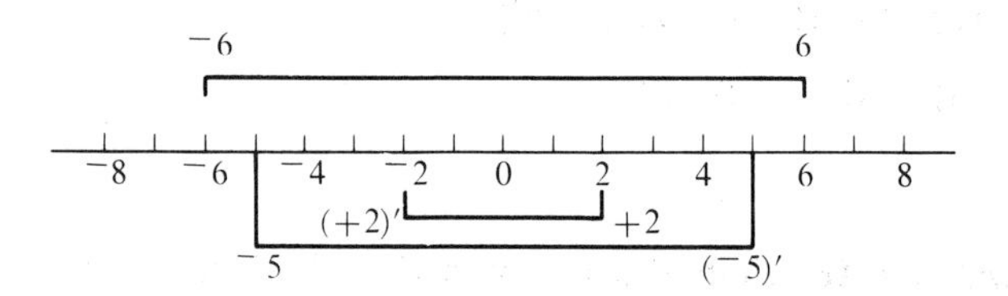

3. (a) No, because ^-n is negative, according to Definition 7.1; and by Definition 7.2, only the positive integers are natural numbers. Yes, by Definition 7.1.

(d)

Proof: Let x be any integer. If x is a natural number, by Definition 7.1 its additive inverse, ^-x, is an integer also. If $x = 0$, its inverse is also 0, an integer. Finally, if x is not a whole number, by Definition 7.1, $x = {}^-m$, the additive inverse of some natural number m; hence $m + {}^-m = 0$. But then by commutativity of addition (from Definition 7.1), $^-m + m = 0$, or $x + m = 0$. But this equation implies that m is the additive inverse of x. So the additive inverse of x is, in fact, a natural number, hence an integer.

4. (a) By Definition 7.1, W is a proper subset of Z. Also, W is equivalent to Z by Definition 2.8, as the following one-to-one correspondence shows:

Z:	0	1	$^-1$	2	$^-2$	$\cdots$	n	^-n	$\cdots$
	$\updownarrow$	$\updownarrow$	$\updownarrow$	$\updownarrow$	$\updownarrow$		$\updownarrow$	$\updownarrow$	
W:	0	1	2	3	4	$\cdots$	$2n - 1$	$2n$	$\cdots$

Hence Definition 2.11 is satisfied.

5. (c)

Proof: By Definition 6.1, $(x')'$ is an additive inverse of x'; so the equation $x' + (x')' = 0$ must hold. But since x' is an additive inverse of x, we must also have, by Definition 6.1, $x + x' = 0$ or $x' + x = 0$. Now, the last equation says that x is an additive inverse of x'. So both x and $(x')'$ are additive inverses of x'. Assuming that each element has only one additive inverse (see footnote 2, Section 7.3), it follows that $(x')' = x$.

EXERCISES 7.3

3. (a) This decomposition was used because $^+3$ is the additive inverse of $^-3$ and $^+4$ that of $^-4$. These facts, in turn, led to the reduction of the right side of the equation to zero.

4. (a) $x + {}^{+}7 = ({}^{-}4 + {}^{-}3) + (3 + 4)$ by substitution and the addition table in W.

$= [{}^{-}4 + ({}^{-}3 + 3)] + 4$ by the generalized associative law of addition (from Definition 7.1).

$= ({}^{-}4 + 0) + 4$ by the additive inverse law (from Definition 7.1).

$= {}^{-}4 + 4$ by the additive identity law (from Definition 7.1).

$= 0$ by the additive inverse law (from Definition 7.1).

5. (a) Since ${}^{+}5 \in N$, by Definition 7.3 (1), $|{}^{+}5| = {}^{+}5$ or 5; since ${}^{-}5$ is negative, by Definition 7.3 (3), $|{}^{-}5| = ({}^{-}5)' = 5$; similarly for the other parts.

(c) (i) It *appears* to be true, after the last four expressions have been simplified by performing the indicated additions.

(ii) The equation $|{}^{-}b| = b$ fails if b is negative: e.g., if $b = {}^{-}1$, then $|{}^{-}b| = |{}^{-}({}^{-}1)| = |({}^{-}1)'| = {}^{+}1$, so $|{}^{-}b| = {}^{+}1$; yet $b = {}^{-}1 \neq |{}^{-}b|$. So it is *not* safe to interpret Definition 7.3 as simply "dropping the sign."

6. (a) (ii) Let $x = {}^{-}2 + {}^{-}6$, and show by means of an appropriate chain (as in Problem 4(a)) that $x + {}^{+}8 = 0$, hence $x = {}^{-}8$.

(b) See Theorem 7.3.

7. For the generalizations, see Theorems 7.4 and 7.5.

8. (b) These equations illustrate the fact, from Definition 7.1, that "If $a \in Z$, then $a + 0 = 0 + a = a$," which is the additive identity law in Z.

9. (a)

Proof: The sum, $x = {}^{-}a + {}^{-}b$, is equal to ${}^{-}(a + b)$ iff $x + {}^{+}(a + b) = 0$. Now, $x + {}^{+}(a + b) = ({}^{-}a + {}^{-}b) + {}^{+}(a + b)$. Since ${}^{-}a$ and ${}^{-}b$ are negative by hypothesis, then $a, b \in W$ by Definitions 7.2 and 7.1. So by addition in W, ${}^{+}(a + b) = {}^{+}a + {}^{+}b$. Substituting in the previous equation, we get $x + {}^{+}(a + b) = ({}^{-}a + {}^{-}b) + ({}^{+}a + {}^{+}b) = ({}^{-}a + {}^{-}b) + ({}^{+}b + {}^{+}a) = [{}^{-}a + ({}^{-}b + {}^{+}b)] + {}^{+}a = ({}^{-}a + 0) + {}^{+}a = {}^{-}a + {}^{+}a = 0$, where each step has an obvious justification from Definition 7.1. Then $x + {}^{+}(a + b) = 0$ implies, from the first sentence, that $x = {}^{-}(a + b)$.

EXERCISES 7.4

2. (a) $({}^{+}3)({}^{+}6)$ is interpreted as the change in temperature after 3 hours $({}^{+}3)$ if the temperature is rising 6° $({}^{+}6)$ each hour; thus $({}^{+}3)({}^{+}6) = {}^{+}18$.

(d) $({}^{-}5)({}^{-}7)$ is interpreted as the change in temperature 5 hours ago $({}^{-}5)$ — as compared with the present — if the temperature has been falling

7° ($^-7$) each hour; since it was 35° higher 5 hours ago than it is now, $(^-5)(^-7) = {}^+35$.

3. (b) $(^-2)(^+3) = x$ has the value $^-6$ iff $x + (^-6)' = 0$ or $x + {}^+6 = 0$. Now $x + {}^+6 = (^-2)(^+3) + (^+2)(^+3)$ by substitution and the addition table in W. But by the distributive law, $(^-2)(^+3) + (^+2)(^+3) = (^-2 + {}^+2)(^+3)$, which reduces to $(0)(^+3) = 0$ by applying the additive inverse law from Definition 7.1 and then Theorem 7.6 (or Theorem 4.16). So $x + {}^+6 = 0$, or $(^-2)(^+3) = {}^-6$, as required.

   (c) Theorem 4.16 applies only to whole numbers; but $(^-2) \notin W$.

4.

Proof:

1. $a = a + 0$	1. add. ident. (from Definition 7.1)
2. $a \cdot a = a(a + 0)$	2. equal. mult. in Z (see footnote 3, Section 7.3)
3. $a(a + 0) = a \cdot a + a \cdot 0$	3. distr. (from Definition 7.1)
4. $a \cdot a = a \cdot a + a \cdot 0$	4. trans. equal. (Steps 2, 3) in Z
5. $(Z,+)$ is a group.	5. Definition 7.1
6. $a \cdot 0 = 0$	6. The group identity, 0, is unique. (Step 4) (See Exercises 6.2, Problem *7)

6. (a) (ii) $(^-4)(^-6) = x$ has the value $^+24$ iff $x + (^+24)' = 0$ or $x + {}^-24 = 0$. Now $x + {}^-24 = (^-4)(^-6) + (^-4)(^+6)$, assuming that $^-24 = (^-4)(^+6)$ has already been proved by the method of Problem 5. Then by the distributive law, $(^-4)(^-6) + (^-4)(^+6) = (^-4)(^-6 + {}^+6)$, which reduces to $(^-4)(0) = 0$ by the method of Problem 3(c). So $x + {}^-24 = 0$, or $(^-4)(^-6) = {}^+24$, as required.

7. (b)

Proof: $(^-a)(^-b) = x$ has the value $^+(ab)$ iff $x + {}^-(ab) = 0$. Now, $x + {}^-(ab) = (^-a)(^-b) + (^+a)(^-b)$ by substitution and Theorem 7.8. But by the distributive law (from Definition 7.1), $(^-a)(^-b) + (^+a)(^-b) = (^-a + {}^+a)(^-b)$, which reduces to $(0)(^-b) = 0$ by applying the additive inverse law from Definition 7.1 and then Theorem 7.6. So $x + {}^-(ab) = 0$, or $(^-a)(^-b) = {}^+(ab)$, as required.

8. (a) No, because, e.g., $(^-1)(^+2) = {}^-(1 \cdot 2) = {}^-2$, by Theorem 7.8, rather than yielding $^+2$, as a multiplicative identity would do.

   (c)

THEOREM 7.10 If $x \in Z$, then $x \cdot (^-1) = x'$.

9. (b) No; for example, $^+1 + (^-2)(^+3) \neq (^+1 + {}^-2)(^+1 + {}^+3)$, since the expressions reduce to $^-5 \neq {}^-4$.

EXERCISES 7.5

2. (first difference) (a) $^{-}9 - {}^{+}6 = {}^{-}15$ because $^{+}6 + {}^{-}15 = {}^{-}9$.

   (b) $^{-}9 - {}^{+}6 = {}^{-}9 + ({}^{+}6)' = {}^{-}9 + {}^{-}6 = {}^{-}15$

3. From New York City to Madrid: $^{-}4 - {}^{-}74 = {}^{+}70$; so Madrid is 70° east of New York City. In the other direction, $^{-}74 - {}^{-}4 = {}^{-}70$; so New York City is 70° west of Madrid.

4. The difference, $^{+}7 - {}^{+}3$, answers the question, "If a person with initial assets of \$3 later finds that his assets have become \$7, how has his financial status changed?" Since it is \$4 better, the difference should be $^{+}4$, according to this model. Similarly, $^{-}3 - {}^{+}7$ may be interpreted as a change from \$7 in assets to \$3 in debts, a loss of \$10; hence $^{-}3 - {}^{+}7 = {}^{-}10$.

5. $^{+}7 - {}^{+}3 = {}^{+}7 + ({}^{+}3)' = {}^{+}7 + {}^{-}3 = {}^{+}4$; $^{-}3 - {}^{+}7 = {}^{-}3 + ({}^{+}7)' = {}^{-}3 + {}^{-}7 = {}^{-}10$

7. (a) $^{+}7$ (c) no quotient in Z. (e) $^{+}28$ (f) 0

   (g) no quotient in Z, or any other number system in which Theorem 4.31 holds.

8. $({}^{-}10)' = {}^{+}10$; $[({}^{+}5)']' = [{}^{-}5]' = {}^{+}5$

9. (a) If the minus signs indicate subtraction and a negative integer, respectively, then $0 - (-5) = 0 - {}^{-}5 = 0 + ({}^{-}5)' = 0 + {}^{+}5 = {}^{+}5$. If they indicate subtraction and taking the additive inverse, respectively, then $0 - (-5) = 0 - (5)' = 0 + [(5)']' = 0 + [{}^{-}5]' = 0 + {}^{+}5 = {}^{+}5$. If both minus signs indicate taking the additive inverse, then $0 - (-5) = 0 + [(5')]' = 0 + [{}^{-}5]' = 0 + {}^{+}5 = {}^{+}5$. If the minus signs indicate taking an inverse and a negative integer, respectively, then $0 - (-5) = 0 + ({}^{-}5)' = 0 + {}^{+}5 = {}^{+}5$. All the results agree.

10. (c) $[(a)']' = a$, which *is* the content of Theorem 7.2.

EXERCISES 7.6

1. It is necessary to check the three properties of Definition 6.12: reflexivity, symmetry, and transitivity. For example, symmetry requires that if $^{-}a = {}^{-}b$, then $^{-}b = {}^{-}a$. But $^{-}a = {}^{-}b$ implies $a = b$ by the fact cited in this exercise. Then from the symmetry of equality in W, $a = b$ implies $b = a$, which implies, in turn, that $^{-}b = {}^{-}a$, as required.

2. (a) Equality for addition (Theorems 4.5 and 4.6), cancellation for addition (Theorem 4.7), equality for multiplication (Theorems 4.13 and 4.14), cancellation for multiplication (Theorem 4.15).

   (c) Yes, for all.

3. (a) $^{+}8 > {}^{+}5$, $^{+}8 > {}^{-}5$, $^{-}8 < {}^{+}5$, $^{-}8 < {}^{-}5$, etc.

   (d) $|^{+}8| > |^{+}5|$, $|^{+}8| > |^{-}5|$, $|^{-}8| > |^{+}5|$, $|^{-}8| > |^{-}5|$, etc.

4. According to the proposed "definition," $4 < 3$, since $4 + x = 3$ for the *integer* $x - {}^{-}1$. But also, from $3 + x = 4$, with $x = {}^{+}1$, we can conclude that $3 < 4$, or $4 > 3$, which violates the trichotomy property.

6. (b) Yes. By Definition 7.2, if a is positive, then $a \in N$. But then the equation $0 + x = a$ is satisfied by $x = a \in N$, so that $0 < a$ or $a > 0$. This argument can be reversed: so the "iff" statement is true.

7. (b) Starting with $1 < 3$, for example, consider in succession $c = {}^{+}2, 0, {}^{-}2$, obtaining $1({}^{+}2) < 3({}^{+}2)$, $1(0) < 3(0)$, and $1({}^{-}2) < 3({}^{-}2)$, respectively. The last two inequalities are false. Problem 8 (Theorem 7.13) provides the correct conclusion.

9. (b) No changes are needed, except to extend to integers in an obvious way the properties of N and W used in the proof: e.g., Definition 4.7 should be extended to Definition 7.7.

EXERCISES 7.7

1.

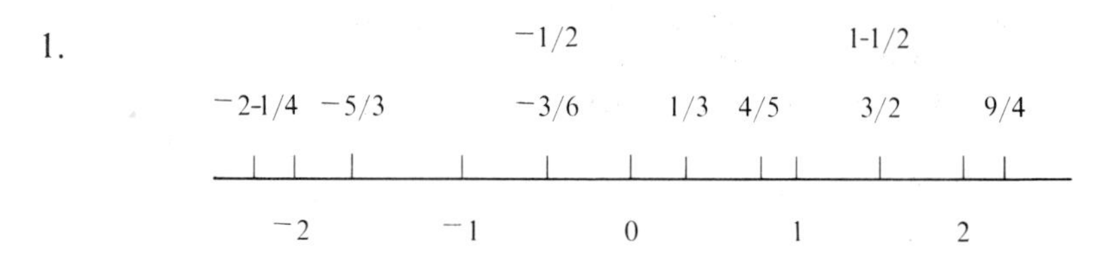

2. (a) $6'' = 1/6$, $0''$ does not exist, $(4/5)'' = 5/4$, $({}^{-}1/6)'' = {}^{-}6/1$ or ${}^{-}6$.

   (b) $6' = {}^{-}6$, $0' = 0$, $(4/5)' = {}^{-}4/5$, $({}^{-}1/6)' = 1/6$

3. (a) If $a/b, c/d \in Q$, then the *quotient*, $(c/d) \div (a/b)$, is a rational number x/y (if x/y exists) such that $(a/b)\cdot(x/y) = c/d$.

   (b) $(3/4) \div (2/3) = 9/8$ because $(2/3)\cdot(9/8) = 2\cdot 9/3\cdot 8 = 18/24 = 3/4$ (by Definition 7.10).

4. (a) $5 \div 4 = (5/1) \div (4/1) = (5/1)\cdot(4/1)'' = (5/1)\cdot(1/4) = 5/4$; $(1/4) \div (3/2) = (1/4)\cdot(3/2)'' = (1/4)\cdot(2/3) = 2/12 = 1/6$

   (b) According to Definition 7.9, $(Q^*, \cdot)$ is a commutative group.

5. $a \div b = (a/1) \div (b/1)$ by the writing agreement of Definition 7.9. Then $(a/1) \div (b/1) = (a/1)\cdot(b/1)''$ by Theorem 6.3 (see Problem 4(b)). But $(b/1)'' = 1/b$. Then $(a/1)\cdot(b/1)'' = (a/1)\cdot(1/b) = a\cdot 1/1\cdot b = a/b$, by the rule for multiplying in Q and the multiplicative identity law in Z.

6. (b) Since $(Q^*,\cdot)$ is a group, but $(Z^*,\cdot)$ is not, Theorem 6.2 applies only to the set Q^*; hence this theorem guarantees closure of division only within the system of nonzero rationals.

7. (a) $2/3 = {}^-8/{}^-12$; $3/{}^-2 = {}^-9/6$; $0/{}^-5 = 0/6$; $8/2 = {}^-4/{}^-1$; $3/3 = {}^-1/{}^-1 = 7/7$; $10/13 = 40/52$; the remaining rationals do not pair up.

 (b) (ii) Equal rational numbers are represented by the same point.

8. (a) $(1/2) + (1/3) = (3/6) + (2/6) = (3 + 2)/6 = 5/6$;

 $(1/2) + (1/3) = (1\cdot 3 + 2\cdot 1)/(2\cdot 3) = (3 + 2)/6 = 5/6$.

 $({}^-3/4) + (1/2) = ({}^-3/4) + (2/4) = ({}^-3 + 2)/4 = {}^-1/4$;

 $({}^-3/4) + (1/2) = [({}^-3)\cdot 2 + 4\cdot 1]/(4\cdot 2) = ({}^-6 + 4)/8 =$

 ${}^-2/8 = {}^-1/4$

 (b) $(1/2) + (1/3) = (1/3) + (1/2) = 5/6$ illustrates commutativity for addition and also the sign law for adding two positive integers; $({}^-1/2) + ({}^-1/3) = {}^-5/6 = {}^-(1/2 + 1/3)$ illustrates Theorem 7.3, extended to rationals; the last four examples illustrate extensions to rationals of Theorem 7.5, the additive identity law, Theorem 7.4, and the additive inverse law, respectively. All the examples illustrate closure of addition in Q.

 (c) $(1/2)\cdot(1/3) = 1\cdot 1/2\cdot 3 = 1/6$; $({}^-3/4)\cdot(1/2) = (-3)\cdot 1/4\cdot 2 = (-3/8)$

9. (b) $(2/3) - (1/2) = 1/6$ because $(1/2) + (1/6) = 2/3$.

 (c) Since $(Q,+)$ is a commutative group, Theorem 6.3 may be used in the form $(c/d) - (a/b) = (c/d) + (a/b)'$.

10. (a) (i) 899

 (ii) No. Definition 7.11 insists on an element of the ordered set S between *each* pair of distinct elements of S— e.g., between 100 and 101. So Definition 7.11 fails for Z.

 (c) No; e.g., 1/11, 1/12, and 3/35 are all between 1/13 and 1/10. An infinite number can be inserted, because the process of choosing midpoints of smaller and smaller intervals may be continued without end. For example, between 0 and 1 are the rationals 1/2 (the midpoint of the interval from 0 to 1), 1/4 (the midpoint of the interval from 0 to 1/2), 1/8, ...

EXERCISES 7.8

1. (a) $7\cdot 60 + 40 + 32\cdot 60^{-1} + 15\cdot 60^{-2}$

 (b) $17\cdot 60^{-1} + 20\cdot 60^{-2} = 17/60 + 20/3600 = 17/60 + 1/180 = 52/180 = 13/45$; $2\cdot 60 + 10 + 12\cdot 60^{-1} + 30\cdot 60^{-2} = 120 + 10 + 12/60 +$

30/3600 = 120 + 10 + 1/5 + 1/120 = 130 + 25/120 = 130 + 5/24 = 3125/24

2. (a) In the Babylonian system, 6, 24, 75, 120, and 450 are regular, since they contain only the prime factors 2, 3, or 5.

3. (a) In the decimal system, 2, 4, 40, and 800 are regular, since they contain only the prime factors 2 or 5.

   (c) The examples in part (b) all have their denominators of the form $2^m \cdot 5^n$. Thus $40 = 2^3 \cdot 5^1$, and 1/40 = .025 (terminating); whereas $450 = 2^1 \cdot 3^2 \cdot 5^2$ and so is not included in part (b), because of the factor 3, and 1/450 = .00222 . . . (nonterminating).

4. (a) (i) Yes, for all parts.

   (iii) No, for all parts except 3/6, which equals 1/2.

   (d) Otherwise the attempt to predict whether the decimal expansion terminates would lead to contradictory conclusions, as in 3/6 and 1/2.

5. (a) $.\bar{4}$, $.\overline{58}$, $5.\bar{8}$, $.58\bar{5}$, $2.1\bar{6}$, $2.1\overline{066}$

   (c) Predict 6 for 5/7; and $5/7 = .\overline{714285}$, whose period *is* 6; predict 5 for 5/6; but $5/6 = .8\bar{3}$, whose period is only 1.

   (d) $b - 1$

6. (a) 9/10, 73/100, 2/25, 5/2, 5/8, 61/20, 7/40, 1/25,000

   (c) The (positive) differences between $1/3 = .\bar{3}$ and the terminating decimals of part (b) become smaller and smaller, approaching zero: they are $.0\bar{3}$, $.00\bar{3}$, $.000\bar{3}$, $.0000\bar{3}$, $.00000\bar{3}$.

   (d) $.\bar{5} = 5/9$, $.5\bar{4} = 49/90$, $.\overline{72} = 72/99 = 8/11$, $1.\overline{05} = 104/99$, $.\overline{624} = 624/999 = 208/333$, $3.05\overline{161} = 76214/24975$, as found by the procedure

$$\begin{aligned} 1000x &= 3051.61161\ldots \\ x &= 3.05161\ldots \\ \hline 999x &= 3048.56000\ldots \end{aligned}$$

   so $x = 3048.56/999 = 304856/99900 = 76214/24975$.

7. (a) $1/4 = .25\bar{0}$ or $.24\bar{9}$, $27/25 = 1.08\bar{0}$ or $1.07\bar{9}$.

   (c) Two: e.g., $1/2 = .5\bar{0}$ or $.4\bar{9}$.

   (d) One: e.g., $2/3 = .\bar{6}$.

   (f)

THEOREM 7.15′ Rationals as Decimals A decimal expression represents a rational number iff it is periodic.

(Theorem 7.15′ is true if "periodic" is understood to include the trivial special cases, $\bar{0}$ and $\bar{9}$.)

8. Only parts (a) and (b), since they are terminating and periodic, respectively. Although part (c) has a pattern, it is not periodic; part (d) does not even have a discernible pattern.

EXERCISES 7.9

1. (b) 169, 32, 0, 169, $^-32$, $^-64$, 256

   (c) x^m is positive if m is even, and negative if m is odd.

2. (a) 1/9, 16/9, 16/9, 9/16, .5625

   (d)

Proof: By Definition 5.1 extended (see Problem 1(a)), $(a/b)^2 = (a/b)\cdot(a/b)$. Then by the product formula for rationals and Definition 5.1 extended, $(a/b)\cdot(a/b) = a\cdot a/b\cdot b = a^2/b^2$.

   (e) 8/27, $^-8/27$, 1/16, 1/16, a^3/b^3, a^m/b^m

3. (a) 7, 0, 1/3, 4/5, 3/2, 1.5

   (d) $5^2 = 25 > 21$ and $4.4^2 = 19.36 < 21$.

   (e) 4.6, because $4.6^2 = 21.16$, the closest to 21 of the ten trial values listed.

4. (a) 1, 5, $^-1$, $^-5$, 3/2, 2, 3

5. (a) All except (iv) and (viii) are true.

   (b) Equations (i) and (ii) are satisfied.

6. (b) $2.2 < \sqrt{5} < 2.3$

   (c) $2.23 < \sqrt{5} < 2.24$

   (e)

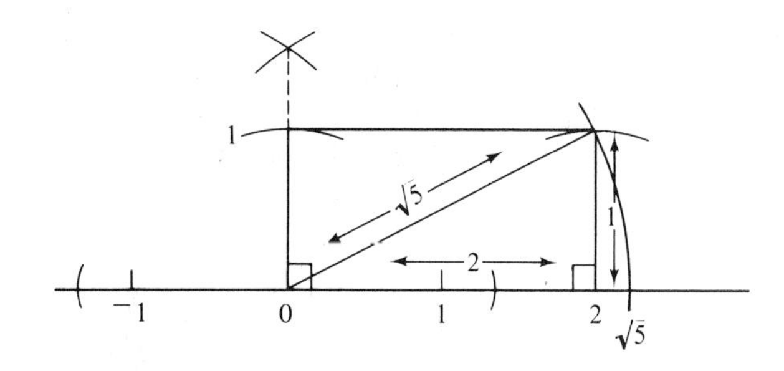

   (g) No, because if $x = \sqrt{^-5}$, Definition 7.14 would require that $x^2 = {^-5}$, an impossibility in view of the laws of signs for multiplying rationals, which are the same as those for integers.

7. (a) $0 < y < 1$, $.4 < y < .5$, $.41 < y < .42$, $.414 < y < .415$, $.4141 < y < .4142$

8. (a) Estimate, $\sqrt{34} \approx 6$;

   divide, $6\overline{)34.00}$;
   5.66

   average, $(6 + 5.66)/2 = 5.83$;

   divide again, $5.83\overline{)34.000000}$;
   5.8319

   average again, $(5.83 + 5.8319)/2 = 5.83095$.

   Hence conclude that $\sqrt{34} \approx 5.831$. (Tables give $\sqrt{34} \approx 5.830952$.)

   (b) $\sqrt{3.4} \approx 1.8439$ (c) $\sqrt{5} \approx 2.2361$ (d) $\sqrt{474} \approx 21.77$

9. (b) By Theorem 7.9, extended to real numbers, the square of any negative real number is positive, so that $x^2 = {}^-4$ is impossible for any real number x.

   (d) r is positive or zero.

   (e) (iii) $\sqrt[n]{r}$ is real and has the same sign as r when n is odd; it is real and positive when n is even and r is positive or zero. If n is even and r is negative, then $x = \sqrt[n]{r}$ is not a real number, because x^n, a product of an even number of negative factors, cannot be negative and so cannot be equal to r, as Definition 7.15 demands.

10. (a) No, because Theorem 7.15 states that all rationals terminate or are periodic.

    (c) Part (a): No, for all cases, by Theorem 7.19; Part (b): Yes, for all cases, since these numbers may be "trapped" within smaller and smaller intervals, which "close down" on a single point of the number line.

    (e) The system W is not closed because, for instance, though $2 \in W$, $\sqrt{2} \notin W$. The same example shows that Z and Q are not closed. Theorem 7.19 confirms these statements. If in $\sqrt[n]{r}$, r is negative and n is even, by the argument of Problem 9(e) (iii), $\sqrt[n]{r}$ is not real; so even the set R is not closed under root extractions.

11. (a) All except $\sqrt{{}^-4}$ and $\sqrt[4]{{}^-\pi/3}$.

GENERAL BIBLIOGRAPHY

Archibald, R. G. *An Introduction to the Theory of Numbers*. Columbus, O.: Charles E. Merrill Publishing Co., 1970.

Bell, E. T. *Men of Mathematics*. New York: McGraw-Hill Book Co., 1945.

Bochner, Salomon. *The Role of Mathematics in the Rise of Science*. Princeton: Princeton University Press, 1966.

Boyer, C. B. *A History of Mathematics*. New York: John Wiley & Sons, Inc., 1968.

Bridgess, M. P. *The Mathematical Theory of the Struggle for Life*. SMSG *Supplementary and Enrichment Series*, 26. Pasadena, Calif.: A. C. Vroman, Inc., 1966.

Courant, Richard and Herbert Robbins. *What Is Mathematics?* New York: Oxford University Press, 1941.

Dantzig, Tobias. *Number, the Language of Science*, 4th ed. New York: The Macmillan Co., 1954.

Eves, Howard. *An Introduction to the History of Mathematics*, 3rd ed. New York: Holt, Rinehart and Winston, 1969.

Gillings, R. J. *Mathematics in the Time of the Pharaohs*. Cambridge, Mass.: The M.I.T. Press, 1972.

Hamming, H. R. *Computers and Society*. New York: McGraw-Hill Book Co., 1972.

Huntley, H. E. *The Divine Proportion*. New York: Dover Publications, Inc., 1970.

Jacobs, H. R. *Mathematics, a Human Endeavor*. San Francisco: W. H. Freeman and Co., 1970.

Kemeny, J. G. et al. *Finite Mathematics with Business Applications*, 2nd ed. Englewood Cliffs: Prentice-Hall, Inc., 1972.

Kline, Morris. *Mathematics in Western Culture*. New York: Oxford University Press, 1953.

Kordemsky, B. A. *The Moscow Puzzles: 359 Mathematical Recreations*. New York: Charles Scribner's Sons, 1972.

Landau, Edmund. *Foundations of Analysis*. New York: Chelsea Publishing Co., 1951.

Lieber, L. R. *The Einstein Theory of Relativity*. Lancaster, Pa.: The Science Press Printing Co., 1936.

Mathematical Thinking in Behavioral Sciences. Readings from the *Scientific American.* San Francisco: W. H. Freeman and Co., 1968.

Mathematics in the Modern World. Readings from the *Scientific American.* San Francisco: W. H. Freeman and Co., 1968.

Montague, H. F. and M. D. Montgomery. *The Significance of Mathematics.* Columbus, O.: Charles E. Merrill Publishing Co., 1961.

Newman, J. R., ed. *The World of Mathematics.* 4 vols. New York: Simon and Schuster, 1956.

Niven, Ivan and H. S. Zuckerman. *An Introduction to the Theory of Numbers*, 3rd ed. New York: John Wiley & Sons, Inc., 1972.

Ogilvy, C. S. *Tomorrow's Math: Unsolved Problems for the Amateur*, 2nd ed. New York: Oxford University Press, 1972.

Polya, George. *How To Solve It.* New York: Doubleday & Co., Inc., 1957.

Sawyer, W. W. *A Mathematician's Delight.* Baltimore: Penguin Books, 1943.

Singh, Jagjit. *Great Ideas in Information Theory, Language and Cybernetics.* New York: Dover Publications, Inc., 1966.

Steinhaus, Hugo. *Mathematical Snapshots*, 3rd American ed. New York: Oxford University Press, 1969.

Tietze, Heinrich. *Famous Problems of Mathematics.* New York: Graylock Press, 1965.

Uspensky, J. V. and M. A. Heaslet. *Elementary Number Theory.* New York: McGraw-Hill Book Co., Inc., 1939.

Williams, J. D. *The Compleat Strategyst.* New York: McGraw-Hill Book Co., Inc., 1954.

DEFINITIONS

THEOREMS

INDEX